Ecology and the Politics
of Scarcity Revisited

Ecology and the Politics of Scarcity Revisited

The Unraveling of the American Dream

William Ophuls

A. Stephen Boyan, Jr.

Foreword by Thomas E. Lovejoy
of The Smithsonian Institution

W. H. Freeman and Company
New York

Library of Congress Catologing-in-Publication Data

Ophuls, William, 1934–
Ecology and the politics of scarcity revisited / William Ophuls,
A. Stephen Boyan, Jr. ; foreword by Thomas E. Lovejoy.
p.   cm.
Includes bibliographical references and index.
ISBN 0–7167–2313–1
1. Environmental policy 2. Environmental protection.
I. Boyan, A. Stephen, 1938–, II. Title.
HC79.E5054 1992
333.7'2—dc20   91–46535
                                                        CIP

Printed in the United States of America

1 2 3 4 5 6 7 8 9 0   VB   9 9 8 7 6 5 4 3 2

To the posterity that has never done anything for us

Men are qualified for civil liberty in exact proportion to their disposition to put moral chains upon their own appetites... Society cannot exist unless a controlling power upon will and appetite be placed somewhere, and the less of it there is within, the more there must be without. It is ordained in the eternal constitution of things, that men of intemperate minds cannot be free. Their passions forego their fetters.

Edmund Burke

Contents

——

List of Text Boxes

Foreword

—

The environmental handwriting is clearly on the wall, but bolder and more insistent than fourteen years ago when the earlier version of this book appeared. Ecological limits mean that it is virtually impossible for all of the 5.4 billion people on earth to achieve the high-consumption habits of U.S. citizens. Yet it is equally impossible for all of the 5.4 billion to embrace the hunter-gatherer life of the remaining pre-agricultural societies. Obviously, something must change in the way we collectively relate to the environment.

Will economic growth come to a halt by design or with a horrifying crunch because we have blithely allowed ourselves to overshoot the planet's carrying capacity? Surely the end is in sight for economic growth based on ever increasing consumption of finite natural resources. We have polluted the clean air and water, thereby changing the atmospheric composition—a certain signal that we have already carried this to the planetary scale.

Perhaps there is an answer in the biological world we so long tended to dismiss and disdain. Individual organisms have two distinct forms of growth. In one the organism simply grows larger and of necessity consumes more resources. In the other, organisms do not grow in size but rather in complexity of structure, and, by analogy, behavior. Perhaps growth in complexity can serve as a model for an ecologically sustainable society.

There is widespread longing for an escape from the environmental squeeze cage through some ingenious technological fix. Environmentalists find that solution suspect because it is close to impossible for any single fix to work, and because most suggestions are of such cosmic scale that they are as likely to create other environmental problems as to solve the ones at hand. It is nonetheless important not to dismiss cavalierly what technology can do to help. Every technological advance in efficient use and conservation of natural resources will make it easier to achieve the transition to a society in balance with its environment.

Even so, it would be foolish to believe that advances in efficiency and conservation and any other marvelous surprises ecotechnology may have in store will in themselves solve the problem. Society faces not only the current situation of human numbers and resource consumption but also frightening rates of increase. A current annual addition of 100 million people leads to projections of human population many billions larger than present. To sustain such population growth with the same set of resources seems beyond the reach of technological fixes, especially while society is undermining the environmental stability of the planet through myriad activities, the most monstrous of which is alteration of atmospheric composition and climatic stability.

The serious question this book explores is whether our institutions, societal values, and economic and political assumptions are up to the adjustment. Or are we without the structures and viewpoints which will allow us to turn away from our dangerous inclination to bicker, dicker, and fritter? Although few would wager that human society currently has the values and structures necessary for survival, the simple point is that there is no choice, and that we must bend our astonishing intellectual capacity toward creating the conditions for change.

Not many readers, I expect, will agree with every analysis this book provides or with its every insight into the workings and assumptions (oft unstated) of human societies and how they relate to the environmental challenge. Nonetheless, the book presents exactly the kind of thinking that must take place if we are to do more than stumble around in an ecopolitical darkness. One hears, at least occasionally, the question of whether democracy is up to the environmental challenge. That is the question explored here, and the answer is positive but guarded, because a democracy of complacency that sees only immediate horizons will most certainly fail. How ironic it would be for democracy, now more widespread than ever in history, to undermine itself by failing to face its greatest challenge.

Over the centuries Thomas Malthus has been criticized each time society has been able by one means or another to postpone a final reckoning. What we really must recognize is that Malthus was only wrong about the date. How much better it is to accommodate to an inevitable reality in advance rather than adjust painfully in retrospect. How much more sensible to learn how best to use and enjoy newly discovered ways to work within our resource capacities, rather than to grasp in desperation for room to maneuver after we have pushed ourselves to the ecological limits, or indeed beyond. This book challenges us to face the environmental music rather than let scarcity force us into an era of unwanted political change and social cacophony.

Thomas E. Lovejoy
Assistant Secretary for External Affairs
Smithsonian Institution
January 1992

Acknowledgments

————

I want to give special thanks to Carol Beyers for her valuable research assistance in the preparation of this manuscript. I also wish to thank Sandy Parker, of the UMBC Geography Department, Margaret McKeon, of the Duke University Political Science Department, for their valuable comments on the text and Paul Hertz, Chairman of the Biology Department at Barnard College, for his contributions to Chapter 1. Thanks also go to Justin Boyan, who helped me with mathematical formulas needed to prepare tables and calculations, as well as to Kitty Boyan, who gave me much appreciated support as I worked on the manuscript.

A. Stephen Boyan, Jr.
January 1992

Preface

___

About a century and a half ago, Victor Hugo described the ethical thought of his day and predicted what would one day become the most challenging task of ethics:

> In the relations of humans with the animals, with the flowers, with the objects of creation, there is a whole great ethic scarcely seen as yet, but which will eventually break through into the light and be the corollary and the complement to human ethics. . . . Doubtless it was first necessary to civilize man in relation to his fellow men. With this one must begin and the various lawmakers of the human spirit have been right to neglect every other care for this one. That task is already much advanced and makes progress daily. But it is also necessary to civilize humans in relation to nature. There, everything remains to be done (*En Voyage, Alpes et Pyrénées,* quoted in Borrelli 1989, p. 39).

I wanted to revise *Ecology and the Politics of Scarcity* because, as the text will make plain, we are continuing to degrade our environment, and we seem unable to stop doing so. "We seem to confront an array of tragic choices: business-as-usual is becoming impossible and intolerable, yet all the immediately available political alternatives appear unworkable, un-palatable, or downright repugnant." The prospect of ecological scarcity thus forces us to consider our ethics. If we are to "civilize humans in relation to nature" we need to think about the larger issues, for "we do not simply live in a world of problems but in a highly problematical world, an *inherently* anti-ecological society. This anti-ecological world will not be healed by acts of [mere] statesmanship or passage of piecemeal legislation. It is a world that is direly in need of far-reaching structural change" (Bookchin 1990, p. 83).

The prospect of ecological scarcity equally obliges us to consider the eternal questions of political philosophy. Dr. Ophuls educated a genera-tion of readers in the connection between ecological scarcity and the questions of political philosophy. He said that "the value of this book lies in the nature and quality of the questions it raises, for I am convinced that they will be central for our era" and "until we have these questions clearly in mind, the . . . answers [to them and our predicament] are bound to elude us." I appreciate Dr. Ophuls's permitting me to update his book and to raise these questions to a new generation of readers. For despite the publication of a great deal of environmental literature, two decades of environmental activism, and billions of dollars spent on environmental programs, the "bottom line" is that the condition of the environment continues to worsen. Clearly we need to do more than what we've been doing. It may be that the future won't take the form that Chapter 8 suggests, but without some fundamental changes, the planet will someday not be a hospitable place for human habitation.

Yet although most environmental indicators are worsening, I revised this book not because I despair, but because I care for the future. It is difficult for me to be optimistic, but it is also easy to overlook the bits of good news that are emerging everywhere. Discouraging as it is that the human population is still increasing prodigiously worldwide, the good news is that eight East Asian and Latin American countries reduced their fertility rates by more than 50% between 1960 and 1987. People even in poor countries are capable of changing their attitudes toward reproduction and their reproductive behavior. On another front, even though modern agricultural practices are still spreading throughout the world, the good news is that some farmers in almost every region are returning to organic or IPM methods and to the use of biological pest controls—and making as much money as they did before. Though the world continues to use ever more energy, the good news (as we will see in Chapter 2) is that some utilities, businesses, and individuals are taking meaningful steps to increase the efficiency with which they use energy. And while the world continues to confuse "growth" with happiness and measures growth by gross national product, the good news is that Germany, France, and Norway, among other countries, are beginning to figure natural resource depletion into their economic analyses.

Other good news is that a change in paradigm is occurring in industrial countries. As John McRuer describes it, the conventional paradigm

is rooted in neoclassical economics. Its vision of earthly paradise is a cultured world of urban luxury, where both recreation and work are high-tech and orderly, and where nature is a manicured garden. Its nightmare is about losing control—fumbling its development strategy and waking to the cold chill of brown-out, angry customers, and missed investment dollars. In a world driven by competition, the Convention, whose ethic is rational humanism, is the paradigm of power in western nations. It [still] drives politics, business, government, the media, and, to some extent, religion.

Its opponents are the "Greens." Their nirvana is a bucolic world of gentle, self-sufficient communities, which use a simple but adequate technology to coexist affectionately with nature. Their nightmare is about technology's power to destroy their wild places—of waking to a world where suburbs desecrate the meadows, where forests have turned to moonscapes, where the song birds are gone, and where the water is befouled with cancerous substances. Their world works by ethics, by cooperation, by beneficence.

McRuer observes that although the Greens may be "wistful, impractical, and in some ways hypocritical" (they may use fuel-efficient cars but jet to the winter sunshine), they are the force behind a "doomsday debate" about the planet's future. The Conventions assume that "human skill can prevail against all—that society can grow forever in wealth and wisdom, and thus in bliss." The Greens "invoke conservation laws of nature, warn that humans are not gods" but are entangled in the fate of natural systems (McRuer 1990, p. 5).

What is interesting is that even as the debate continues and the condition of the planet worsens, the Conventions themselves are beginning to support the concept of "sustainable development." Advocates of sustainable development—for example, the World Resources Institute's Global Possible Conference (1984)—still have high hopes for growth and technological solutions (hopes that, as we shall see, are at least in some respects a delusion) but also recognize that a stable birthrate, a highly efficient use of energy with a shift to renewables, and a reliance on nature's income rather than its capital (quoted in Corson, 1990, p. 322) (which Greens have traditionally identified as key aspects of a "steady state") are essential.

The fundamentals of "sustainable development" do not yet enter into the everyday decisions of most corporations, but again, here and there, they are beginning to do so. For example, the chemical industry has quietly asked for the help and cooperation of environmental scientists in determining how best to reduce its toxic air emissions. We will see how some utility executives and state utility regulators are recognizing the wisdom of efficiency measures and a shift toward renewables. The 3M Corporation has hired an environmental vice-president; in 12 years the company has reduced its waste generation by over 50 percent and is still coming up with new ways to reduce pollution on- and off-site. Encouragingly, it has saved money in the process. The McDonald's Corporation is radically shifting the way it serves food, substituting recycled and recyclable materials for new styrofoam containers. The Borden chemical company has altered its procedures to reduce the organic chemicals in its waste water by 93%. It too has saved money in the process.

The concept of sustainable development is even beginning to take root in the developing world. And this despite the fact that industrial countries are selling developing countries environmentally harmful products, such as pesticides, and transferring their own environmental costs to those developing nations by relocating industry there to avoid being subject to their own country's pollution controls. The Environmental Policy Institute and the National Wildlife Federation have published a booklet on environmentally responsible projects in the developing world. They include windbreak planting, watershed restoration, integrated pest management, environmental reserves, habitat protection, energy efficiency and conservation, fish cultures, agroforestry—even family planning in Zimbabwe. In Thailand, just one man launched a private nonprofit corporation whose efforts resulted in 70% of the population practicing birth control and led to a reduction of the birth rate in that country from 3.2% to 1.6% in just 15 years! A change in thinking takes time to be translated into concrete action. Sometimes that action is tentative; sometimes outside pressure or events must push it forward; sometimes one energetic individual with a vision can do so. But when thought gets translated into the first action, each step makes the next step easier and reinforces the previous shift in perspective.

Even more encouraging is the change in perspective among young people. In schools across the United States, environmental education courses are springing up. Among the objectives of these courses is to teach youngsters such concepts as the "web of life," to show them that "you can't do just one thing," and to involve them personally in planting, recycling, and cleaning up debris in their communities. The fact that these courses are being taught from elementary schools "on up" itself suggests changing attitudes among today's educational elites. Moreover, those who teach these courses are often surprised at how readily children accept ecological concepts—how, indeed, children will encourage their parents to launch recycling programs or will educate their parents about other environmental issues. One observer has suggested that this should come as no surprise at all. Just as today's 40- and 50-year-olds grew up with a vague fear of the bomb and worried at times that their elders would blow up the world, he suggests that 10-year-olds today are vaguely fearful of environmental deterioration and worry that their elders are polluting the world (Montagna 1991). If this is so, and these children are adopting a change in paradigm, suggestions dismissed today as politically unrealistic will, in some modified form, become tomorrow's political imperatives. Environmental impact will no longer be the nuisance of generating required paperwork but will be an important underpinning for political decisions.

The point of these examples is not to kindle a naive faith that the future will take care of itself or that human skill in some way will solve all. On the contrary, the text makes it clear that some ecological over-shoot is already predetermined—that human suffering is already bound to increase because of our propensity to do too little, too late. Rather, the point of these examples is to demonstrate that we are slowly learning that all our behavior has consequences. In civilizing ourselves in our relations to our fellow beings, we learned that when we treat people badly, it may seem as though we get away with it, but in the long run, people will "get us back." Now we are learning that when we treat nature badly, it may seem as though we get away with it, but in the long run, nature also will get us back. Nature is beginning to get us back. The sun that has given us life is now giving us cancer. The rain that has given us drink is now bearing poisons.

The examples I have mentioned show that some are learning an ethic for our relationship to nature. In many quarters, the task of civilizing humans in relation to nature has begun. Perhaps the only outcome will be more rear-guard actions against the planetary deterioration fueled by the growth god. But despite the troubles about us, the new ethic is making progress daily, and I believe that all is not lost.

A. Stephen Boyan, Jr.
January 1992

Ecology and the Politics
of Scarcity Revisited

Introduction

The reality and gravity of the environmental crisis can no longer be denied. What *had* been a somewhat remote controversy among specialists and the committed few over the limits to growth (for example, Meadows et al. 1972 vs. Cole et al. 1973) was brought home forcefully to the common person during the energy crisis of 1973–1974. We began to understand in our bones that, whatever the causes of this particular crisis, there might not always be enough material or energy to support even current levels of consumption, much less the higher levels many aspire to. Again, in the summer of 1988, with drought baking the soil from east to west, with 100° heat in cities across the country, and with coastal beaches befouled by garbage, raw sewage, and medical wastes, Americans again had a sense of foreboding about what we're doing to our planet. No one, for example, seriously asserts any longer that ecological concern is a mere fad, which after a brief pirouette in the media limelight will cede its place to the newest crisis. Nor are there many who still maintain that those concerned with environmental issues are perpetrating a political hoax designed to siphon money and public support away from disadvantaged minority interests. Whatever the excesses of some who have espoused the cause of environmentalism, the crisis is real, and it challenges our institutions and values in a most profound way—more profoundly even than some of the most ardent environmentalists are willing to admit. This book is about that challenge.

Of course, both theory and common sense have always told us that infinite material growth and unlimited population increases on a finite

1

planet are impossible. But somehow, at least in this country, the problem has not seemed all that pressing, and for every expert who said it was, another could be found to say it was not. Rather, said the latter, it was a problem we could safely leave to our grandchildren—and a good thing *that* was, because bringing an end to material growth would demand that we make agonizing economic, social, and political choices. Thus, despite its undeniable reality, the environmental crisis remains controversial.

However, few really disagree with the ultimate implications. At least seven major studies concluded, in the 1970s, that population and material growth cannot continue forever on a finite planet and that conducting "business as usual" will fail to meet basic human needs (Corson, 1990, p. 15, citing Meadows et al. 1982). These studies called for a "steady-state" society characterized by frugality in the consumption of resources and by deliberate setting of limits to maintain the balance between humanity and nature. In the 1980s five more studies came to similar conclusions. Some, including the World Commission on Environment and Development, called for "sustainable development." Its report, *Our Common Future* (1987), was endorsed by the United Nations Environmental Program. Like the concept of a steady state, sustainable development requires stable population, high energy efficiency, a resource transition based on utilizing nature's "income" without depleting its "capital," and an economic and political transition. However, it allows for more growth than "steady state" advocates, and advocates of sustainable development are generally more optimistic about whether the necessary changes can be made.

Looking at the studies as a whole, the major controversy concerns the time scale. The so-called optimists believe that (1) the current situation in general is not quite so bad as the doomsayers make out, (2) continued scientific and technological ingenuity will keep the ecological wolf from the door indefinitely, and (3) there are many social negative-feedback mechanisms (such as the economic marketplace, the impact of media-propagated information on values, and the political process itself) that will promote gradual human adjustment to physical limits when and if it becomes necessary. Thus, to put it crudely, business as usual can continue for the foreseeable future, and to look beyond that is borrowing trouble. The so-called pessimists believe, to the contrary, that (1) the situation is more urgent than most are willing to admit, (2) limits on our scientific and technological ingenuity and on our ability to apply it to the problems confronting us are already discernible, and (3) the negative-feedback mechanisms on which the optimists would rely have already begun to fail. Thus time is short, and far-reaching action by the current generation is imperative to avoid overwhelming the earth's capacity to support us in dignity. Failure to act soon and effectively could lead us into the apocalyptic collapse—wars, plague, and famine—predicted by the early demographer and political economist Thomas Malthus, whose famous essay on the dangers of overpopulation (1798) was the first explicit statement of the environmental limits on human activity.

We count ourselves among the pessimists. Thus the first purpose of this book is to make clear the nature of the crisis and why it is pressing. The second and more important is to draw out in full measure the political, social, and economic implications of the crisis, for even some of the more prominent environmentalists appear not to have understood their import. At least in their public statements, they maintain that a sufficient quantity of reform—fairly radical reform, to be sure—would rescue us from our ecological predicament. To the extent that more radical changes are urged, the language used is often vague. Concrete political and social arrangements are rarely discussed, and really fundamental changes in our way of life or our constitutional arrangements are, one could gather, virtually unthinkable.

This book argues, to the contrary, that the external reality of ecological scarcity has cut the ground out from under our own political system, making merely reformist policies of ecological management all but useless. At best, reforms can postpone the inevitable for a few decades at the probable cost of increasing the severity of the eventual day of reckoning. In brief, liberal democracy as we know it—that is, our theory or "paradigm" of politics (see Box 1)—is doomed by ecological scarcity; we need a completely new political philosophy and set of political institutions. Moreover, it appears that the basic principles of modern industrial civilization are also incompatible with ecological scarcity and that the whole ideology of modernity growing out of the Enlightenment, especially such central tenets as individualism, may no longer be viable.

This conclusion may strike many as extreme. Despite overwhelming historical evidence for the rapid mortality of all merely political structures, we tend to think of the set of political values and institutions that we inherit, whether monarchy by divine right or liberal democracy, as eternal, immutable, and, above all, *right*. They are not. Political paradigms are, in fact, extraordinarily fragile creations. They

1

Paradigms and Political Theories

The political theories and institutions by which people govern themselves have a high degree of intellectual, emotional, moral, and practical coherence. A political society is characterized by definite institutional arrangements, both explicit and tacit standards for political behavior, and widely shared understandings on such issues as what makes political power legitimate and how constituted authority ought to treat members of society (especially how the norms of the political association are to be enforced). We can speak of this ensemble of institutions, practices, and beliefs as the political "paradigm" of the society (Wolin 1968, 1969).

Because political paradigms have the same kind of internal consistency as scientific theories, the process of political change is analogous to scientific change. Most scientific inquiry—so-called normal science—aims at routine puzzle solving under the conceptual umbrella of a fundamental scientific theory or paradigm (Kuhn 1970), like the famous DNA or double-helix model of gene replication in molecular biology. As long as such basic (and partly metaphysical) theories are successful in solving the puzzles thrown up by nature, allowing normal science to make apparent progress, all is well. However, once the puzzles can no longer be solved and disturbing anomalies resist all efforts to incorporate them into normal theory, then the community of scientists sharing this paradigm is ripe for revolution. Scientists begin to cast around outside the framework of the old paradigm for answers to the crucial anomalies; from this episode of "extraordinary" science emerges a new paradigm that overthrows the old, just as one regime replaces another in a political revolution.

Putting this in political terms, every society undergoes stresses. New classes, new economic relationships, and new religious or racial patterns emerge. But political associations are conservative. Retooling the paradigm is "unthinkable" and is likely to be resisted to the bitter

may of course persist long after the conditions that made them viable have vanished, but unhappy the people who live during the long period of decay or the swifter decline into revolutionary turmoil. However, our predicament is not hopeless. We can adapt ourselves to ecological scarcity and preserve most of what is worth preserving in our current political

end; before it considers radical change, a political society will exhaust all possibilities for reform by normal politics—that is, reform within its basic constitutional structure. If the puzzle is more or less solved by such reform, then the political paradigm carries on as before, slightly changed. An example is the extension of suffrage to the working class in England. By contrast, the same set of political "facts," rising political consciousness and assertiveness among non-elites, could not be accommodated by the political paradigm of Czarist Russia. Reform efforts were unsuccessful, and the new political facts therefore constituted an anomaly that led to political crisis and eventually revolution as the only solution.

There are, of course, significant differences in the way the scientific and political communities respond to anomaly. First, although it is rare for the ability of the leaders of the scientific community to be impugned, this is one of the most characteristic responses of the political association to impending crisis. As part of its effort to cope with change through normal politics—provided the paradigm allows for it, whether by election or routinized *coup d'état*—it will throw one set of leaders out in hopes that the next lot will solve the puzzle better. This is often successful. However, a genuine anomaly cannot be solved in this fashion, and once it is clear that changes in leadership will not produce a solution, the political community can no longer avoid confronting its crisis. Second, for a great variety of reasons, political communities are much more long-suffering than scientific communities. Unlike scientists, who make radical efforts to replace a suspect theory as soon as possible, members of a political community can tolerate gross anomalies—for example, the disparity between the theoretical and the actual status of American blacks during the century following the Civil War—for generations. However, when the "facts" that constitute anomaly will not go away and can no longer be ignored or borne, then revolutionary (but not necessarily violent) change becomes inescapable.

The crisis of ecological scarcity constitutes just such a gross and ineluctable political anomaly.

and civilizational order. But we must not delay. Events are pressing on us, and our options are being rapidly and sharply eroded: Already we face an array of potentially tragic choices. To see clearly how and why these choices are indeed forced on us, we must commence by examining basic concepts.

Ecology

This work is an ecological critique of American political institutions and their underlying philosophy.* What is this "ecology" upon which the argument rests? *Webster's Third New International Dictionary* gives three meanings:

1. A branch of science concerned with the interrelationship of organisms and their environments.
2. The totality or pattern of relations between organisms and their environment.
3. Human ecology, [that is,] a branch of sociology that studies the relationship between a human community and its environment; specifically, the study of the spatial and temporal interrelationships between humans and their economic, social, and political organization.

The first definition describes the work of the professional ecologist, who uses laboratory or field observation and experimentation to understand the laws governing the interactions of organisms with their living and nonliving environment. The second definition indicates a more general use of the word--for example, one can speak as readily of "the ecology of a peasant community" as of "the ecology of a mountain pine." Thus one would indeed expect human ecology to concern itself with the totality of the relationship between a human community and its environment. Unfortunately, as the second part of the third definition reveals, the purview of human ecology has in practice been rather limited, so that at present there exists no genuine science of human ecology in the full sense. It is such a science that environmentalists wish to create. Meanwhile, they are trying to broaden the meaning of the term *human ecology* to embrace the totality of people's relationships with their physical and living environment, and it is in this sense that we shall use the word *ecology,* except where the context makes it clear that the reference is to the science of ecology described in the first definition.

---

* Of course, the environmental crisis is global and civilizational in character, but beginning with the American case offers a number of advantages, such as familiarity to the reader and ready availability of information. Also, American society epitomizes the modern way of life in most respects, and if it can be shown that modernity will no longer work here, then modernity can be presumed to be in trouble elsewhere. Indeed, as we shall see in Chapter 7, very little modification is needed to make the argument apply to other developed countries and internationally.

There is etymological justification for this broad use of the term. The root of the prefix *eco* is the Greek word *oikos,* which means "household." Thus ecology logically is the science or study of the household of the human race in its totality. Interestingly enough, the original meaning of the term *economics,* a word also derived from *oikos,* was "a science or art of managing a house or household," whereas economy was "the management of a group, community, or establishment with a view to ensuring its maintenance or productiveness." Today economics has become "a social science that studies the production, distribution, and consumption of commodities." Thus, from the science of management of the human household in all its dimensions, economics has narrowed itself to an exclusive focus on the problems of a particular subsystem of ecology—the money economy—and treats this subsystem as though it were autonomous.

Of course, professional ecologists are often equally guilty of the narrowmindedness that comes from overspecialization. Indeed, economist and ecologist alike are victims of the almost vicious degree of specialization characteristic of the modern world. The science of human ecology now evolving is an effort to bridge the gap between specialties and make possible the rational management of the whole human household. This effort will require us to become, in effect, specialists in the general. Its spirit is well reflected in the redefinition of ecology offered by Paul Sears, dean of American professional ecologists:

> It may clear matters somewhat to modify the usual definition of ecology as the science of interrelation between life and environment. Actually, it is a way of approaching this vast field of experience by drawing upon the *best information available* from whatever source it may come [Sears 1971].

To be human ecologists of the kind Sears envisions, we must integrate the better part of all human knowledge—clearly an impossible goal. Yet, it must be attempted. We must hope that, although any individual work in human ecology must fall short of the ideal, there will emerge a body of works that complement each other and give us the global understanding we need to find our way out of the environmental crisis. What follows is the work of one human ecologist and one political scientist who happen to be concerned principally with the political aspects of managing the human household and who therefore have drawn on ecology, other natural sciences, engineering and technology, the social sciences, and even the humanities to construct a human–ecological critique of the American political economy.

Politics

Much of the ensuing argument will appear not to be about politics at all as it is usually defined by the man or woman in the street or the academic specialist in politics. The difficulty arises in large part from a narrow definition of *politics*. The word is used to mean either the winning and losing of elections and other political battles, for which a more appropriate word is *politicking,* or the organization and administration of units of government, thereby excluding economic and social phenomena as well as religion and many other matters that were once considered part and parcel of politics. This pinched understanding of politics reflects the impoverished, fragmented view of reality resulting from excessive academic specialization, which has created a gap like the one between economy and ecology, with similarly perverse consequences. The basic political problem is the survival of the community; two of the basic political tasks are the provision of food and other biological necessities and the establishment of conditions favorable for reproduction. Neither of these can be accomplished except in the human household provided by nature, and in this sense politics *must* rest on an ecological foundation.

The model for such a comprehensive view can be found in the political theories of the classical world. As any reader of Plato's *Republic* or Aristotle's *Politics* knows, for the ancient philosophers politics was all-inclusive: Religion, poetry, education, and marriage were just as much political matters as war, the regulation of property, and the distribution of administrative office. Aristotle's famous description of the human as the "political animal" graphically conveys our uniqueness in being responsible for organizing our own communal life. Aristotle said that men without politics would be either gods or beasts. Beasts are ruled by instinct and natural necessity; their "government" is genetically given. Likewise, the spontaneous, infallible right action of the gods is ordained as part of Creation; free of all mortal necessities, gods need no artificial government. Only humans—half beast, half god—struggle to govern themselves with no certain guide and no assurance of success. For Aristotle, as well as for Plato and other major political theorists, "politics" concerns this struggle to live in community on the earth, and it therefore extends to many things besides government narrowly defined. Aristotle asks how this political animal can design and create institutions that will assure the survival of the city of man and some measure of the good life within it. It is just such a broad conception of politics that informs this book: Is the way we organize our communal life and rule ourselves compatible with ecological imperatives and other natural laws?

Ecology is about to engulf economics and politics in that how we run our lives will be increasingly determined by ecological imperatives. For example, one definition of the term *politics* that is prevalent among academic specialists is "the authoritative allocation of values." But as Woodhouse (1972) points out, what happens in the beds and on the sleeping mats of the world should therefore be considered politics, for no single thing is likely to determine the general world allocation of values in the years to come more "authoritatively" than the reproductive behavior of millions of anonymous human beings.

Thus, whether we like it or not, embracing the larger conception of politics that characterized early political theorists is becoming virtually inevitable, and this alone is sufficient justification for carefully delineating the nature of the laws of human ecology that our politics must henceforth reflect. Moreover, practically speaking, the cogency of many of the arguments in the second half of this book depends largely on the existence of the kinds of ecological imperatives documented in the first half. Thus, expanding our conception of politics to include these ecological imperatives is an important first step toward coming to terms with ecological scarcity.

Scarcity

The habitual condition of civilized people is one of scarcity. Goods have never been available in such abundance as to exhaust people's wants; more often than not, even their basic needs have gone unmet. The existence of scarcity has momentous consequences, of which one of the most important is the utter inevitability of politics. The philosopher David Hume argued that if all goods were free, as air and water are, any person could get as much as he or she wanted without harming others. People would thus willingly share the earth's goods in common "as [do] man and wife." However, without a common abundance of goods, "selfishness and the confined generosity of man, along with the scanty provision nature has made for his wants," inevitably produce conflict; thus a system of justice that will restrain and regulate the human passions is a universal necessity (Hume 1739, III-2-11). The institution of government, whether it takes the form of primitive tabu or parliamentary democracy, therefore has its origins in the necessity to distribute scarce resources in an orderly fashion. It follows that assumptions about scarcity are absolutely central to any economic and political doctrine and that the relative scarcity or abundance of goods has a substantial and direct impact on the character of political, social, and economic institutions.

This understanding, however, has been undermined over the past three centuries, during which abnormal abundance has shaped all our attitudes and institutions. The philosophes of the Enlightenment, dazzled by the rapid progress of science and technology and the beginnings of the Industrial Revolution, envisioned the elevation of the common person to the economic nobility as the frontiers of scarcity were gradually pushed back. The bonanza of the New World and other founts of virgin resources, the take-off and rapid-growth stages of science and technology, the availability of "free" ecological resources such as air and water to absorb the waste products of industrial activities, and other, lesser factors allowed this process to unfold with apparent inexorability. Karl Marx, who documented and criticized the horrors and inhumanities of the Industrial Revolution, nevertheless celebrated its coming because the enormous productive forces unleashed by the bourgeois overthrow of feudalism could be used to abolish scarcity. With scarcity abolished, poverty, inequality, injustice, and all the other flowers of evil rooted in scarcity would simply wither away; not even a state would be needed, he thought, because everything would be a free good, like Hume's air and water, that humans could share together without conflict.

Marx's utopian assessment of the possibilities of material growth was shared or came to be shared by almost all in the West, though in a less extreme form and with considerable difference of opinion on how the march to utopia should be organized. For example, the works of the political philosopher John Locke and of the economist Adam Smith, the two men who gave bourgeois political economy its fundamental direction, are shot through with the assumption that there is always going to be more: more land in the colonies, more wealth to be dug from the ground, and so on. Thus virtually all the philosophies, values, and institutions typical of modern society are the luxuriant fruit of an era of apparently endless abundance. The return of scarcity in any guise therefore represents a serious challenge to the modern way of life.

Worse, scarcity appears to be returning in a new and more daunting form that we call *ecological scarcity*. Instead of simple Malthusian overpopulation and famine (as though that weren't enough), we must now also worry about shortages of the vast array of energy and mineral resources necessary to keep the engines of industrial production running, about pollution and other limits of tolerance in natural systems, about such physical constraints as the laws of thermodynamics, about complex problems of planning and administration, and about a host of other factors Malthus never dreamed of. Ecological scarcity is thus an ensemble of separate but interacting limits and constraints on human action, and it appears to pose problems far surpassing those presented to our ancestors by scarcity in its classical form.

The nature and difficulty of the challenge we confront are apparent in the ironic fact that the very things Hume used to illustrate the state of infinite abundance—air and water—have become scarce goods that must be allocated by political decisions. The profundity of the challenge is also apparent in the economist Kenneth Boulding's use of the concept "spaceman economy" to describe the consequences of ecological scarcity. According to Boulding, because our overpopulated globe is coming increasingly to resemble a spaceship of finite dimensions, with neither mines nor sewers, our welfare depends not upon increasing the rate of consumption or the number of consumers, both of which are potentially fatal, but on the extent to which we can wring from minimum resources the maximum richness and amenity for a reasonable population. A good, perhaps even an affluent, life is possible, but "it will have to be combined with a curious parsimony"; in fact, "far from scarcity disappearing, it will be the most dominant aspect of the society; every grain of sand will have to be treasured, and the waste and profligacy of our own day will seem so horrible that our descendants will hardly be able to bear to think about us" (Boulding 1966). There is, of course, no historical precedent for such a society. What is ultimately required by the crisis of ecological scarcity is the invention of a new mode of civilization, for nothing less seems likely to meet the challenge.

Political Theory

It is not the aim of this book to prescribe the form of post-industrial civilization but rather to document the existence of ecological scarcity, show how it will come to dominate our political life, and then make plain the inability of our current political culture and machinery to cope with its challenges. From this analysis, a range of possible answers to the crisis will emerge. For example, if individualism is shown to be problematic in an era of ecological scarcity, then the answer must lie somewhere toward the communal end of the political spectrum. Also, certain general dilemmas that confront us—for example, the political price attached to continued technological growth will be made explicit.

In brief, then, this work is a prologue to a political theory of the steady state. Yet although it stops well short of formulating a genuine political theory of the steady state, it is directly concerned with the great issues that have dominated traditional thought about politics. Our essential purpose is to show how the perennial, but dormant, questions of political philosophy have been revived by ecological scarcity. We shall see, for example, that the political problems related to the task of environmental management have to do primarily with the ends of political association, rather than with the political means needed to achieve agreed-upon goals. The questions that arise from the ensuing analysis are essentially value questions: What is the common interest? Under current conditions, is liberal democracy a suitable and desirable vehicle for achieving it? What, indeed, is the good life for men and women? In other words, we confront the same kinds of questions that Aristotle, in common with the other great theorists of politics, asks. We are obliged by the environmental crisis to enlarge our conception of politics to its classical dimensions. To use a famous capsule description of politics, the questions about *who gets what, when, how, and why* must be reexamined and answered anew by our generation. Our goal in this book is to set the agenda for such a philosophical reexamination of our politics.

However, we do not approach this task as does a traditional political philosopher. Past theorists seeking guidance for human action have grounded their ideas on revelation or induction. Either, like Plato, they have appealed to some *a priori* metaphysical principle from which the shape of the desirable political order can be deduced, or, like Aristotle, they have examined human behavior over time to see whether certain kinds of political institutions are more effective than others in producing a happy and virtuous people. Of course, many theorists have mixed these approaches, and some have introduced other considerations. In almost all cases, however, humanity's linkage to nature has counted for little. By contrast, like Malthus, we start with humanity's dependence on nature and the basic human problems of biological survival.

To be sure, most political and social thinkers have acknowledged humanity's ultimate dependence on nature, and a few have devoted some attention to the specific effects that environmental constraints have had on people. In Book One of *The Politics,* Aristotle discusses scarcity and other ecological limits, implying that because of them slavery may be necessary for civilized life. Plato in Book Two of *The Republic* and Rousseau in *The Second Discourse* also display a subtle awareness of the impact the evolving process of getting one's daily bread can have on social institutions. Nevertheless, with the major exception of Malthus, political and social theorists have tended to take the biological existence of men and women as given. This is no longer possible.

Nor is it possible any longer to ignore humanity's impact on the environment. Of course, concern about this impact and the consequent damage to human welfare also has a long history (Glacken 1956). Over two thousand years ago, Plato in Greece and Mencius in China both worried about the destruction of habitat caused by overgrazing and deforestation. The early Christian writer Tertullian called wars, plagues, famines, and earthquakes blessings because they "serve to prune away the luxuriant growth of the human race" (Hardin 1969, p. 18), and Aristotle found the poverty caused by population growth to be the parent of revolution and crime: "If no restriction is imposed on the rate of reproduction...poverty is the inevitable result; and poverty produces, in its turn, civic dissension and wrong doing" (Barker 1952, p. 59). Clearly, certain of the environmental problems we face today have been with human beings since the very beginning of civilization.

The character of these problems, however, has changed markedly over the centuries. In ancient times, humanity's impact on the environment was local; by the eighteenth century, worldwide effects were becoming apparent; writers of the nineteenth century remarked on the extent of this impact and its cumulative effects; and observers in this century have focused on the acceleration of change. Accumulating quantitative impact has thus brought about a qualitative difference in our relation to the physical world: We are now the prime agent of change in the biosphere and are capable of destroying the environment that supports us. The radically different conditions prevailing today virtually force us to be ecological theorists, grounding our analysis on the basic problems of human survival on a finite and vulnerable planet endowed with limited resources.

A second contrast between this work and traditional political theory is that, again like Malthus, our effort throughout is to identify the critical limits to and constraints on human action. We wish to discover what is possible—or, alternatively, what we are forced to do—rather than what is desirable. In other words, values come last in this supposedly philosophical analysis.* This is not because we disdain the eternal questions of value, but because a value-neutral approach is called for on very practical grounds.

---

* As will be explained shortly, values will be crucial to creating a steady-state society, but values that are widely accepted today as immutable may have to change so that we can either avert or endure harmful changes in the natural world.

First of all, philosophical, ethical, and spiritual arguments seem to appeal only to the converted. Hard-headed scientists, technologists, bureaucrats, and businesspeople—the men and women who make the basic decisions that shape our futures—do not often pay much attention to such arguments. If one is to argue constructively with the people who incarnate our cultural and political norms, one must argue the case in their own terms. This requires that one adopt a fundamentally empirical and scientific or agnostic approach, putting aside the question of values, at least temporarily, to find instead what is possible given the natural laws that govern our planet.

Second, one of the most important reasons for focusing on limits and constraints is the nature of our predicament. Although the human species has never enjoyed total freedom of choice, at some times and places a relative abundance of everything needed for the maintenance of life and the construction of culture has made the latitude of choice correspondingly large. By contrast, people cast adrift in a lifeboat with short supplies, say, or the trapped inhabitants of a besieged town, face many painful dilemmas; if they wish to survive, they must impose stringent limits on their behavior. Similarly, by its very nature a spaceship imposes a certain type of social design on those embarked. As our circumstances come to resemble those of space travelers, we may expect knowledge of this social design to tell us a great deal about what we must do—in other words, to plot the relatively narrow range within which the values and the moral requirements can lie. Nature's dictates become our policies if we wish to survive.

Nevertheless, questions of value are inescapable. There being no agreed-upon prime value—not even survival—that dominates all others, solving every problem of public and private morality neccessitates trade-offs between desired goods. To illustrate briefly, even if ecologists could predict with absolute certainty that a continuation of current trends would produce massive death and other catastrophes by A.D. 2000, people might still decide, in a spirit of profligate fatalism, to doom posterity rather than forgo current enjoyment. Moreover, we shall not face totally forced choices. There are a number of possible solutions to the lifeboat problem and an even larger number to the spaceship problem, so the outcome will be the result of a complex interplay between limits and constraints, our present and future capacity to evade or manipulate limits, and our values. In brief, science can only define the limits to political and social vision; it cannot prescribe the contents. Where science ends, wisdom necessarily begins, and we hope this book will help prepare the reader for making decisions and judgments at that point.

The Steady State

Many of those who examine our ecological predicament tend to agree that we are headed toward a *steady-state society.** Although the concept must be refined further, a steady-state society is one that has achieved a basic, long-term balance between the demands of a population and the environment that supplies its wants. Implicit in this definition are the preservation of a healthy biosphere, the careful husbanding of resources, self-imposed limitations on consumption, long-term goals to guide short-term choices, and a general attitude of trusteeship toward future generations. Useful analogies include living off annual income instead of eating up capital and managing the earth as we would a perpetual-yield forest, so that it continues to thrive and replenish itself "for as long as the grass shall grow and the sun shall shine."

The rest of the book will help make these abstractions somewhat more concrete. For one thing, from an analysis of our current errors we can infer at least some aspects of the steady-state society, even though a full and systematic description is beyond our present ability. However, it is important to understand from the outset that the exact nature of the balance at any time depends on technological capacities and social choice, and as choices and capacities change, organic growth can occur. For this reason, the steady state is by no means a state of stagnation; it is a dynamic equilibrium affording ample scope for continued artistic, intellectual, moral, scientific, and spiritual growth.

Indeed, without substantial human growth in every dimension, the steady-state society can never be realized. Devising an ecological technology or a new set of political institutions for the steady state is the lesser part of the problem, for its core is ethical, moral, and spiritual. This idea is well expressed in a metaphor suggested by George Perkins Marsh, a major figure in the history of the American conservation movement and the greatest pioneer of human ecology after Malthus. Marsh's major work (1864) depicts the human race as a heedless cottager tearing down his earthly abode for kindling in order to keep a lively but evanescent fire blazing in the hearth. By inference, the men and women of a steady-state

---

* *Stationary-state society* and *equilibrium society* are alternative terms. The former is the traditional economic label for a state of zero growth. Because it tends to imply a condition of rigor mortis, it is not entirely suitable as a description of what is in store for us. Some believe that even *steady-state* is too static and prefer *equilibrium* or a *sustainable society.* Advocates of a sustainable society believe that "growth" can continue to occur but must conform to natural limits. Rightly understood, however, *steady-state* is appropriate and we shall normally use it.

society would take excellent care to see that their earthly household was preserved intact, knowing that posterity (if not they themselves) would have use for it. Through frugality and good stewardship, they would seek ways to be warm that would nevertheless allow them to pass the cottage—improved if possible, but at all costs undamaged—down to their children. The ultimate goal, then, is as much an ethical ideal (the good stewardship enjoined by the Biblical parable) as a concrete set of political, economic, and social arrangements.

We shall return to this point at the end of our analysis. Meanwhile, let us in Part I explore ecology and the ecological limits and constraints now beginning to press down on us. Next, in Part II, we shall go on to examine the political challenges. Perhaps then we shall understand better why we need not only a new theory of political economy but probably a new theology as well.

I

Ecological Scarcity and the Limits to Growth

1

The Science of Ecology

The Synthetic Science

Ecology's synthetic nature distinguishes it from the more reductionist branches of science. On the grandest scale, ecologists try to understand the process of life in the context of the chemical, geological, and meteorological environment by assembling the isolated knowledge of specialists into a single, ordered system. Indeed, the subject matter of ecology is so large that simple experimentation is often not feasible. Hence ecologists often conduct observational studies on a functional unit called the ecosystem (the community of organisms living in a specified locale, along with the nonbiological factors in the environment—air, water, rock, and so on—that support them, as well as the ensemble of interactions among all these components).

Understanding the process of life requires seeing ecosystems in dynamic and historical terms. Contemporary ecosystems have developed from particular origins and are undergoing both short-term changes and long-term evolution. Ernest Haeckel, one of the founders of ecology, defined this science as "the body of knowledge concerning the economy of nature," and Charles Elton described his ecological work as "scientific natural history [concerned with] the sociology and economics of animals" (Kormondy 1969, pp. vi–ix). In sum, systems ecologists try to reveal the general principles that govern the operation of the whole system called the biosphere, the part of the planetary system that contains or influences life.

From this description, it should be obvious that because humanity inhabits the biosphere, ecology must also be concerned with human activities. Today there can be no valid distinction between ecology and human ecology. Nevertheless, ecologists have until recently concerned themselves with the economy of nature in pristine environments relatively undefiled by human intervention. In this way they have succeeded in discovering some of the general principles that govern the economy of nature. Using their findings, we shall give a synoptic overview of the basic principles of ecology. However, our discussion is framed in terms of human rather than scientific ecology. That is, those things that have special relevance to human action are emphasized in this description of general ecological principles and of the basic structure of the natural life-support system of our planetary spaceship.

Interdependence and Emergent Properties

A fundamental principle of ecology is that an ecosystem is more than the sum of its individual parts; that is, just as the properties of water are not predictable from the individual properties of oxygen and hydrogen, so the emergent properties of ecosystems are not predictable solely from the properties of the living entities and nonliving matter of which they are composed. Each ecosystem on Earth must be understood in terms of the *interactions* of its components. This principle requires ecology to be a synthetic and process-oriented science.

Flowing immediately from this first principle is the fact of interdependence. Every phenomenon within any ecosystem one chooses to examine can be shown to be related to every other phenomenon within it. Moreover, there are rarely any simple relationships; every effect is also a cause in the web of natural interdependency. Of course, not all relationships are equally important or equally sensitive, and many are indirect. Certain kinds of important interrelationship are intuitively obvious even to the casual observer. We all know that there are predators and prey, that microbes can cause disease, and that worms inherit the bodies of those who are buried. However, the casual observer is unaware of the numerous other interrelationships in nature, many of critical importance. The number of living components alone and the variety and complexity of their couplings are bewildering. For example, no one has ever made a complete census of all the organisms that inhabit so simple an ecosystem as a pond.

The fact of interrelationship is so pervasive that it bridges the classic dichotomy between the living and the nonliving components of an ecosystem: "The living and nonliving parts of ecosystems are so interwoven into the fabric of nature that it is difficult to separate them" (E. P. Odum 1971, p. 10). The evolution of animals, for example, did not take place in a static physical environment to which life then adapted. Rather, early life modified the physical environment, gradually transforming an extremely inhospitable environment into one suitable for the organisms we know today. Both air and soil are the products of living systems, and their maintenance depends on the work of minute organisms. As a result of their ordinary metabolic processes, tiny plants have respired the oxygen in our atmosphere and created soils out of rock and dead organic material. If we humans leave them relatively undisturbed, they will continue to supply us with the breath of life, keep our soils viable, and purify our waters.

The fact of interdependence makes the concept of community one of the most important in ecology. Diverse organisms live together and engage in complex reciprocal interactions. The implications of this fact are profound, for people are inevitably major players in the communities they occupy. The principle also shows that many of our practices are misguided. For example, ecologists have often found that the most effective and safest way of controlling a so-called pest is not to attack the pest organism directly but to modify the community so that the pest is naturally controlled within the network of interdependencies that constitute the community. This gives rise to the catch phrases "Everything is connected to everything else" and "You can never do just one thing." Let us examine some specific examples to see why human intervention often produces unexpected and unintended effects on ecosystems.

Unintended Consequences, the Price of Intervention

Because every effect is also a cause, changing one factor in a well-adapted and smoothly functioning ecosystem is likely to unleash a chain of second-, third-, and fourth-order consequences. For example, when an organism from one ecosystem is transferred (either unintentionally or by design) to another, it is likely at first to "run amok," for the new ecosystem has no history of dealing with such an organism and therefore no mechanism to control its spread. It becomes an "instant pathogen," like the measles that decimated Eskimos and South Sea Islanders following their first contacts with Western civilization. Examples of introduced pests—the Japanese beetle, Dutch elm blight, the gypsy moth, and the rabbit in Australia—are also well known.

Just as adding organisms to ecosystems spells danger, subtracting them may cause the whole web of interdependencies in the ecosystem to unravel. Many different types of human activities (habitat destruction, unregulated hunting and fishing, fire, pollution, and others that are less obvious, such as noise) can have the effect of eliminating a key species in an ecosystem, which in turn causes related components to decline or collapse. Insecticides, especially DDT, provide the classic illustrations of what happens when we use drastic measures to accomplish what appear to be simple goals—when we try to do "just one thing." In a remote jungle village in Borneo, health workers sprayed the walls of the villagers' huts with DDT to control the mosquitoes that spread malaria. And control the mosquitoes it did. However, the lizards that patrolled the walls of the huts inevitably absorbed large quantities of DDT (both from coming in contact with the sprayed walls and from eating poisoned prey), and they died. This had the unfortunate effects of killing the village cats, which ate the moribund and poisonous lizards, and leaving the straw-loving caterpillars (hitherto kept in check by the lizards) that inhabited the thatched roof free to gorge without limit. The end result was a plague of rats, the population of which exploded in the near absence of cats, and destruction of the roofs of the villagers' huts (Anon. 1968).

This might be simply an entertaining story were it not, in effect, a model of what pesticides and other chemicals are doing to the global life-support system. Like DDT, many of these chemicals are poisonous to a broad spectrum of life forms. And like DDT, they are persistent. Because they are synthetic rather than natural compounds, no organisms have evolved an ability to metabolize them; hence they accumulate to dangerous levels in ecosystems. In addition, as with DDT, the phenomenon of "biological magnification" concentrates poisonous substances approximately tenfold with each step up the food chain, because each grazer or predator must eat many times its own weight in smaller organisms. The release of even modest quantities of chemicals can therefore become lethal to the carnivores (eagles and pelicans, for example) at the top of the food chain. We would be wise to remember that humans also feed at the top of our food chains.

People who protest that the extinction of a few carnivore species is a small price to pay for protecting our crops against the ravages of pests are missing the point. The Bermuda petrel, say, could disappear, and the richness of the biosphere on a worldwide scale would hardly be diminished. Indeed, the extinction of many top-carnivore species as a result of chemical poisoning would not jeopardize our survival directly. However, their disappearance is an indicator of an ecological sickness, just as sugar in the urine, though not dangerous in itself, indicates diabetes.

The long-term consequences of such ecological illness are potentially grave. It has been experimentally shown, for example, that many synthetic chemicals in widespread use affect the species composition of plankton in the ocean, a change that might have serious implications not only for the structure of oceanic food chains (of which plankton are the base) but also for the role of plankton in the ocean's governance of the cycles that rule the biosphere, such as the reduction of atmospheric carbon dioxide. One of the most worrisome aspects of this kind of problem is that, owing to the inherent lag in biological systems, the peak concentrations, and therefore the full impact, of chemicals that are released is not experienced until some time in the future. The possibility therefore exists that we have already done irreparable damage but that we will find out only when it is too late.

Moreover, trying to make simple modifications to ecosystems frequently turns out to be futile as well as self-destructive. Chemical insecticides again provide a model. Plant-eating insects have a long history of adapting to chemical warfare, because the principal defense of plants against being eaten is to make their vulnerable parts unpalatable or poisonous. Thus, although an application of insecticide may kill all but a few of a given pest population, those hardy few live to reproduce, and they pass to their numerous progeny the resistant genes that enabled them to survive the poison. Before too many generations have passed, virtually an entire population of pests has become resistant to the chemical being used, and the war must be escalated with vastly increased dosages or new chemical weapons.[*] In the long term, this is a no-win strategy, for we cannot expect to stay ahead indefinitely in chemical war with insects.[†]

---

[*] The same kind of problem is being encountered in our war against microbes. Bacterial resistance to the common antibiotics is increasing rapidly, to the alarm of the World Health Organization (Dixon 1974).

[†] By 1980 more than 400 insects, ticks, and mites had developed pesticide resistance, along with more than 100 bacteria and viruses. Chemical manufacturers put more than 1000 new pesticides on the market each year in an attempt to overcome pest resistance. Since the 1940s, crop losses to insects have doubled even as farmers increased their use of pesticides tenfold (Mott 1988, pp.20-29). According to the National Academy of Sciences, "alternative agriculture," defined as farming without chemicals or with "low-input" pesticide applications, has become more productive than chemical farming in some cases because of the expense of applying more drastic pesticide applications (1989, pp.4-6) (see Chapter 2). Nevertheless, substantial organizational and government obstacles prevent widespread adoption of these techniques.

Indeed, it is likely to leave us worse off, because the insecticide may destroy many of the natural controls on the pest's population. Insectivorous birds and insects, which ordinarily exert control over the pest, are killed off. (Predators, because they have much smaller populations and reproduce more slowly than plant-eating insects, are much less likely to evolve resistance to the pesticides.) Thus, even if they were not ecologically destructive, single-purpose technological solutions would probably not succeed in a natural environment characterized by an all-pervasive interdependence.

The human ecological problem is that all the activities we call development tend to involve relatively single-purpose additions to or subtractions from natural ecosystems. Cases abound in which the negative unintended effects of intervention outweigh the intended primary effects (Farvar and Milton 1968). Consider the dams and irrigation projects that have spread schistosomiasis (bilharzia, an extremely debilitating parasitic disease) or that have led to loss of productive land through salinization or erosion (see Chapter 2). In order to understand more clearly how "everything is connected to everything else" and why human action can therefore boomerang ecologically, let us examine the economy of nature more closely.

Homeostatic Stability

A major characteristic of undisturbed natural systems is that they are in a state of homeostasis—that is, they include mechanisms of self-maintenance and self-regulation that, in some cases, produce a relatively stable balance or dynamic equilibrium. Even certain kinds of disruption, such as fire and flood, that one might consider destructive of natural ecosystems, actually contribute to the renewal of those systems. Fires and floods, after all, have been around for so long that plant and animal communities have become highly adapted to these stresses. Thus, apart from local disruption due to volcanism and earthquakes, the status of natural ecosystems is threatened only by long-term changes in climate and geology and by the actions of humans.

The interactions of the biological components tend to prevent any one species from changing the character of the ecosystem. If the population of one species starts to grow, then the population that preys on it responds by growing also. Even top predators, who need not fear being eaten themselves, are subject to parasites and disease as well as other "density-dependent" causes of mortality. And if for some reason too many predators are alive, the number of prey is soon reduced to the point where some predators starve and a balance is restored.

Ecologists conceive of the biosphere as an open system in a steady state that is driven by the fairly constant input of energy from the sun and in which a finite stock of materials is constantly recycled. What characterizes this steady state and how is it maintained?

The Life Cycle

Sunlight is the energy source for photosynthesis in plants ranging from microscopic phytoplankton to giant trees. In photosynthesis, carbon dioxide, water, and other inorganic chemicals are combined to create the carbohydrates that plants require for their own metabolism and growth and upon which animals feed. At the same time, plants also respire the oxygen that animals use to metabolize their food. Because plants take nutrients in raw inorganic form from the environment and convert them to the organic form required to support the higher levels on the food chain, the plants are called *producers*. The producers are consumed by organisms at the next level of the food chain, the herbivores, who are called the *primary consumers*. They in turn are eaten by the carnivores, the *secondary consumers*. Except for the carbon dioxide respired by the consumers, which is available for the producers, the flow of materials so far described is unidirectional. If it continued, the transfer of energy would not be cyclical. Soon producers would exhaust their supplies of chemical nutrients and cease producing; the rest of the food chain would then collapse. But there is another major group of organisms whose role is to take all organic debris—dead producers and consumers, feces, and detritus such as fallen leaves—and break it down into its inorganic components. These components can then be reused by the producers, and the whole system stays in operation. These organisms—bacteria, fungi, and insects—are called *decomposers*.

In reality, of course, things are much more complex than this simple schema conveys. In nature food webs usually replace simple food chains, and there are even some plants that eat insects. Nevertheless, the schema conveys the essence of the major cycle of life.

It is useful to note that the absence of complexity in human agricultural fields is responsible for pest problems. A monoculture of corn or any other crop is a highly simplified ecosystem containing large numbers of one highly succulent species of plant. Responding to the banquet spread before them, insects multiply rapidly and become pests. When farmers attempt to kill the pests, they destroy the natural controls on the pest population, thus simplifying the system still further and exacerbating the problem.

Biogeochemical Cycles

In addition to the basic life cycle, there are many material cycles of critical importance to the operation of the biosphere. In the well-known water cycle, water evaporates from the oceans and other bodies of water to be transported by the atmosphere (via energy from the sun) over the land, where it precipitates to be used by plants and animals and evaporate once again, or run off eventually to the sea, thus closing the cycle. The nitrogen cycle, on the other hand, is a major biogeochemical cycle little appreciated by the nonspecialist. Plants need nitrogen compounds to grow, but the nitrogen gas in the atmosphere is biologically inert and cannot be used directly by the plants. Various bacteria and algae, some living free in the soil and others associated with the roots or leaves of plants, are capable of "fixing" the nitrogen from the atmosphere into compounds that plants can absorb and use. Other microorganisms decompose fallen organic material and break down its nitrogen into a form suitable for plant nutrition. The plant may then be eaten by an animal; the nitrogen is either excreted or returned to the soil when the animal dies, thus closing the cycle.

All such natural cycles are of the same character: The materials necessary to maintain the processes of life are used and then recycled such that they can be used again. Unlike the energy in sunlight, these materials have no cosmic inputs; the biosphere works with the finite quantity of each element that is present on the planet. In contrast to human activities, a well-functioning ecosystem reuses materials with great efficiency. Indeed, there is almost no such thing as waste in nature, for by and large one organism's waste is another's food. Without this cycling of materials, organisms would long ago have drowned in their own wastes.

Many of the natural cycles interlock in critical ways. For example, the soil is the home of the decomposers that provide terrestrial plants not only with nitrogen but also with all other necessary nutrients. Damage to the soil disrupts many different cycles. In an interdependent ecosystem, some components are more sensitive to disruption than others because their role is large or critical, and comparatively minor disruptions in such components can be amplified into major consequences for the system as a whole. In this connection, it is impossible to overemphasize the utter dependence of all life on the tiny creatures that play essential roles in all natural cycles. Just as the nitrifying and denitrifying bacteria play a critical role in the nitrogen cycle, so the phytoplankton in the oceans are indispensable for the homeostatic maintenance of the oxygen and carbon cycles. The survival of these critical components and the integrity of the basic cycles are crucial to long-term human welfare.

The instructive contrast with human practices should be apparent. In earlier eras, human detritus was principally organic, and it was, like humanity itself, scattered thinly over the surface of the earth. More recently, humans have come to generate large quantities of nonorganic waste, and their organic wastes are highly concentrated in limited areas. Local ecosystems cannot absorb all that is asked of them. Furthermore, organic waste no longer tends to be deposited near where the raw materials were obtained, for some regions are agricultural mines that produce food for people living far away, whose waste is not returned to the soil but dumped into the ocean. Farmers try to make up for this loss by applying artificial fertilizer, which not only requires a large amount of energy to manufacture but also creates water pollution when it runs off into lakes and rivers, becoming yet another waste out of place in natural cycles. Despite the prevalence of cycles in the biosphere, not all natural processes are perfectly cyclical. Although, in general, the biosphere is characterized by closed cycling in a steady state, the cycle of materials is not completely closed except over the longest spans of time. Indeed, some processes are irreversibly linear (erosion is an obvious example), and ecosystems do not reuse chemicals with total efficiency. Thus materials eventually find their way into a "sink," typically the abyss of the sea. After many years, the process of mountain building may return these materials to the active part of the biosphere and they may thus reenter the cycle, but in the short run their path is unidirectional. Similarly, fossil fuels are the residue of nonrecycled organic material deposited during past geological ages. They are a stored source of solar or photosynthetic energy that we are using up and that can be replenished, if at all, only in some future geological era.

The Limits of Ecosystems

Why are human-made wastes a problem for ecosystems? After all, if ecosystems are self-maintaining and self-regulating, might one not infer that they can repair themselves after experiencing major environmental stresses? Unfortunately, this is true only within certain natural limits. Over the course of evolution, species become adapted to conditions their ancestors faced in the past, and by analogy, ecosystems can survive certain stresses to which they are adapted. Thus ecosystems can cope with fire, but not with large doses of radiation, synthetic chemicals (for which no decomposer exists), and the other novel stresses with which human activity taxes them. Given enough time, organisms might evolve to meet the new conditions, and ecosystems would restructure themselves until they were once again in a state of dynamic homeostasis. Unfortunately, these processes require time on a geological scale, and human-made stresses come too fast for the processes of evolution to keep pace. Some wastes (such as sewage) are biodegradable, and some human-made stresses (such as heat) are natural in the sense that they are not completely new to nature; yet these too can cause environmental disruption if the rate at which they enter an ecosystem is greater than the rate at which they can be absorbed. Under such conditions, the system is driven away from homeostatic stability toward disruption.

This process is illustrated by the phenomenon called eutrophication, the overenrichment of an aquatic ecosystem by an excess of nutrients. In a freshwater lake, algae (the producers) grow and are grazed on by herbivorous invertebrates and fish (the primary consumers), who in turn are eaten by carnivores (the secondary consumers). The fish release waste and die, providing food for the decomposers, who generate the inorganic products the algae need as food. The system is essentially closed and in balance. Furthermore, it is capable of handling normal environmental stresses. In response to the light of the summer sun, the algae multiply rapidly, temporarily depleting the supply of inorganic nutrients. But the fish react by eating more algae and increasing their own population, which creates additional organic waste that the decomposers turn into nutrients for the algae. A new level of seasonal stability is attained. Having had to adapt to this stress annually for eons, the ecosystem has developed an appropriate self-regulating response.

However, when a lake is artificially enriched with inorganic phosphorus compounds (previously common in household detergents), the algae, which are ordinarily limited by the low level of phosphorus in the lake, multiply very rapidly. If the additional level of nutrients is sufficient, the algae can become so dense that the water is impenetrable. The fish cannot respond (reproductively) fast enough to contain the algal population explosion. As algae complete their life cycle and start to die in large numbers, the amount of dead organic matter becomes so great that the bacterial decomposers begin to deplete the oxygen content of the water. The lowered oxygen availability results in the death of many fish, further adding to the decomposers' load. But the algae continue to grow because they no longer depend on the decomposers to supply them with nutrients and because the primary herbivores (the fish) have died off. And so on, in a vicious circle. Sudden enrichment has destroyed the negative-feedback mechanism governing algal growth, resulting in excessive production that the rest of the ecosystem cannot tolerate, given the existence of other limits such as oxygen content.

Note that two kinds of intrinsic limits operate in this case. First, there is the physical limit set by the oxygen supply when the nutrients that ordinarily limit algal growth are available in excess. Second, the system is limited by the biological lag built into its self-regulating mechanisms. If fish could reproduce as rapidly as algae (and if other factors, such as oxygen, needed by the fish as well as the decomposers, were not limiting), then any algal bloom could be contained. But because ecosystems are adapted to much more modest levels of stress than humans are capable of inflicting, they typically cannot respond to sudden or massive stress rapidly enough to prevent the system from being overwhelmed.

Our understanding of biogeochemical cycles underscores the ecological axiom that "everything must go somewhere." The law of the conservation of matter states that matter can be transformed but cannot be created or destroyed. The self-maintaining cycles circulate materials throughout the biosphere as long as the sun continues to provide the energy required. Waste is essentially a human phenomenon; it obstructs or destroys the natural cycles unless it is introduced into ecosystems in disposable form, in acceptable amounts, and at manageable rates, all of which depend on the natural limits of the ecosystem. Merely dumping wastes does not solve our waste-disposal problem in any but the most temporary fashion, for the consequence of pollution is a decaying or dying ecosystem. The biosphere is in effect our biological capital, from which we draw income in the form of food, water, and breath (to mention only the most fundamental requisites of life), so human health and survival are directly related to the health of the biosphere.

The Price of Intervention

Human intervention in nature is inevitable because we, like all other organisms, have an impact on the ecosystems in which we live. The ancestral humans were hunter–gatherers without fire or even the crudest technology. With the subsequent development of technology, our impact on the environment was magnified, and unfortunately that impact has had a detrimental effect on nearly every ecosystem we inhabit.

The issue of ecosystem disturbance has become an economic one: Because every intervention in nature to solve a problem or obtain a benefit simultaneously creates new problems and generates environmental costs, we must make certain that the trade-off between benefits and costs is truly in their favor. As indicated in the Introduction, environmental disruption has a very long history, but the magnitude of human intervention has grown enormously, particularly in the modern era. In earlier days, the benefits of environmental intervention probably far outweighed the costs. Through ignorance, we damaged ecosystems more than necessary to get those benefits, but our numbers and the low level of technology we possessed made it impossible for us to tear asunder the fabric of nature. This is no longer true. Nature has indeed been destroyed, at least to the point where humans may find it difficult or impossible to survive in many areas. What are some of the trade-offs built into the economy of nature?

Ecological Succession

Nature is not static. The biosphere is a dynamic and open steady-state system; ecosystems always contain the latent capacity to change in response to external or internal perturbations. Because the history of the earth consists of long periods of geological and climatological stability interspersed with much shorter episodes of relatively rapid change, at most times the major physical and chemical parameters of the biosphere have been relatively unchanging, and nature appears to have been relatively stable. This appearance is deceptive, however, for natural undisturbed systems can be observed to change even on human time scales.

If we could watch for a long time a bare boulder that had rolled down from a mountain peak to the valley below, we would see regular changes occurring. Devoid of life at first, the boulder would eventually be colonized by lichen. Aided by such mechanical processes as weathering, the lichen would change the chemical composition of the rock's surface until moss could grow in one or two areas. After many years, the moss would have created conditions that enabled other plants to grow, and insects as well as microorganisms of various kinds would have long since found a home on the boulder. After many more years, large areas of the boulder would have been transformed by this biological activity, and one day a pine seed would be able to take root in the newly created soil. By this time, many of the early inhabitants of the boulder would have been displaced: The "pioneer" species that created conditions favorable for other types of organisms thus did themselves out of home and job. During the pine's growth, it would further transform the boulder, perhaps splitting it into smaller fragments. Thus through a combination of physical and biological processes, a large, hard piece of rock was transformed into life-supporting soil. Ecologists refer to such transformations of a habitat as ecological succession.

This brief and simple example illustrates a number of important ecological principles. As we stated earlier, living things do not merely adjust to their environment, they modify it. And lichens and mosses repeat on a much smaller scale the task performed in the primitive oceans by the microscopic creatures that created the atmosphere as we know it.

Ecological succession moves nature from non-life to life and from simple life forms and communities to more complex and diversified ones that exhibit a higher degree of negentropy, or biophysical organization.

The second law of thermodynamics, one of the fundamental physical principles of the universe, tells us that entropy, or disorder, always tends to increase, impelling all systems toward a state of uniform, random chaos. What, then, accounts for evolution and ecological succession, for the incredible negentropic diversity and order of nature?

The answer lies in the capacity of life to trap the energy of the sun and perform work that decreases entropy before the energy is degraded to dispersed, random heat. Why life exists at all remains a mystery; yet the existence of life is a fact, and once it gets started, the nature of the transformations it causes is quite clear.

By studying the orderly succession of plant and animal communities in diverse habitats, ecologists have derived the dynamics of the process. An abiotic environment is colonized, and an extremely simple ecosystem characteristic of a young or pioneer stage of succession is created. Gradually the first colonizers transform the environment, until a slightly more complex pioneer ecosystem is established. This in turn creates the preconditions for more and more complex ecosystems, until the most complex ecosystem the local climate and physical environment are able to sustain comes into being. This final stage, which represents the temporary end of development, is called the climax ecosystem. It will endure essentially unchanged as long as it is undisturbed by fire, human intervention, or other unusual stresses, or until major geological or climatic changes cause the process of succession to continue.

Pioneer and climax ecosystems differ sharply in almost every important attribute by which ecologists describe ecosystems. The essential differences are listed in the following table (after E. P. Odum 1971). This table can be summarized by saying that in creating a more complex and diversified ecosystem, nature replaces opportunistic species with more established species. Plants characteristic of pioneer stages (like weeds, which epitomize pioneer species) grow prodigiously, and the amount of vegetable matter produced, compared to the total amount of the biomass, is very large. In mature stages, plants grow much more slowly and have longer lives; although total productivity (that is, photosynthesis) is actually higher in mature stages, net community production (annual "yield") is lower than in the pioneer stages, because energy is invested not in a new "crop" but in the maintenance and the slow, organic growth of the already existing biomass. Thus, whereas in the pioneer stage energy is used in a fairly straightforward way to grow more plants, so that a relatively empty habitat can be populated, in the mature system the

greater part of the energy is used to enhance the persistence of the
existing community, which already occupies all the territory available.

Pioneer State	Climax State
Few species	Many species
One or few species dominate	Relative equality of species
Quantity growth	Quality growth
Few symbioses	Many symbioses
Short, simple life cycles	Long, complex life cycles
Mineral cycles relatively open and linear	Mineral cycles circular and closed
Rapid growth	Feedback control/homeostasis
Relatively inefficient use of energy	Efficient use of energy
Low degree of order (high entropy)	High degree of structured, complex order (negentropy)

Humans, the Breakers of Climaxes

Where do humans fit into this scheme? Unfortunately, ever since they
acquired technology in the form of fire, humans have disrupted ecological
succession and climax communities. Humans have lived as breakers of
climaxes, which contain the stored wealth of the ages in their plants,
animals, and soil. Instead of living on the income (the production) of the
biological capital inherited by the species that populate such ecosystems,
humans have invaded the capital itself. One of the first and most important human interventions was the use of fire: Early humans found that
burned-over areas produced a new growth of succulent grass that at-
tracted an abundance of game. However, the agricultural revolution
resulted in the greatest simplification of natural ecosystems, as described
by the cultural ecologist Roy Rappaport:

> Man's favored cultigens...are seldom if ever notable for hardiness and
> self-sufficiency. Some are ill-adapted to their surroundings, some cannot
> even propagate themselves without assistance and some are able to survive
> only if they are constantly protected from the competition of the natural
> pioneers that promptly invade the simplified ecosystems man has con-
> structed. Indeed, in man's quest for higher plant yields he has devised some
> of the most delicate and unstable ecosystems ever to have appeared on the

face of the earth. The ultimate in human–dominated associations are fields planted in one high-yielding variety of a single species. It is apparent that in the ecosystems dominated by man the trend of what can be called successive anthropocentric stages is exactly the reverse of the trend in natural ecosystems. The anthropocentric trend is in the direction of simplicity rather than complexity, of fragility rather than stability (1971, p. 13).

It is not cultivation alone that simplifies ecosystems. The sheep rancher does not want bison eating the grass that could be used to feed more sheep, so the bison must go. So must the mountain lion, the wolf, the coyote, the eagle, and any other predator that might cut into production. Ecological poisons such as DDT and radiation also simplify ecosystems, because they tend to kill the organisms high on the food chain, leaving behind large numbers of a few resistant species.* Overfishing and overhunting have the same effect. So does nearly every form of human activity. The dilemma is clear. Humans must have productive ecosystems in order to survive, but high productivity requires simple and even dangerously fragile ecosystems. Further, since the biosphere is highly intergrated, other ecosystems are also simplified, natural cycles disrupted, materials lost, and the whole system of the biosphere rendered less stable. If every forest is cut down, what will perform the forests' flood-retaining and oxygen-making functions? If all marshes and estuaries along our shores are developed, because that is the most "productive" use for them, what will take the place of the oxygen-producing plankton supported by the large quantities of organic matter washed by the tides from estuaries to the continental shelves? And what will replace the fish that depend on them?

In short, humans do not live by food and fiber alone. However, although maximization of production as we have traditionally defined it would totally compromise our life-support system, discreet cropping of climax ecosystems is possible only in the hunting–gathering mode of human existence. Humans in a technological society must therefore strike a balance between production from their environment and protection of their environment. Stated another way, humans must be prepared to optimize their level of production, taking into consideration the contribution of nonproductive elements of nature, such as wilderness, to

---

* One goal of current biotechnological research is to breed crops immune to poisons so that farmers can apply broad-spectrum sprays and kill all life, except for the immune crop, in their fields (*The Washington Post,* May 17,1988, p. C1).

their well-being. The maximization of productivity, as narrowly defined by economists, is eventually fatal for the system as a whole. A fundamental principle of human ecology thus emerges: "The optimum for quality is always less than the maximum quantity that can be sustained" (E. P. Odum 1971, p. 510).

Variability of the Climax

Even in areas where climate and other general factors are the same, microclimates and habitats differ as a result of local variations in topography and soil composition: Some low-lying areas are boggy, some are on higher-than-average ground, some get more sunlight or are more exposed to wind. In accordance with these variations in microclimate and habitat, species are unevenly distributed. We therefore find not one uniform climax everywhere, but a general climax state (sometimes called a polyclimax) that is a mosaic of a large number of "edaphic" climaxes, adaptations of the basic climax to special local conditions. Thus even within the overall biogeochemical determinism of nature, there is a surprising degree of pluralism.

Sometimes ecosystems never attain the true climax state. Areas that are subjected to periodic natural stresses, such as fire inundation or frequent hurricanes, may never reach the climax state that is theoretically attainable. Instead, the community adapts to the stress and displays what is known as a cyclic climax. Again, special local conditions overrule the general direction of nature.

In some cases humans are a source of stress, and the result is called an anthropogenic subclimax. Many grasslands, for example, were produced by deliberate burning and are thus an anthropogenic subclimax. Asian paddy land, which has been cultivated for millennia, is another human-made subclimax, one that mimics a natural marsh. Anthropogenic subclimaxes are inevitably less natural than the true climax, but the best of them attain quite high levels of maturity and yet also furnish humans quite a high yield of useful products. Even today, some productive areas of the English countryside are at this high level of ecological maturity. Unfortunately, this type of subclimax is rare compared to those subclimaxes, such as once-fertile lands turned into desert, that result from humanity's destruction of habitat.

Exploiting Ecosystems for Production

The likelihood that an anthropogenic climax will succeed depends on local conditions. Intensive agriculture of the kind we practice drives our fields back to the pioneer stage, where our cultigens, which are adaptations of natural weeds, flourish. In temperate areas, climate, the condition of the soil, and the nature of the broken climax all permit this to occur without immediately harmful effects, provided that it is done in the right way (by not grossly simplifying ecosystems with broad-spectrum insecticides, by not mining the soil, by minimizing monocultures, and so on).

Intensive agricultural exploitation of mature tropical ecosystems, on the other hand, can produce total collapse. For example, in a tropical rain forest, almost all the nutrients are tied up in the biomass; very few are contained in the soil. Thus when a plot of forest is cleared for cultivation, much of the ecosystem's biological capital is hauled away to lumber mills (or worse, burned) and so is lost to that ecosystem. Most of the animals flee to still-undisturbed areas, dispersing another portion of the biological capital. At most a few years (typically four or five) of profitable exploitation are possible before rain leaches out the remaining nutrients and the sun bakes the resulting clay into concrete. As a result of ill-advised exploitation, this highly diverse, mature, and stable ecosystem may now be an essentially abiotic (lifeless) environment.* Moreover, because the soil is no longer stabilized by plants, the heavy rains cause accelerated erosion, which can effectively prevent life from making a significant comeback for many generations, if ever. (The erosion also disrupts the dynamics of the river ecosystems into which the soil may be washed). Thus ancient and apparently stable ecosystems (for example, the Amazon basin forests) are paradoxically quite vulnerable to human intervention, for they can be driven by overexploitation to the point of total and irreversible collapse. And what is lost is not merely future production but all the invisible contributions such forests make to the general health of the biosphere. Hence, in addition to the local trade-off between production and protection, humans must confront this kind of global ecological paradox in an era when demands for food are rising, in the tropics above all.

Ecologists propose a basic agricultural strategy reflecting the fact that the safest landscape is one that contains all the variety of nature—crops, forests, lakes, marshes, and so forth—as well as a mixture of communities of different ecological ages. (They point out that this would also be the most pleasant landscape to inhabit.) Margalef (1968, p. 49) calls for a "balanced mosaic, or rather a honeycomb, of exploited and protected areas"—in short, compartmentalization.

---

* According to the World Resources Institute, 40 to 50 million acres of tropical forest are being destroyed each year. (1990-91, p. 102).

2

Some Consequences of Destroying Tropical Forests

The destruction of tropical forests may be bringing about a new age of extinction, similar to the Cretaceous Period some 65 million years ago. For example, only 1% of Brazil's Atlantic forests remain; apart from unknown numbers of plants species found nowhere else, they are the only home of 20 species and subspecies of monkeys, about 10% of the world's total. Some of these are in immediate danger of extinction. Likewise, only 10% of Madagascar is still covered with natural vegetation, and 7000 species of plants exist nowhere else in the world. The same is true of lemurs, a major group of primates. Worldwide, 2 or 3 species are becoming extinct each day. All of this is not merely of aesthetic or ethical interest. Wild plants often interbreed with humanity's cultivated crops, making the crops stronger or more resistant to disease. One-fourth of our medicines come from chemicals found in plants. And five-sixths of the estimated number of plants species in the world are still unknown and unclassified by the world's biologists. Most of these are in the tropics.

Tropical forest destruction also affects humans directly. At one time, 3 to 6 million people lived—sustainably—in the Amazon rain forests; today only 500,000 people remain. Needless to say, the people burning the rain forests—cattle ranchers and peasants—are not the same people whose habitats and lives are being wiped out.

Another solution is the development and utilization of compromise ecosystems; these are essentially "good" anthropogenic subclimaxes, systems that combine the virtues of production and protection. Rice paddies and fish ponds, which are cultivated analogues of natural marshes and estuaries, are examples of successful compromise ecosystems. Such ecosystems are highly productive and more could be developed, especially in the tropics. Detritus agriculture (mushrooms) and pisciculture (the breeding, rearing, and transplantation of fish, as in carp ponds) also offer opportunities for the invention of productive compromise systems. Yet another kind of compromise is tree cropping for food, especially in tropical areas, where it has long been practiced (though not always wisely) for cash crops such as coffee and cocoa. Also, native tropical gardening techniques are ecologically very sophisticated, and research might disclose ways to make them more productive. *

The basic strategy of all these compromise systems is to study the nature of the climax and then, instead of breaking it completely, to mimic it closely or to insert humans into the process as careful parasites that preserve the host while siphoning off as much food as possible. However, compromise systems will not work everywhere. Some ecosystems are too vulnerable or difficult to manage. Furthermore, the productivity of many compromise systems is not high enough to feed great masses of people living in cities. Thus for large parts of the globe, we must attempt to strike a balance between production and protection by comprehensive zoning, retaining some areas as a source of biological capital from which we draw interest in the form of security and well-being, while subjecting other areas to intensive (but, one hopes, less destructive) cultivation in order to obtain the food and fiber a large population needs. (Ecologists generally insist that the oceans must remain essentially protected areas; rational cropping of naturally occurring fish, and compromise systems of mariculture, such as the growing of oysters, are all that we should expect from the oceans if we wish to avoid the risks that would be involved in exploiting this crucial regulator of natural cycles.)

In general, then, ecologists urge a move toward an overall global anthropogenic subclimax that would give the optimal trade-off between production and protection, as well as a considerable degree of amenity in the form of a varied and pleasant landscape. Ecological succession provides a model for the transformation of our agriculture: from rapid-growth, high-production, pioneering stages to a relatively stable and mature subclimax that is optimal considering all of our needs and that is

---

* As the drawbacks of modern industrial-intensive agriculture have become more and more apparent, particularly as a means of increasing food production in tropical areas, agronomists and ecologists have rediscovered some of the virtues of traditional agriculture. Though unproductive in terms of the market economy, many old agricultural practices turn out to be superior in terms of productivity per acre as a function of energy input and long-term environmental compatibility (Armillas 1971, Rappaport 1971, Thurston 1969). Moreover, as we will note in the next chapter, traditional agriculture, when modified by modern organic farming techniques, *can* be productive in terms of the market economy. Even in an advanced country like the United States, organic farming was found to be competitive with industrial-intensive agriculture (National Research Council 1989).

characterized by constructive symbiosis rather than warfare between humans and nature.

Life Is Energy

Whether an ecologically optimal agriculture will also serve to feed large numbers of people will depend largely on how well it adapts itself to the character of the energy flows that make the economy of nature operate. Energy is the currency of nature's economy; the biomass and stock of materials are its inventory, or capital. Natural systems trap incoming solar energy and make use of it for production; the more mature the ecosystem, the more efficiently this energy is utilized and the more is produced. Thus energy determines productivity. This seems paradoxical. If our ecologically immature agricultural systems are so inefficient in terms of energy use, why are they so productive? Part of the reason is that our cultigens are domesticated weeds, adapted by nature and then by us for rapid growth, which gives a high yield when cropped. Moreover, the total amount of radiation is so large and the total area in production is so vast that enormous quantities are produced despite the inefficiency of the process. However, to answer this question fully, it is necessary to look at the energetics of food chains and at the concept of energy subsidy in ecosystems.

An examination of food chains soon reveals that very large numbers of producers are required to maintain a much smaller number of herbivores, which in turn maintain a still smaller number of carnivores. There are several reasons for this, but the principal one is the loss of energy in food chains due to the tendency of energy to degrade into nonuseful forms, as ordained by the second law of thermodynamics. Natural processes are typically of rather low thermodynamic efficiency from a mechanical engineer's point of view. Photosynthesis, for example, is only about 1% efficient in terms of energy fixation or the amount of protoplasm created by producers in proportion to incoming solar radiation. * Also, producers have a high metabolic rate, so they tend to burn up a lot of what they produce just to stay alive, and herbivores consume energy maintaining themselves and grazing the producers. Thus, at each step in the food chain, energy dribbles away in the form of waste and heat. In the typical

---

* This is not so inefficient as it might seem. Remember that enormous quantities of solar energy are used to warm the globe, evaporate water, blow clouds around, and do all the things that keep natural cycles going. To use ecological language, most of the incoming energy is used for maintenance.

food chain, the energy available declines by a factor of ten at each trophic (feeding) level, although the ratio can vary. Thus the efficiency of agriculture in feeding people depends a great deal on where food is taken from the chain. When cereal is fed to pigs or cows raised for human consumption, only about a tenth as many people can be fed as when humans themselves live as cereal and vegetable eaters. There is thus a trade-off in agriculture between the quantity and the "quality" of our diet. Low at best, the efficiency of agriculture in supporting human beings is even lower if they wish to eat high off the hog. *

However, the crucial factor in agricultural productivity is the amount of energy subsidy that humans provide to crops. Natural energy subsidies exist, such as the tidal flushing that brings nutrients to estuaries. But energy subsidy is largely the work of humans, often unintentional and unwelcome, such as the eutrophication of a lake), but more often deliberate. All agriculture, no matter how primitive, depends on an input of energy in the form of labor or materials. In traditional agricultural systems, this subsidy primarily takes the form of human labor, with some help from draft animals. The total amount of the subsidy is typically rather small compared to the yield (returns of up to 50 to 1 are possible, although 15 to 1 is more common). However, gross productivity is often (but not always) comparatively low. By contrast, modern technological agriculture is quite productive but not very efficient, for it is little more than a biological machine for turning fuels into food. In fact, when all the subsidies supporting industrial agriculture are taken into account, it clearly spends more energy than it produces; in other words, *the energy yield of industrial agriculture is negative.*

The dependence of industrial agriculture on energy has important implications. The United States produces about three times as much food per hectare as India, but this increased production requires ten times the input of energy. (The comparison is conservative if anything, because the hotter the environment, the higher the metabolic rate and thus the energy loss of a plant. All other factors being equal, temperate zones produce higher yields than tropical zones.) This gives some insight into the problems of raising agricultural productivity in developing areas. For them to achieve anything like the productivity enjoyed by the developed

---

* If the world's food supply were distributed equally, and it were consumed in the form of grain, enough food exists for 6 billion people to be fed an adequate diet. On the other hand, only 4 billion people can be adequately fed if 10% of their calorie intake is from animal sources, and only 2.5 billion people can be fed if 30% of their calorie intake is from animal sources (Corson, 1990, p. 73).

temperate zones, the energy subsidy must be increased at least ten times. In the final analysis, food, like every other aspect of life, is a matter of energy.

The Essential Message of Ecology

To recapitulate, although it is possible in principle to exploit nature rationally and reasonably for human ends, humans have not done so. Because they have not been content with the portion naturally allotted them, humans have invaded the biological and ecological capital built up over evolutionary time. Moreover, as a result of human ignorance of nature's workings, they have done so in a peculiarly destructive fashion. With our new ecological understanding, we can see that linear, single-purpose exploitation of nature is not in harmony with the patterns in the biosphere and must be abandoned. Instead, we must learn to work with nature and to accept the basic ecological trade-offs between protection and production, optimum and maximum, quality and quantity. This will require major changes in our lives, for the essential message of ecology is *limitation:* There is only so much the biosphere can take and only so much it can give, and this may be less than we want. In the next chapter, we shall explore the consequences for human action that follow inescapably from the existence of fundamental biospheric limits.

2

Population, Food, Mineral Resources, and Energy

In the last three centuries, human beings, by invading the biological capital built up over the course of evolution, have created economic wealth for unprecedented numbers of people. In the last four decades, the additional wealth created in each decade on the average equalled that added between the beginning of civilization and 1950 (Brown 1990, 1990b, p. 3)! Yet during the past 20 years, studies based on computer projections of trends in resource use, environmental impacts, and population growth have shown that the growth in wealth has started to decelerate. The studies differ in detail, but all agree that population and material growth cannot continue indefinitely on a finite planet. All agree that environmental, population, and resource stresses are being inflicted on us now; they disagree only on how drastically and how soon we must respond in order to avert human disaster. The Club of Rome study reported in *The Limits to Growth* (Meadows et al. 1972) appeared to show that an immediate rather than an eventual transition to the steady state was necessary and that the actions required to cope with the problems of transition violated both conventional wisdom and our current values. More important, this was the first study of a holistic nature. Past ecological warnings had tended to focus on or become identified with one particular limit to growth, such as pollution (Carson 1962) or population (Ehrlich 1968). They were therefore vulnerable to the counterargument that the problem could be solved by using resources from other sectors. What *The Limits to Growth*

purported to show, by identifying all the important relationships among various sectors and linking them in a computer model designed to reveal the resultant of all these interactions, was that this strategy of borrowing from Peter to pay Paul would not work much longer. *The Global 2000 Report to the President,* published in 1980, used a similar approach and came to the same conclusion.

Paradoxically, however, these studies tended to intensify debate about particular limits to growth, for the critics charged that they were based on excessively Malthusian and pessimistic assumptions about many of the alleged limits that largely determined their outcomes. It was asserted that with more "reasonable" or optimistic assumptions about population dynamics, resource availability, pollution-control technology, and the like, quite different conclusions could be reached. In the United States, a new administration characterized *The Global 2000 Report* as projecting "doom and gloom" and cut the budgets of the agencies that had prepared it. Not surprisingly, however, budget cutting did not make the problems cited in the report go away. The World Resources Institute report, *The Global Possible,* published in 1985, reiterated the major trends found in *The Global 2000 Report* and suggested priorities for action in such areas as curbing population growth rates, reducing poverty, protecting forests and other habitats, and curbing pollution. So did *Our Common Future,* commissioned by the United Nations, World Commission on Environment and Development (1987). Moreover, every year since 1984, the

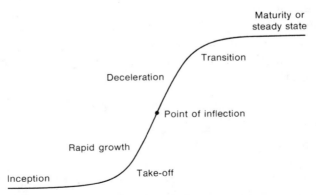

FIGURE 2-1 The logistic curve, or sigmoid curve, and its features. Because humanity lives on a planet endowed with finite resources, it is certain that, like every living population and process, it must obey the law of growth depicted by a sigmoid curve.

Worldwatch Institute has issued an annual report on *The State of the World*. Year after year, it has recorded the continuing deterioration of the earth's physical condition and the threat that this deterioration is becoming unmanageable, leading to economic decline and social disruption.

What follows is a synoptic review of the various limits (primarily but not exclusively physical) that have been identified by qualified specialists. The argument makes four general points: (1) There are indeed demonstrable limits to the demands humanity can place on its environment. (2) Although technology can help us "juggle" limits in accordance with human preferences, outright repeal of the limits is impossible. (3) Manipulating the limits technologically entails costs that we may not be able, or wish, to bear. (4) Time is of the essence if we wish to cope with limits effectively and humanely. In sum, an era of ecological scarcity has dawned. The argument focuses on a few key factors, such as pollution, that are truly critical for the system as a whole, and it makes as explicit as possible the interactions among separate limits, for ecological scarcity is not simply a series of discrete problems. It is an ensemble of problems and their interactions—a "problematique"—and can be understood in no other fashion.

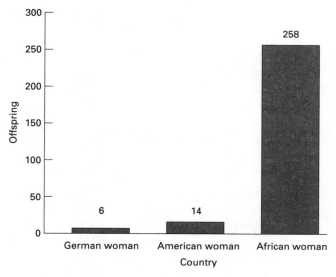

FIGURE 2-2   Number of offspring in three generations.

Population and Food

No aspect of ecological scarcity has received more attention over a long period than the "population explosion." The primordial limit on population is food, so let us consider the supply of arable land and other basic factors: water, the state of agricultural technology, and above all the costs and consequences of feeding people—especially the large numbers of additional people who will be born in the decades to come—to see how many human beings the earth can reasonably support.

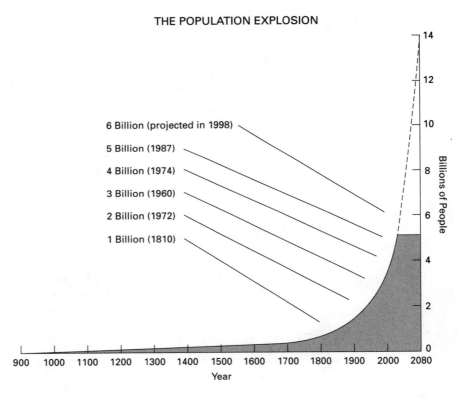

THE POPULATION EXPLOSION

6 Billion (projected in 1998)
5 Billion (1987)
4 Billion (1974)
3 Billion (1960)
2 Billion (1972)
1 Billion (1810)

Billions of People

Year

FIGURE 2-3   It took from the beginning of time to about the year 1810 for the human population to reach 1 billion people. Just over 100 years later, the next billion people were added. By 1987 the earth was home to 5 billion human beings. Currently, population is growing by more than 93 million each year. Human numbers are projected to exceed 6 billion before the year 2000, and unless the availability and use of contraceptives increase dramatically, the human population could reach 14 billion by the end of the next century.

The Inevitability of More People

In 1977, when the first edition of this book was written, approximately 4 billion people inhabited the earth. In 1990 the figure had risen to 5.2 billion. The population of the earth is exploding. 10,000 people are added to the earth every hour, 240,000 every day, *about 1 million every 4 days.*

If the current rate of world population growth continues, there will be 10 billion people on the globe by 2025 and 16 billion by 2100 (Haub 1988). Of course, these projections are only as good as the assumptions on which they are based. The United Nations Population Fund projections assume that human fertility will decline to replacement levels by the middle of the twenty-first century and that world population will stabilize at about 14 billion people by 2100. That number could go higher or lower, depending on the extent of birth control use within the next few decades. Even so, most experts regard a world population of 12 to 16 billion people as foreordained. *

The primary reason for this inevitability is the phenomenon of demographic momentum. For example, even though the United States attained replacement-level fertility in the 1970s, the population still grew. In fact, assuming that we maintained a fertility rate of 1.8, the level achieved in the 1970s, the population would continue to grow from its current 215 million to at least 292 million in 2080. This is because rapid growth in decades past has bequeathed us a young population—that is, a population distribution with disproportionate numbers of young people who have yet to replace themselves. Thus, even assuming a constant low fertility rate, the U.S. birthrate would continue to exceed the death rate for many more years.

But the United States has *not* maintained the fertility levels of the 1970s. In 1990 the fertility level had reached 2.0 and had not yet peaked. If the fertility rate went as high as 2.2, as the Census Bureau assumes in its "high-fertility" projections, the United States population would reach not 292 million but 421 million by 2080, and it would increase by 20 million people per decade after that. Even these numbers might be too low. This would be especially true if life spans were extended, as some

---

* The staggering magnitude of "billions" is often not appreciated. The mathematician John Allen Paulos brings home the reality of these numbers by asking audiences how long it will take for a million seconds to pass. The correct answer, he tells them, is about 11.5 days. He then asks how long it will take to get to a billion seconds? The correct answer is 32 years.

medical scientists hope, for then the death rate might remain below the birth rate for additional years before growth ceased.

Of course, the U.S. population problem is of trivial dimensions compared to that of underdeveloped countries, where a replacement level of fertility is far from being achieved. Most have population growth rates of about 2.1% per annum, resulting in truly explosive growth—a doubling every 33 years. Many hope that such rapid growth is a temporary phase and that the so-called demographic transition (a drop in fertility in response to lowered death rates and economic improvement, similar to that experienced by the now-rich countries in the course of their development) will begin to slow the population explosion. However, the last 20 years have produced little evidence for such demographic transition. A few countries have lowered their birthrates with little industrial development (Sri Lanka, Costa Rica, and China); other

3

Halting Population Growth

Halting population growth in the developing world is a monumental task. The average family size in these nations is now 4.8 children. Because these countries have large numbers of young people, family size must drop below replacement levels (that is, to fewer than 2 children) and must remain there for some time. Only China has approached this goal, and that by coerced draconian measures.

The obstacles are many. Children are valued for important economic reasons in developing countries: to work in the fields and to support parents in their old age. High fertility is considered both a mark of virtue and proof of a man's virility. Women in most places are uneducated and lack equal rights. There is some evidence that where women gain education and equal rights, they concentrate on improving the health and sanitation of their homes and become receptive to contraception. This occurs because their children are more likely to survive and because they gain sources of status other than children. Although such signs are encouraging, women are only slowly becoming educated or achieving equal rights, and, when they do, there is no evidence that most will voluntarily reduce their family size below replacement levels.

countries failed to lower their birthrates despite some industrial development (Brazil and Mexico). Also, statisticians have found little connection between GNP and lowered birthrates in developing countries. Although a demographic transition might yet develop in some countries, most developing countries are growing as rapidly as ever. It seems likely, in the absence of catastrophic famine, disease, or other Malthusian checks, that demographic momentum in countries with extremely young populations virtually guarantees a world population of 12 to 15 billion within the period 2040–2070. Can we feed this many people?

How Many Can Be Fed?

A number of studies have tried to establish how many people the earth can support. One concludes that we could feed a world population of 50 billion. But to arrive at this figure, a number of totally unrealistic assumptions are needed. First, all potentially arable land, regardless of fertility or suitability, would have to be dedicated to food production. Humans could inhabit only wasteland totally unsuited to agriculture; suburbs and cities would have to be uprooted and turned into farmland; the bulk of the population would perforce be housed in Arctic regions; and so forth. Second, all land, even infertile land that had been pressed into production, would have to yield food at a level of production that has not been attained on our most fertile soil with the best means of which we are capable—in other words, a level that plant physiologists dream about but that has no real prospect of being achieved outside a laboratory. Third, all the possible side effects of intensively farming every conceivable acre at the highest technological level must be ignored, including the pollution caused by fertilizer runoff, the enormous flow of energy required for such colossal and high-technology agriculture, the climatic effects of turning all forests into farmland, the enormous demand on water resources, and so forth. Thus that 57 billion is an optimistic limit based purely on the potentiality of plant physiology, not on the realities of agriculture.

The Food and Agricultural Organization (FAO) published a study in 1983 that made more realistic assumptions about land use, soil quality, water supply, and the application of agricultural inputs (basic fertilizers, simple conservation measures, improved plant varieties, and so on). It projected that most countries of Asia and South America could feed their populations in the year 2000 from their own cultivated lands, whereas most African and Middle Eastern countries could not. But even this study was optimistic, because it assumed that all cultivable land would be brought into production, that the land would be used to grow only food crops, and that the food would be distributed fairly to all economic groups. In reality, world grain output per person has *declined* since 1984. In Africa and Latin America, the food produced per person has declined dramatically in the past decade, and those regions, as well as Asia, began the period with high rates of malnutrition (see Table 2-1). Presently, one-fifth of the world's population does not consume enough calories to lead an active working life. Even worse, 40 to 60 million people die in developing countries each year from hunger and diseases related to hunger. The nutritional shortfall worldwide is such that even with perfectly equal distribution of the current world food production (theoretically possible but politically inconceivable), everyone would be somewhat malnourished. Thus we do not really provide for even the existing 5.2 billion human beings (see Box 4).

Recent developments in agricultural technology (the so-called Green Revolution) now being applied to underdeveloped countries have averted a catastrophic global famine thus far. But the Green Revolution is not immune to drought (a perennial problem in many less developed countries) and as we shall see, it may have achieved the maximum agricultural productivity of which it is capable in many parts of the world. Besides, as even its proponents admit, the Green Revolution cannot solve the problem of overpopulation; it simply buys us some time to bring population growth under control. Moreover, the world already confronts severe ecological problems because of its current mode of agricultural production, and the Green Revolution intensifies these problems. (It also has a number of painful social side effects—for example, making poor farmers worse off while making rich farmers better off.) Thus the time

TABLE 2-1   Decline in Annual Grain Production Per Person in Latin America, Asia, and Africa, Peak Year and 1990

	Year of Peak Production	Kg Per Person in Peak Year	Kg Per Person in 1990	Percent Change	Percent Malnourished 1985–87
Latin America	1981	250	210	-16	14
Asia	1984	227	217	- 4	22
Africa	1967	169	121	-28	32

Source: Brown 1991, p. 13; WRI 1990, p. 87.

4

A Mere Distribution Problem?

In discussing world hunger, some claim that there is no population problem, only a food distribution problem. But this statement is misleading. If food were distributed equally, and everyone ate the way Americans do, less than *half* the present world population could be fed on the record harvests of 1985 and 1986 (Erlich 1990). Only if food were distributed equally and everyone got their needed calories from vegetarian sources would the food supply be sufficient to feed everyone. If everyone ate like the average Latin American and consumed a mere 10% of their calories from animal sources, only 4 billion people could be fed (Corson 1990, p. 73).

The average American consumes about 30% of his or her calories from animal sources. And there is no indication that many of them are ready to stop eating their steaks, cheeses, and chickens to benefit of the world's hungry (assuming that their abstinence *would* mean more food for the hungry). And what if they *were* willing? World population is increasing at a rate of 1 million people every 4 days. So at best, a strict vegetarian diet by everyone, along with equal distribution of grain, would provide a mere temporary respite from the hunger problem. Clearly, the urgency of the population problem is merely camouflaged by the assertion that it is only a matter of finding more equitable ways to distribute food.

being bought now is at the expense of the future. A more detailed examination of the difficulties, dilemmas, and contradictions in current and projected agricultural practices will show why feeding adequately even double the present number of humans will overstrain the earth's ecological and energetic resources to the breaking point and is likely to be out of the question.

The Basic Agricultural Predicament: Limited Land

The fundamental fact about agriculture is that it requires land, and good agricultural land is in diminishing supply. First, virtually all good agricultural land in the world is already in use; it is this good land that provides us with almost all our food. The best 50% of the land in use probably supplies 80% or more of the total agricultural output. The marginal lands now in use thus make only a modest contribution (although it is often critical for dietary quality). Second, much of the presently good agricultural land on the planet is becoming marginal while some marginal agricultural land is becoming altogether useless to agriculture.

There are several reasons for the loss of agricultural land. One is the sheer overrunning of agricultural land by population pressures—for the homes, industries, and roads that people demand. This is happening everywhere, but especially in East Asia, where population pressures claim a half-million hectares of agricultural land per year (Brown and Young 1990, p. 65).

Another problem is soil erosion. Whereas traditional agriculture conserved topsoil by such practices as rotating crops and permitting land to remain fallow for periods of years, many farmers today grow the same crop each year on the same land. They minimize periods when the land is plowed and left "unproductive." These practices deprive the soil of the organic matter it needs to rebuild the soil structure; as a result, some of the topsoil is lost to wind and water erosion. One estimate is that 25 billion more tons of topsoil are lost to erosion each year than are formed, resulting in the loss of 9 to 21 million tons of grain production annually (Brown and Young 1990, p. 61).

A third problem is salinization and waterlogging. Salinization occurs when land is irrigated. Some of the diverted water evaporates, leaving behind minute amounts of salt. Eventually these salts accumulate in the top few inches of soil; salts reduce the yields of the crops and ultimately kill them. Waterlogging occurs when irrigated lands are not properly drained. The water table therefore rises, eventually reaching the root zones of the crops. The roots in this water cannot get the oxygen they need to survive. Salinization and waterlogging today are serious problems that are degrading 24% of the world's croplands. Each year, 1.1 to 3.6 million tons of crop output are lost to salinization and waterlogging.

Finally, deforestation causes the loss of cropland. Seventeen million hectares of the earth's tree cover are destroyed each year, an area the size of Austria (Brown 1991, p. 7). Some deforestation may produce temporary gains in croplands, but on the massive scale on which deforestation is occurring, it destroys more than it creates. Deforestation increases water runoff and affects rainfall cycles and rainfall distribution. It therefore causes flooding in some places and desertification in others. The desertification brought about by deforestation (and by the overgrazing of livestock) affects almost one-third of the earth's land surface. Desertification results in the loss of 6 million hectares of agricultural land each year.* In addition, deforestation indirectly adds to the problem of soil erosion. People in deforested areas are forced to burn cow dung and crop residues for fuel, thus depriving the soil of fertilizer and organic matter that builds up the soil structure and holds it. Putting these and other factors together, Lester Brown observes that "each year, the world's farmers must try to feed 88 million more people with 24 billion fewer tons of topsoil" (Brown 1989, p. 29)

Although there appears to be other land in the world that could be developed, bringing into production any sizable quantity of new land would require enormous amounts of capital, vast expenditures of energy, and above all, ecological expertise beyond any we now possess. And production gains are likely to be ephemeral. The preceding chapter addressed the ecological futility of trying to clear and farm tropical forest lands. Yet numerous countries are allowing the irreversible destruction of tropical forests at a rate of 500,000 trees per *hour* (Newsweek 1989, p. 59), reaping a few years of harvest but leaving a legacy of serious ecological problems and potential climatological consequences. Even where the soils of forest lands are capable of supporting some kind of more intensive cultivation, cutting down forests carries the risks of erosion, flooding, and desiccation of climate, the result of ignoring the ecologist's warning that the so-called nonproductive parts of an ecosystem perform invaluable protective functions and are vital to production on even the most suitable land.

Some novel kinds of ecologically sophisticated exploitation of unused land—tree culture, pisciculture, modernized versions of native gardening techniques, and so on—can provide a useful supplement (of vital protein especially) to a basic cereal diet, but except in a few favored areas, they are not the answer to the present and future food needs of humanity.

---

* Air pollution, though it does not destroy cropland, also causes crop losses. By one estimate, 1 million tons of grain are lost just to ozone pollution each year (Brown and Young 1990, p. 63). Other pollution damage to crops, (such as that caused by acid rain, rising ultraviolet radiation levels due to the depletion of the ozone layer in the stratosphere, and global warming) have not been reliably calculated yet. But the evidence suggests that each of these factors will accelerate crop losses.

...And Limited Water

Irrigation has opened up many arid and semiarid lands to agriculture throughout human history. In this century, irrigated land has risen at an astonishing rate, from 50 million hectares to 250 million (Postel 1990, p. 40). But irrigation also has limits. Since 1978, irrigated area per capita has been decreasing worldwide. We have already noted some of the environmental costs of irrigation—the intensified erosion, waterlogging, and salinization that force some irrigated land out of production. Furthermore, in developed countries, irrigated runoff is becoming contaminated with fertilizer and pesticide residues, polluting the sources of the irrigated water itself, as well as human drinking water. In the United States, over 50 pesticides contaminate groundwater in 32 American states (Corson 1990, p. 163). In third-world countries, water-development projects have spread disease, destroyed traditional agriculture upon which poor people rely, ruined fisheries, and destroyed many species by eliminating their habitats. Major outbreaks of schistosomiasis have resulted from a parasite that lives in reservoirs and irrigation systems. Today at least 200 million people have the disease; another 600 million are in danger of getting it (Corson 1990, p. 162). Poor people who farmed or fished in floodwaters lost their way of life as dams upstream stopped the flooding on which they depended. (These dams do not normally become substitute fisheries, for their deep waters are usually sterile, providing insufficient nutrients for fish).

The environmental consequences of irrigation projects, however serious, are only part of the reason why such projects will be limited in the future. Another problem is that both water and irrigable land are getting scarce. The cost of irrigation is therefore becoming too expensive to justify in many parts of the world. Africa, where food needs are greatest, is the worst case; the cost per hectare of irrigating land on that continent has risen to between $10,000 and $20,000. This cost is prohibitive for agriculture, which partially explains why Africa's irrigated area increased by only 5% in the last decade (Postel 1990, p. 42).

The most fundamental limit on irrigation is the decreasing availability of good water. Rivers and lakes worldwide receive enormous quantities of industrial discharges, agricultural runoff, and municipal sewage. A quarter of the world's population—1.2 billion people—do not have access to safe drinking water. Because of the scarcity of good water, agriculture and growing numbers of people compete for it. In northern and eastern Africa, in parts of the Middle East, in China, and even, occasionally, in the western United States, where getting usable water is a serious problem, growing populations often require that water be shifted from agriculture to drinking purposes.

Water tables are falling in many parts of the world. Indeed, in some areas water has already become too expensive to pump, forcing land to be abandoned. In many parts of the world, including one-fifth of the irrigated area in the United States and large parts of China and India, farmers are pumping water out faster than it is being "recharged"—a process that cannot continue indefinitely. Gigantic aqueducts to transport water very long distances have been built to overcome these shortages. Such systems require enormous quantities of energy to build and maintain, and even more grandiose ones are being considered. But realistic prospects for major gains in irrigated areas of the world are poor. In Thailand, the Philippines, and parts of India, some potential exists for enlarging irrigated areas, but in other parts of India, wells are going dry. China and the United States have actually declined in irrigated area since the 1970s, and because pumping is outpacing replenishment, that decline is expected to continue. Even general adoption of drip irrigation and other techniques that conserve water (but boost energy costs) will not alter this picture substantially. In fact, the proliferation of waterweeds, which reduce water flow and increase evaporative loss, more than outpaces improvements in water supply in many areas.

Despite high costs, desalination is a possible answer for some local areas, but it requires such great expenditures of energy for production and transportation that its widespread use is unlikely. Moreover, the hot brines that are an inevitable by-product will create a pollution problem of very large dimensions. Such remote possibilities as towing icebergs from the Antarctic are occasionally advanced as solutions to this very serious limit to growth, but even the most favorable estimates relegate them to a minor, local role. It is just barely conceivable, for example, that icebergs will one day supply water to the Atacama Desert in California, but the ecological, economic, and practical barriers to providing North America with agricultural water from this source are immense.

Finally, global warming may make the problem of limited water even more severe. (For a discussion of the physical mechanism of global warming, see Chapter 3). One possible effect of global warming will be to shift the areas where water is available. Dams, flood control, hydropower, reservoir and other water management projects built where water is now available will become less productive or even useless where rainfall and snowpack have shifted elsewhere. New water management projects will need to be built, at enormous cost, in the new wet locales that materialize. Global warming is also predicted to increase evapotranspiration in crops. If this occurs, increased irrigation of existing irrigated cropland and the irrigation of cropland now rain-fed will be required to achieve the same crop productivity. Although a number of

now-irrigated areas may benefit from increased rainfall levels, some
scientific models suggest that much of the newly available water may be
in the form of intensified monsoons. If so, occasional flooding, rather than
a stable source of soil moisture for crops, will be the result.

Little Help from the Sea

Nor can we expect the sea to provide basic subsistence for added billions.
For one thing, the oceans are for the most part biological deserts. All
major fisheries depend on the few areas where large quantities of nutrients are brought to the surface by upwelling. For another, experts
believe that we are already close to the maximum sustainable yield of the
sea: 100 million tons of fish per year (Corson 1990, p. 142). And even this
harvest may be unsustainable in the long run. The top millimeter of the
ocean is critical to its productivity and general ecological health, and we

5

Mariculture: A Ray of Hope?

Although some species are being overfished or are becoming too polluted to consume, the farming of fish offers an alternative source of
protein. In 1980, worldwide production of seafood and edible seaweed
was 8.7 million metric tons, nearly one-eighth of the total harvest of
such resources. Third-world countries produced 74% of this harvest
(Corson 1990, p. 148). Aquaculture has the potential for good growth
in third-world countries, because it is usually labor-intensive and can
be practiced not only in bays and the sea but in freshwater ponds and
semi-freshwater estuaries as well. It also can be extremely productive,
for it is potentially much more efficient than terrestrial agriculture at
converting solar energy into animal protein.

    However, aquaculture may never realize its full potential. First, it re-
quires scientific knowledge and capital; the requirements of water
purity, location, spacing and so on, are extremely demanding; and the
techniques of managing aquatic ecosystems are intricate. Second, pollution and coastal development must be well controlled for mariculture
to work. So far, unfortunately, such control is weak and is losing ground
to both pollution and development.

are polluting it with a multiplicity of land-generated wastes—municipal, industrial, and agricultural—as well as with oil and plastics from both land and our ships at sea. Much of the increase in fishing since 1983 has been of species used to feed animals and to make fish meal, rather than of fish which people consume. Yields of some major fisheries have leveled off or are declining. Once a fishery collapses, as did the Peruvian anchovy and the Alaskan crab fisheries, harvests may never recover to former levels. Many species of whales are already "extinct" for practical purposes, because for more than a century, the few remaining animals have failed to reproduce at a rate adequate to satisfy commercial demand.

People are also damaging salt marshes, mangroves and coral reefs in many parts of the world. Salt marshes and mangroves are temperate and tropical wetlands, respectively, and they are far more biologically productive than the open ocean. Mangroves, especially, spawn and support enormous quantities of finfish and shellfish, and at the same time they filter out human-made pollution. But they are being converted to shrimp ponds and rice fields, and the salt-tolerant mangrove trees are being cut down for building materials and firewood. Coral reefs are starving, and in many cases dying, throughout the world (Booth 1990 p. A8). Some die because rivers filled with sediment flow into them, preventing sunlight from reaching the reefs. Others die because fishers dynamite reefs to kill the fish that hide in them. Mine tailings, pesticides, cyanide, and other pollutants kill still other reefs. But a new and ominous reason why the coral reefs are dying is abnormally warm seas. The 1980s were the warmest decade in the last 100 years, and 1990 was the hottest year in modern records. The dying of the reefs may be an early biological effect of possible global warming. *

At the same time, in more temperate climates people are destroying wetlands for development. Estuaries, where commercially valuable fish spend part of their lives, are becoming reservoirs of human contaminants. Contrary to a popular assumption, the estuaries into which polluted rivers flow do not flush these pollutants out to the sea. Rather, the

---

* The destruction of the ozone layer is also affecting the seas in a worrisome way. In 1988, an ozone hole over Antarctica resulted in a 15% overall decline in ozone levels. Surface phytoplankton levels in that area also decreased by 15 to 20% (Roberts 1989, p. 288-289). If the decrease in ozone levels *caused* the decrease in phytoplankton, then ozone depletion harms the oceanic food chain. In addition, a decrease in phytoplankton decreases the ocean's ability to absorb atmospheric carbon and accelerates global warming.

pollutants concentrate in the estuaries, endangering marine life and making the fish (especially shellfish) dangerous for human consumption.

National and international efforts have been made to control ocean pollution. Certain treaties limit ocean dumping from ships; another gives nations control over the seas within 200 miles of their coasts, theoretically enabling them to regulate and protect the resources within that territory. The United States and some other countries have laws restricting development in wetlands and protecting clean water. Yet as a whole, these efforts have failed. Some nations do not agree to or comply with international controls. Many also fail to enforce their clean water laws or adequately fund clean-up projects to reverse the pollution from internal sources. In the United States, for example, most bays are badly polluted. Americans dumped 3.3 trillion gallons of sewage into marine waters in 1980, and since then the amount of dumping has increased. Industries add 5 trillion additional gallons of industrial waste to domestic waters each year (Corson 1990, p. 146). Urban runoff, in the form of solid wastes, lawn pesticides, toxic chemicals, oil, and heavy metals, adds billions of gallons more—to say nothing about fertilizers, herbicides, and pesticides from agriculture and fallout from atmospheric pollutants. All of this occurs in spite of national, state, and local laws enacted to deal with the problem. So far efforts to curb overfishing have likewise been half-hearted and largely unsuccessful in the United States and throughout the world.

Diminishing Returns from Increased Energy Input and Intensification of Agriculture

Experts agree that the only possible answer to the problem of feeding double the present world population lies not in the opening up of new frontiers either on the land or in the oceans, but in the preservation and more intensive exploitation of lands that are now farmed and that are the most suitable for intensive agriculture. Even if we overlook its ecological consequences, intensification of production requires a vast input of energy. Modern intensive agriculture is essentially a technique for converting fossil fuels and minerals into grain and fiber.* From 1950 to 1984, world

---

* For example, even though the atmospheric nitrogen for making nitrogenous fertilizer is itself free (and essentially inexhaustible), the process of converting it into chemical forms that plants can use is costly in terms of materials and energy: Every ton of nitrogenous fertilizer requires one ton of steel and five tons of coal to manufacture (McHale 1970, p. 12). Some of the minerals essential to intensive agriculture may come to be in short supply.

fertilizer use grew from 14 million tons to 146 million tons, a fivefold increase per capita. This increase offset a one-third decline in grain area per person (Brown 1989, p. 52). But the *rate* of increase in fertilizer use has declined since 1984. As a result, world grain yield has declined from 343 kilograms per person in that year to 329 kilograms per person in 1990 (Brown 1991, p. 13). This is not to say that agricultural production declined during those years; it increased. But while grain production grew by 1% annually from 1984 to 1990, population grew by 2% (Brown 1991, p. 14). In addition, the rate of increase in yields of the staple crops may have peaked. First, the miracle strains that were the basis of the Green Revolution are already in widespread use. Second, some of these miracle grains may have reached their biological limits of productivity. Rice yields, for example, have not increased since 1984 (Brown 1991, p. 12). When the upper limit of photosynthetic efficiency of a plant is reached, further application of fertilizer does not increase yields. In a finite environment, all biological processes grow in a manner reflected by an S-shaped curve, at the end of which they do not grow further (see Figure 2-1).

Some people hope that biotechnology research will result in dramatic increases in world food production. Biotechnologists have already achieved some impressive gains. For example, they have produced hormones that increase milk production and have developed vaccines and drugs to increase livestock productivity. Biotechnology promises to create some foods in the laboratory. Biotechnologists may be able to breed natural enemies of plant pests or render existing pests harmless. They may also be able to breed plants more tolerant of heat, drought, frost, pests, and salt. These techniques increase the finite environment for the growing of food—in terms of both area and growing season. Biotechnology thus may be able to offset the losses of croplands currently in use.

But biotechnological progress in some areas has been controversial or slow. Disagreement exists over the safety of foods to which residues of hormones or drugs cling. Plant genetics has not been easy to understand. Genetic manipulation has had unpredictable results. In the few cases where gene transfers have been successfully introduced to crops, it has taken as much time to obtain a new variety as conventional breeding takes. Furthermore, the biotechnology research agenda so far has been dominated by large corporations. Their primary objective has been to produce herbicide-resistant crops so that pesticides can be spread pervasively over fields, thereby reducing tillage. This contributes to soil erosion and has other dangerous consequences (see Box 6)—and in any event does little to help farmers in developing countries, where tilling by

6

Agribusiness Biotechnology

The biotechnology research priority for agribusiness interests is to produce herbicide-resistant crops. For example, Calgene CEO, a biotechnology company, is testing 1.5 million "transgenic" cotton plants in 12 states. (Rauber 1991, p.33) This firm developed transgenic cotton to be resistant to the herbicide bromoxynil. Bromoxynil, when applied to cotton fields, kills the weeds that compete with cotton. Unfortunately, it also kills the cotton. Rhone-Poulenc, the manufacturer of bromoxynil and a major funding source of Calgene's research, expects to double its sales of the herbicide if Calgene's "transgenic cotton" works.

Bromoxynil, however, has other dangers. It is extremely toxic to fish—a thousand times more toxic than other pesticides that have caused massive fish kills. In addition, applying small amounts of the pesticide to the skin of rats causes deformed offspring. The National Wildlife Federation opposed Calgene's testing project, calling it a "deadly wrong ... quick fix" (Rauber 1991, p. 34).

But similar projects abound. The Monsanto Corporation is trying to develop cotton plants resistant to Roundup, a broad-spectrum weed killer that the EPA regards as a "possible human carcinogen" (quoted in Rauber 1991, p. 34). The U.S. Forest Service has spent $2.8 million to fund research on herbicide-resistant trees. The U.S. Department of Agriculture has begun tests on biotechnology-engineered potatoes that are resistant to bromoxynil and to 2,4-D, the active ingredient in Agent Orange. The fact that these pesticides are harmful to people and the environment does not seem to diminish agribusiness or government interest in such projects.

hand is cheaper. Finally, even biotechnology cannot produce plants with yields that exceed the limits of photosynthetic efficiency.

This suggests that the best prospect for increasing food yields is to use agricultural inputs more efficiently. Such techniques as drip and interval irrigation can both reduce damage to cropland and conserve water. Another possibility is for humans to stop deforestation, the overgrazing of livestock, and overcultivation in order to prevent desertification. It is also possible to change to organic farming methods, which can reduce energy subsidies and conserve soil structure, although these methods may require greater labor intensity (one reason for the high yield of Japanese agriculture is some use of labor-intensive, essentially horticultural farming techniques). Biotechnologists can possibly develop pest-resistant or nitrogen-fixing plants. Farmers can apply biological controls to pests and develop hydroponics to increase agricultural yields. The governments of the world can implement land reform to encourage small-scale farming. They can also employ controls to prevent agricultural land from being turned into homes and factories. People in industrial countries can eat less meat, permitting land now used for grazing to be used for farming, and make it possible to use grains to feed people rather than animals. Mariculture has good potential. In those parts of the world where yields are still low (Africa and South America), energy subsidies can still be productively applied to increase yields.

Unfortunately, most of these possibilities are unlikely to be implemented in a timely way, as we shall see. But if they were applied, along with other techniques to improve the efficiency of agricultural production and distribution, some estimate that the earth could probably support 8 billion people living on a cereal diet.

Why Even 8 Billion Cannot Be Fed: Nutritional Limits

It should be obvious that even this figure, which the population will reach in a few years, is an unrealistic, purely theoretical maximum. Arable land is not evenly or equitably distributed, and a study by the United Nations Environmental Programme (UNEP) concludes that desertification, salinization, waterlogging, and erosion may remove as much land from production as is added each year (quoted in WRI 1990-91, p. 88). Moreover, global calculations of this nature tend to sidestep the problem of dietary quality. Humans do not live by the calories in grain alone; they must also have protein as well as other dietary supplements to survive. As we explained in the last chapter, animal protein can be produced in large quantities only by diverting grain from people to animals at a considerable thermodynamic loss: One cow eats for five to ten people. Even vegetable protein of the right quality is much harder than mere calories to produce in quantity. Soybeans, for example, have so far resisted the best efforts of plant geneticists to produce "miracle" strains, and output per acre is still low. Global production of root crops has actually been declining since 1984 (WRI 1990-91, p. 84). The nutrition problem is compounded by the tenacity and nutritional "blindness" of cultural food preferences and habits. For example, Balinese regard rice as the only food fit for humans, so as population pressure increases, marginal land that might reasonably produce something else is unreasonably forced into rice production. In addition, humans need fiber and other inedible agricultural products, so some of the world's arable land must be withheld from food production to satisfy these needs.

...And Ecological Limits:
Monocultures, Crop Losses, and Pollution

Even assuming that we can find sufficient energy, at acceptable costs and environmental consequences, to surmount the nutrition problem, the ecological limits to energy-intensive agriculture will make it unlikely that the world can feed 8 billion people. The Green Revolution aims at universalizing the methods of temperate-zone industrial agriculture. As we saw in the last chapter, these methods are profoundly anti-ecological. Moreover, they are particularly unsuited to the tropics (Janzen 1973). Some aspects of this issue, such as the side effects of irrigation projects, have already been treated in sufficient detail, but the liabilities of monoculture and agricultural pollution need to be explored further.

Even though the practice of monoculture may be justified by economic considerations, and an individual farmer can usually get away with it most of the time, it is on general principles always ecologically unsound. The drive toward total replacement of traditional crop varieties with genetically uniform cultigens—that is, with the so-called miracle strains of the Green Revolution—raises the specter of regional monocultures, with every farmer producing exactly the same species and variety of inbred (and therefore vulnerable) crop. This is an open invitation to pest infestation and areawide plant disease, as the Irish potato famine and recent U.S. experiences with corn blight have shown. (Nor are crop losses confined to the fields. In the tropics, post-harvest losses of up to 30 percent of the stored crop occur, and some of the miracle strains seem to be more susceptible to storage pests than the traditional varieties.) To make matters worse, in order to combat diseases and pests or to increase yields further, plant physiologists need genetically diverse raw materials for cross-breeding. However, the natural sources of genetic diversity, the many traditional varieties typically grown by peasants, are disappearing at an alarming rate as farmers everywhere rush to take up modern methods and the new genetically uniform strains. Thus the price to be paid for higher production will be exactly what the laws of ecology predict, extreme instability. (This effect could be worse without the technological oversight capacity of the industrial world).

TABLE 2-2   Basic Natural Resources Per Person with Projected Population Growth

Resource	1990 (hectares)	2000	Percent Decline in One Decade
Grain land	0.13	0.11	15
Irrigated land	0.045	0.04	11
Forest land	0.79	0.64	19
Grazing land	0.61	0.50	18

Source: Adapted from Brown 1991, p. 17.

The liabilities of using pesticides also must be considered. In the United States, agricultural use of pesticides went from 1 million pounds per year in the 1950s to 815 million pounds per year in 1987. The National Academy of Sciences has concluded that pesticides in common American foods may be responsible for as many as 20,000 cancer cases annually (Weisskopf 1987, p. A33).* Agricultural pollution is not confined to pesticides. Intensified use of fertilizer also has its side effects. To squeeze the last increment of possible yield from a crop, fertilizer application must be increased disproportionately. However, at more than moderate levels of application, much of the fertilizer runs off into the water supply, causing eutrophication in lakes and rivers and polluting the groundwater supply with nitrates. These nitrates can disable or even kill livestock and infants (Corson 1990, p. 164). Moreover, intensification of industrial agriculture creates non-farm pollution. For example, animal wastes are increasingly serious pollutants now that animals are factory raised. Also, fertilizer and other inputs must be manufactured and transported; agricultural products must be transported and processed. All of this

---

* The report may have underestimated the number of these cancer cases. First, it focused on only 28 of the 53 pesticides deemed carcinogenic or potentially carcinogenic. Second, it did not consider possible synergistic effects of the many pesticides we consume. Third, it did not estimate the effect of pesticides we consume in our drinking water (Weisskopf 1987, p. A33).

generates pollution. Finally, when food is prepared and consumed in the cities, an organic-waste disposal problem is created. Just as with every industrial process, therefore, pollution is the inevitable concomitant of food production; the greater the agricultural production, the greater the agriculture-caused pollution we can anticipate.

Emerging Conflicts over Resources

To an increasing extent, agriculture is in ecological competition for basic resources with its suppliers in the energy, mineral-extraction, and manufacturing industries. For example, every additional mine, fuel-processing plant, power plant, factory, and city dweller reduces the amount of water available for food production. Also, as populations grow, as urbanization increases, and as industry develops, productive farm land is necessarily taken over for habitation and other nonagricultural uses. We have already noted that in east Asia (including China, Japan, Taiwan, and South Korea), half a million hectares of agricultural land are lost to non-farm uses each year. The problem also exists in the United States, with its sprawling pattern of land use. Highways and new suburbs eat up more than a million acres of prime farmland each year; in another few decades, even the United States may no longer have much surplus agricultural capacity. Most developing countries do not consume their farmlands at such a rate, but the trend is evident there too, and the sad fact is that it may make perfect economic sense to take land out of production for so-called higher uses, even in a country that suffers from a severe food shortage, just as it makes perfect economic sense for Latin American countries to export food to the United States despite the fact that 55 million people in that region are malnourished.

The Effects of Weather and Climate on Agriculture

Weather is a very important ecological constraint on agriculture. Like death and other painful realities, the vulnerability of agriculture to the vagaries of the weather is simply a brute fact of life on our planet. Yet the actual and potential impact of weather on agricultural production is rarely taken into sufficient account by the "experts." For instance, normal weather fluctuations alone render absurd any global calculation of how many people the earth can support "on the average," for there is no such thing as average weather except in the statistical records. During the minor periods of drought that are climatically normal, production can easily drop 30 to 40%. In 1987, for example, there was enough food in storage to feed the world for 102 days (Brown 1991, p. 15). But in 1988, because of drought conditions in North America, global cereal consumption far exceeded worldwide production, drawing down world cereal stocks to the lowest level of the 1980s (Brown 1991, p. 15). By 1990, carryover stocks of food had been reduced to 62 days, 2 days over the point at which agriculture prices become extremely volatile and can double or more in price within months. If such a drought were to occur when food stocks were close to this 60-day "threshold of price instability," millions of people now barely surviving could die from spot or regional food shortages. Thus, even assuming that we can indeed feed 8 billion "on the average," the number that we could expect to keep alive over the long term, in good years and bad, might well be less than that.

The problem goes beyond the year-to-year variations in weather; climate changes can also occur. Major volcanic eruptions, although rare, can release particulates and gases that block the sun's rays and can therefore cause declines in crop production. And many manufactured pollutants have a similar effect. Of more concern is global warming, which will have unpredictable consequences for agriculture. Though some crops will grow faster with higher CO_2 levels in their environment, and though some may grow in higher latitudes than they can grow in today, increased crop yields resulting from these factors could be outweighed by adverse changes elsewhere. For example, global warming models predict that moisture levels in mid-level latitudes will decrease. At the same time, warmer temperatures require that more water be supplied to plants because of higher evapotranspiration rates. Lower crop yields may result, therefore, in places such as the United States. Where crops are now grown with irrigation or with barely adequate rainfall, changes in rain patterns could make it impossible to continue to farm in those areas. In tropical latitudes, rainfall changes are expected to shorten the growing season, which may also reduce crop yields.

The regional consequences of global warming under currently understood climate models is uncertain at best. But to the extent that global warming causes changes in food production in the major food-producing areas, it will make it difficult to manage food crises. If global warming, as predicted, also decreases agricultural productivity in food-deficit regions, famine will result. Basic changes in climatic regime have occurred in the past, and even comparatively minor shifts have had major effects on agricultural production.

In sum, the assumption that our weather and climate are constants upon which we can rely absolutely is unwarranted. Because normal weather fluctuations adversely affect agricultural productivity, a world just able to feed a given number of people in "average" years, may be unable to feed that number in a "bad" year. Moreover, natural and human activities can change the regional or global climate, affecting the number of people that can be fed over the long term.

No Technological Panaceas

Surely, many will say, technological answers to most of these problems can be found. Actually, even apart from some of the general limits to indefinite technological innovation (to be considered in the next chapter), the limits discussed above strongly suggest that we have almost exhausted the possibilities of our current form of industrial agriculture. If we push our technology to an extreme, we shall make enormous demands on our energy resources and create serious pollution problems. Only a jump to a new level of agricultural technology can alter this assessment. Unfortunately, such a new technology is nowhere in sight.

We discussed earlier the potential of biotechnology to improve food production. Biotechnology has the potential of genetically altering plants, introducing nitrogen fixation into cereal crops, making crops more pest-resistant or tolerant of adverse growing conditions, and improving photosynthetic efficiency. Some of these possibilities are more promising than others; so far, for example, introducing nitrogen fixation into cereal crops has been elusive. And some techniques, even if they were to be successful, might have drawbacks. Assuming, for example, that biotechnologists can increase photosynthetic capability in plants, more nutrients would need to be supplied to the plant. Current agricultural techniques cause the loss of topsoil containing these needed nutrients. For this and other reasons, most experts believe the shift to biotechnology will necessitate the use of much more chemical fertilizer (Russell 1987), which may have unacceptable economic costs in underdeveloped countries and unacceptable environmental costs worldwide. Biotechnology can create substitute controls on agricultural pests, reducing the need for chemical pesticides. Even here, however, risks abound. New organisms introduced into the environment can themselves become pests; more than 120 species of intentionally introduced crop plants, for example, have become weed pests in the United States (Corson 1990, p. 87).

In short, biotechnology does not appear to be a panacea for our food problem. It may increase food production, but by itself, will do so at some cost: higher energy requirements, lower ecological stability, possible side effects or pollution, and the like. More promising would be a shift to low-input and organic agriculture. Low-input techniques take advantage of some biotechnology (such as biological controls of pests and the planting of pest-resistant plants) to reduce or eliminate the use of pesticides. Organic farmers, in addition, recycle animal manures for fertilizer, rotate crops, plant companion crops, intercrop (plant a row of legumes between two rows of, say, wheat; the legume fixes nitrogen for itself and the wheat), and engage in conservation tillage to maintain and improve soil productivity. Low-input farmers use "fertigation," the application of fertilizer and water around the roots of individual plants, to reduce the need for fertilizer and water and also to reduce the pollution caused by conventional techniques. Examples of low-input and organic agriculture already exist. In China, application of a fungus and parasitic wasp achieved 80–90% control of a major corn pest; IPM techniques on cotton reduced pesticide use by 90% while cotton yields increased (Corson 1990, p. 85). A parasitic wasp has controlled the cassava mealybug, which was wiping out the cassava plant in Africa. The cassava plant is the primary staple of more than 200 million Africans (Postel 1987, p. 37).

In the United States, some farmers have chosen to go completely organic. Eleven cases were cited in a 1989 National Research Council Report, which found that these farmers generally derived "significant sustained economic and environmental benefits" (WRI 1990-91, p. 98; Commoner 1990, p. 98). Organic farming techniques meet the ecologist's ideal for agriculture described in the preceding chapter, "a relatively stable and mature subclimax that is optimum considering all of man's needs and that is characterized by constructive symbiosis rather than warfare between man and nature." But only 5% of American farmers have adopted alternative agriculture, and worldwide, the trend is in favor of conventional modern farming techniques. One measure of this is that world pesticide use is increasing at a rate of more than 12% each year.

One reason for this preference is that, over the short run, modern agricultural techniques increase agricultural productivity.* The world now lives on a very thin margin of food supplies in a time of rapidly increasing population; the problem in most areas is to increase food production as fast as possible. Moreover, although the NRC report concluded that alternative agriculture can be as productive as conventional farming, productivity is briefly lower during a transition period

---

* Actually, researchers are learning that this statement is not always true. When farmers plant trees and other plants along with their crops, crop yields can be higher than in monocultures, as prescribed by the Green Revolution. Trees pump nutrients from deep soil to the surface, making them available to crops. Moreover, they lessen wind and water erosion of the soil and hold rainwater. Shrubs are similarly helpful. A mixed plot in Nigeria not only produced higher crop yields than a nearby monoculture plot, but the shrubs also yielded fuelwood for local people (Matthews 1990, p. 41).

7

Ecological Farming

A mode of agricultural production that gets only a one-to-five or, worse, a one-to-ten return on its energy input may make economic sense in the short run, but it is ecological nonsense in the long run, unless energy is superabundant and ecologically harmless to tap, which is not the case. Moreover, despite what technologists and spokespeople for agribusiness say, there is a real possibility of breaking the vicious circle of technological addiction in agriculture and shifting back toward an agriculture based on dilute but renewable and nonpolluting solar energy but informed by a high degree of scientific understanding and biological sophistication. With care, very high yields could be obtained for millennia from such an agricultural technology. Some of the principles and techniques of ecological farming were suggested in the preceding chapter. Ironically, many of them resemble earlier farming techniques that we have scorned as primitive and inefficient: combined forestry and grazing, controlled cropping of game animals (game ranching) instead of cattle raising in tropical areas, fish ponds that turn wastes into protein, mixed farming instead of monocultures, crop rotation and the use of both animal manure and "green fertilizer," substitution of labor for herbicides and pesticides, and so forth. Especially if they are brought up to date via modern science, these techniques are highly productive (on a per-acre basis they can outproduce industrial agriculture), *but only when human labor is carefully and patiently applied.* Thus farming that is both productive and ecologically sound seems very likely to be small-hold, horticultural, possibly peasant-style agriculture finely adapted to local conditions (especially in the tropics). It should be obvious that many of the developing countries are well poised to make the transition to this modernized version of traditional agriculture. Except for the excessive use of insecticides and chemical fertilizers in some areas, the agriculture of China, Taiwan, Korea, Ceylon, Egypt, and some others is already close to this mode—and has high per-acre yields to show for it. In the United States and most other developed countries, on the other hand, either agriculture will have to "dedevelop" to make the necessary changes, or agribusinesses will have to adapt alternative techniques to large operations. Superior Farming Company and Steven Pavich and Sons, two of California's largest table grape companies, have successfully done so, but resistance elsewhere is high (WRI 1990–91, p. 98).

(typically, for three years). In those countries where the Green Revolution has only recently been adopted, changing to alternative techniques may therefore seem insufferable. In developed countries, where farmers have largely put the farmer's life of toil behind them, fertilizer and pesticide companies not only argue (misleadingly) that crop losses are greater with organic than with conventional techniques, but also stress that alternative farming is more labor intensive, thereby increasing agribusiness's resistance to change.

In sum, while ecological farming can work, it is unlikely to be rapidly embraced and widely adopted. Furthermore, advances in industrial agricultural technology and even food technology will certainly occur, but they come with costs and risks. There is no reason to believe that possible changes in agriculture will be rapid or thorough enough to feed 8 billion people even as well as we feed 5 billion now (surely not a standard we ought to feel proud of). The prospect of feeding 10, 14, or even 16 billion people seems unrelievedly dismal.

Mineral Resources

People eat food. Modern industrial civilization "eats" mineral and energy resources and would collapse if these items essential to its diet were not available in sufficient abundance. Apart from minerals used directly, as are mica in electronics, copper in wires, and nickel in jet engines, minerals are used as alloys, such as steel and bronze, or as essential components of important materials, such as glass, cement and concrete. The value of nonfuel minerals to the United States economy in 1988 was 418 billion dollars. Are there enough minerals to satisfy the present and anticipated requirements of our industrial society? As with the question of the limits to growth in general, there are two basic schools of thought on the availability of minerals.

Mineral Availability: Ecologist versus Technofixers

According to one school of thought, mineral resources are essentially finite, we are using them up at a rapid rate, and we shall run out of many of the raw materials our industrial civilization needs rather abruptly in the near future. The second school holds that supplies of resources are not at all finite; on the contrary, *provided energy is abundant,* they can be almost indefinitely expanded by means of technological innovation and substitution. Spurred by the price mechanism, these will suffice to keep supplies well ahead of increasing demand, until demand levels off, gradually and gently, to an eternally sustainable level. Let us employ the term *ecological* for the former school of thought and *technofixer* * for the latter.

A review of the debate between the two schools will show that, just as for agriculture, an answer to the question turns upon ecological scarcity, rather than resource scarcity in itself. That is, for the next four to five decades the critical question is not whether we are capable of expanding mineral resources to keep up with rapidly growing demands but whether we can bear very much longer the costs of doing so, in terms of energy use, land devastation, and pollution. Let us begin by examining the dynamics of resource use with a fixed stock of a hypothetical resource, applying different kinds of assumptions about supply. Then, following this explanation of the basis of the ecological position, we shall look at the counterarguments of the technofixer school.

The Nature of Exponential Growth

A quantity grows exponentially when its rate of increase during a period of time is a fixed percentage of the changing size of the quantity. Thus a population with access to unlimited supplies of the necessities of life grows exponentially: The increase in population with each new generation means that there are more breeders to produce a further increase in the next generation, and so on. Similarly, money earning compound interest grows exponentially because it yields a certain percentage of both the original amount and the accumulated interest. (By contrast, money invested at simple interest yields a return on only the original amount; there is no compounding, so growth is linear.) With exponential growth, at the end of some period that depends on the percentage rate of increase, the original quantity will be doubled. Naturally, raising the percentage rate of increase shortens the doubling period. At a steady 5%, a quantity doubles about every 14 years; at 10 percent, there are only 7 years between doublings. † Figure 2-4 illustrates the increase with time of any quantity that grows exponentially. (Note that Figure 2-4 is simply the bottom half of the familiar sigmoid curve.) However, the significance of

---

* We are indebted to Margaret McKean of Duke University for suggesting this term instead of economic, which was used in the first edition. Among other things, she points out that the ecological view pays attention to ecological variables but that it is not "non-economic" and should not be so characterized.

† Dividing the percentage rate of increase into 70 gives (roughly) the doubling time in years. Of course, a quantity growing linearly doubles too, but the time between doublings gets longer and longer as the series continues.

continued exponential growth may be conveyed more vividly by two parables.

One tells of a greedy merchant who asked the king to pay for his services with grains of wheat on the squares of a chessboard—one for the first square, two for the second, four for the third, and so on to the sixty-fourth square. Unfortunately for the merchant, the king was not mathematically naive. A quick calculation showed the king that the amount of wheat required for the final square would be about 8.5×10^{18} grains, an astronomically large amount, and he quickly put the greedy and treacherous merchant to death. This story illustrates the levels that an exponentially growing quantity, starting from insignificant amounts, can reach after comparatively few doublings. As one looks higher on the exponential curve (Figure 2-4), the absolute amount of increase at each doubling soon becomes staggering. We have already noted some of the very large demands further doublings of the world population would place on the earth. It should now be obvious why the sigmoid curve is universally observed in nature: An exponential growth curve must eventually bend over and become a flat line, for a mere 20 or 30 doublings of almost any population or quantity suffice to produce amounts that cannot be sustained or produced on a finite earth.

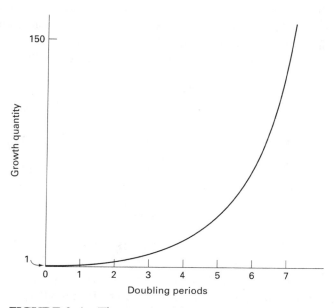

FIGURE 2-4   The exponential curve, which describes the size of any exponentially growing quantity after a given number of doubling periods.

The second parable, popularized during the public debate over *The Limits to Growth,* concerns a lily pad growing in a fish pond. According to the parable, the lily doubles in size each day and if the lily keeps growing, in 30 days it will cover the pond, killing the fish. At first the lily is a tiny speck in the pond, and even after many doublings it remains very small relative to the pond. The farmer who owns the pond is not concerned. He believes that he has plenty of time to cut back the lily and save his fish. Even on the morning of the twenty-eighth day, when the lily covers only one quarter of the pond, he does not realize that by the next morning the lily pad will occupy half of the pond's surface and that he will then have precisely one day to cut back the now rapidly burgeoning lily. This story illustrates the insidious nature of exponential growth when there is a finite limit. Until the very last stages of the progression, the limit appears so far away that it may seem like a problem for our great-grandchildren, not us, to worry about or, at the very least, a problem that we can handle in due course once it becomes pressing. The catch is that realization of the consequences of unchecked growth often comes late. Once the problem has become pressing, it may require heroic measures to check the momentum of rapid growth and prevent a crash.

Growth versus Resources: Running Out in Theory and Practice

In light of this general understanding of the nature and power of exponential growth, let us see what happens to a hypothetical mineral resource of finite dimensions as demand grows exponentially at 3.5% per annum, so that doubling of the absolute demand occurs every 20 years.

Doubling	Remaining Stock (tons)	Current Demand (tons per annum)	Static Reserve (years)
Start	50,000	100	500
1st	47,000	200	235
2nd	41,000	400	103
3rd	29,000	800	36
4th	5,000	1,600	3

Thus, as demand doubles, redoubles, and doubles again, absolute demand rises from a modest 100 tons per annum to 16 times that amount by the fourth doubling, and a resource that appeared at the start to be adequate for half a millennium begins to be depleted very rapidly until, finally and abruptly, the stock is exhausted. The static reserve—the number of years of useful life at current levels of demand—is therefore a very poor indicator of how long a resource is likely to last if demand is growing. Much more important is the exponential reserve—that is, how long the reserve can be expected to last when probable future demands on the resource are taken into account. In this case, the exponential reserve is 83 years, less than 20% of the 500-year static reserve.

Suppose now that we wish to take into account new discoveries and new technologies that will expand our hypothetical resource. Let us say that instead of 50,000 tons total reserve we really have 500,000 tons, or 10 times as much. This allows usage to go on for a few more years, as continuing our tabulation under the new assumptions shows:

Doubling	Remaining Stock	Current Demand	Static Reserve
4th	455,000	1,600	284
5th	407,000	3,200	127
6th	311,000	6,400	49
7th	119,000	12,800	9

The stock will now last for about 147 years. Thus multiplying the total available stock by 10 (1000%) extends the time to exhaustion by less than 80%.

However, a mere tenfold increase in the stock may still be too pessimistic for some, so, as an allowance for resource expansion through any conceivable combination of technological innovation and discovery of new resources, let us assume that we start with 100 times our original stock. This does in fact allow growth to continue for a little longer:

Doubling	Remaining Stock	Current Demand	Static Reserve
7th	4,619,000	12,800	361
8th	4,235,000	25,600	165
9th	3,467,000	51,200	68
10th	1,931,000	102,400	19

But alas, the impact of even such a fantastic increase in our hypothetical resource is dismayingly small. A 10,000% increase in the original stock gives a mere 217 years exponential reserve at a 3.5% growth rate. This is hardly more than double (261%) the original 83-year exponential reserve. Note also how staggeringly large the absolute level of demand becomes as growth continues.

It is obvious from this hypothetical example that once absolute demand attains a substantial level, continued growth in demand begins to consume the remaining resources at an extremely rapid rate. However, the example contains a more important lesson about the insidiousness of exponential growth: even if foresight leads to adoption of a no-growth policy when about half of the stock is used up, there is very little effect on the time to exhaustion. For example, in the last tabulation, if at the time of the ninth doubling we forbid any further growth and simply continue to use up the resource at the annual rate then attained, the stock will last only an additional 68 years. In other words, assuming a 3.5% growth rate until then, the total life of the stock will be 248 years—only 31 years (14%) longer than if we simply allow the growth curve to run its course. This is so even though *nearly 70% of the original stock remains* when we prudently switch to a no-growth policy. Thus we have established three general principles:

1. Given steady exponential growth, the absolute size of the stock of any resource has very little effect on the time it takes to exhaust the resource.
2. Given already high absolute demand on a particular resource, the rate of growth in demand thereafter has almost no effect on the time it takes to exhaust the resource.
3. The time for concern about the potential exhaustion of a resource comes when no more than about 10 percent of the total has been used up.

Since demand for minerals has increased exponentially over the last two centuries and has now reached levels that are very high indeed, the concern of the ecological school appears to be amply warranted, and real-world statistics provide further support for pessimism. Table 2-3 lists identified reserves of major minerals.*

The case is clear. More than half the static reserves are less than 50 years, the average growth rate is about 2.3 percent, the doubling time is

---

* As the name implies, "identified" resources are those that are definitely known to exist and are believed to be recoverable with current technology or under prevailing conditions. That these resources exist does not necessarily mean that extraction facilities are in place or operative.

36 years, and the exponential reserve figures indicate that about half of these major minerals will be exhausted in less than 25 years at current growth rates. It is true that in the major industrial countries, the rate of increase in the per capita consumption of these raw materials has been decreasing since the 1970s. But the level of consumption remains extremely high: The United States alone is estimated to have "consumed more minerals from 1940 to 1976 than did all of humanity up to 1940" (quoted in Young 1991, p. 41). Humanity today consumes twice the copper and steel it consumed in 1950; it consumes seven times the amount of aluminum. As we have seen, even if the people of the world agreed now, at these high rates of usage, not to increase the rate of consumption of mineral resources, that action would have little effect on the time it takes to exhaust the resource. The figures show that lead, mercury, tin, and zinc would still be exhausted in less than 25 years at current rates of consumption. Thus the predictions of the ecological school appear to be valid, unless some combination of new discoveries

TABLE 2-3 Identified Reserves of Important Minerals

Mineral	Static Reserve (years)	Growth Rate of Demand (percent)	Doubling Time (years)	Exponential Reserve (years)
Aluminum	224	4.0	18	99
Copper	41	2.7	26	23
Iron	167	2.4	29	66
Lead	22	1.8	39	11
Mercury	22	1.4	50	★
Nickel	65	3.0	23	36
Tin	21	1.0	70	16
Zinc	21	2.0	35	17

★Data unavailable.
Source: WRI, "STARS" DATA BASE 1990-91, condensing data taken from the U.S. Bureau of Mines and the World Bureau of Metal Statistics. Growth rate taken from U.S. Bureau of Mines, *Mineral Facts and Problems*, 1985.

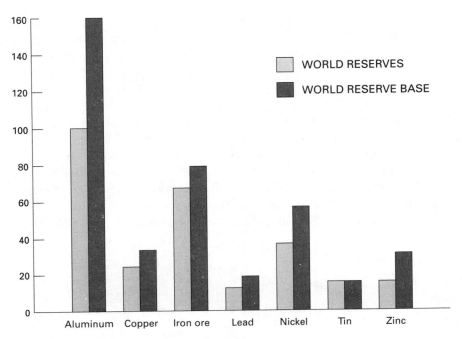

Figure 2-5 Years remaining of metal reserves.
Source: Adapted from WRI, 1990–91.

and technological innovation can indeed expand mineral resources more or less indefinitely. Let us, then, examine the various possibilities that the technofixer school would rely on.

The Limits to Discovery and Substitution

Most mineral geologists and mining engineers are on the whole pessimistic about our finding substantial new supplies of ore. Sophisticated forecasting techniques have been developed to estimate the total size of any given resource under various geological and economic assumptions, and expert predictions of total potential reserves tend to agree rather substantially (that is, the highest estimate rarely exceeds three or four times the lowest). Furthermore, geologists believe that the major metallogenic provinces (areas with a high concentration of ore deposits) are well known and for the most part well explored. Ore deposits remain to be found, but we have already skimmed off the cream, and much more sophisticated and expensive techniques must be used to ferret out the better hidden and less extensive deposits. Going deeper into the earth is no answer; the probability of ore formation decreases with depth, while the costs of extraction rise very steeply. The ocean, too, has been vastly overrated as a potential source of minerals. The available quantities are less than popularly imagined, and the difficulties and cost of extraction are enormous. The probability of really major discoveries has thus declined sharply and will continue to decline with each passing year. It is of course possible that new developments, such as the current "earth sciences revolution," may lead to some unexpected discoveries that, unlike manganese nodules on the ocean floor, readily are exploitable. However, as our hypothetical example clearly shows, given the present very high levels of absolute demand, even major discoveries would have only a modest effect on either the static or the exponential reserve. Doubling the stock merely extends the time to exhaustion by one brief doubling period, and even a tenfold increase has little impact.

In fact, no assumed basis of optimism really makes much difference, as is clearly illustrated by Table 2-4, which takes into account not only resources that could be economically extracted but also those that are marginally economic and subeconomic. Comparing Table 2-4 with Table 2-3 shows that most static-reserve figures do not improve spectacularly (only aluminum shows even a tenfold increase); exponential-reserve figures, with the exception of aluminum, differ hardly at all (about half of the mineral resources will be exhausted in 33 years, and this is assuming that the now subeconomic portion can be mined). In sum, the optimism of the technofixer school about the impact of future discoveries appears to be unwarranted: Very little remains to be discovered about current and projected demand, and the technological difficulties and economic costs of finding and extracting hypothetical and speculative resources are far from trivial.

Even if future discoveries enlarge the static reserves beyond these projections, that good fortune will buy us only a few more decades at best. Will substitution be the answer to impending shortages? Here again there are difficulties that the technofixer school has not confronted. First, the resource problem is one of confronting *general* scarcity, not simply of coping with the exhaustion of one or two particular minerals. Without substantial new discoveries, we shall have to invent a new metallurgy that can do without copper, lead, nickel, tin, zinc, and probably mercury—and in fairly short order. Second, although it is possible, for example, to substitute aluminum for copper

Table 2-4 Maximum Reserves of Important Minerals, Assuming the Growth Rates Listed in Table 2-3

Mineral	Static Reserve (years)	Exponential Reserve (years)
Aluminum	2338	160
Copper	66	33
Iron	236	78
Lead	37	18
Mercury	42	*
Nickel	144	56
Tin	21	16
Zinc	42	30

*Data unavailable.
Source: WRI, "STARS" DATA BASE 1990-91, condensing data taken from the U.S. Bureau of Mines and the World Bureau of Metal Statistics. Growth rate taken from U.S. Bureau of Mines, *Mineral Facts and Problems,* 1985.

in most electrical uses, this transfer of demand would simply cause aluminum to be used up even more rapidly than projected. Third, even good substitutes (such as aluminum for copper) are on the whole less efficient than the material they substitute for, and more energy is therefore required to perform a given function.* Fourth, some metals have properties that are unique. For example, mercury is the only metal that is liquid at normal temperature and pressure; it is essentially irreplaceable in temperature- and pressure-control equipment. Other metals that would be hard to replace include cobalt for magnets, silver in many photographic uses, and the platinum group for industrial catalysts.

* Fortunately, this is not always true. For example, the substitution of fiber optics for copper in some applications results in a net *improvement* in the efficiency of electrical transmission.

Moreover, many metals serve the same function in metallurgy as vitamins in our diets; only small quantities are needed, but they are essential (for example, the manganese in steel). Needless to say, this will add to the difficulty of inventing new metallurgies. Fifth, plastics are often proposed as metal substitutes, but plastics are petrochemicals, and before long we will run out of petroleum (see Table 2-5). Also, the production of plastics involves serious pollution-control problems. Finally, many proposed technological solutions to environmental problems, such as the breeder reactor and thermonuclear fusion, require large quantities of very particular types of materials of a very high quality, and for many of these major resources, such as helium and lithium, there seems to be no satisfactory substitute. It appears, then, that substitution will not help us much in overcoming the emerging problem of mineral scarcity. It can alleviate particular shortages and buy time, but it is not a solution, either alone or in combination with new discoveries.

Limits to Extracting Minerals from Seawater and Ever-Thinner Ores

Technologists (especially those connected with the nuclear industry) and most economists usually see energy as the answer to all mineral-resource problems. They say that if abundant and cheap energy is available, we can, if worse comes to worst, literally burn up the rocks to obtain our minerals (and even the fuel to extract them). Resource scarcity is thus not seen as a problem, for nuclear energy will provide us with cheap, abundant, nonpolluting energy. Fortunately, this seductive argument ignores a number of crucial problems, even if the availability of energy is granted.

First, only six metals are abundant in the earth's crust (*abundant* is defined as present in a quantity exceeding 0.01% of the crust by weight)—iron, aluminum, manganese, magnesium, chromium, and titanium. All the rest of the metals are geologically scarce. It is a characteristic of the former that, as the standard for economically minable ore is lowered, the quantity of lower-grade ore increases. In other words, there is a more or less gradual continuum from the very richest ore bodies of these abundant metals to the almost infinite quantities of very low-grade "ore" contained in ordinary rock. Therefore, the theory goes, all it takes to keep extracting the metal is energy and the technical ability to mine and refine ore of successively lower grades. However, the lower the grade of the ore, the greater the economic cost (which rises exponentially with decline in grade), the greater the volume of rock that must be processed and disposed of per unit of useful output, the higher the energy cost per ton of metal, and above all, the greater the ecological cost due to pollution and increased demand for water. All of these factors are serious limitations on our ability to continue extracting the abundant metals, despite our technical ability to do so.

Second, the scarce metals are characterized by a sharp discontinuity between their concentration in currently minable ore and their average concentration in the earth's crust. Once concentrated ore deposits are mined out, lowering the grade slightly does not increase availability. Only a leap to ore that may be leaner *by 3 to 5 orders of magnitude*—that is, ordinary rock—will produce more metal. Because the economic and ecological costs, the volume of rock that must be moved, and the quantity of energy required for continued production of a scarce metal from ever-thinner ores all increase exponentially with decline in ore grade, the side effects from burning the rocks to get mercury, zinc, tin, and the other scarce metals will be staggering. For example, to obtain a mere 400 tons of zinc (a tiny fraction of the annual world demand of over 7,114,000 metric tons,) it would be necessary to process 5 million tons of ordinary rock with perfect efficiency, for the crustal abundance of zinc is only 0.0082% by weight.* Thus, in order to satisfy even a fraction of current demand for these metals under such a regime, truly astounding amounts of material would have to be obtained, processed, and disposed of. The economic, energetic, and environmental costs would be of even more astounding dimensions: current problems with stripmining would seem trivial by comparison. It thus appears extremely doubtful that we shall ever obtain zinc and the other scarce metals in any quantity from either ordinary rock or seawater—and if we were to do so, these metals would cost considerably more than gold does today.

Third, the harm to the environment that will result from efforts to extract and process more metals from the earth may necessitate halting such mining well before a metal "runs out." Even today, mining causes extensive harm to the environment. Millions of hectares of land are

---

* The problem of extracting minerals from seawater, which has also been proposed, is essentially similar but is even more intractable, because the concentration of most minerals in seawater is vastly less than their concentration in the crust. Using zinc as an example again, we find that to get the same 400 tons, it would be necessary to process 9,000 billion gallons of seawater (equivalent to the combined annual flows of the Hudson and Delaware Rivers) with 100% efficiency (Cloud 1969, p. 140).

devastated each year, and millions of trees are destroyed. Billions of tons of solid waste are produced, to say nothing of vast air and water pollution. In the United States, 9 million hectares are covered by current and abandoned coal and metal mines—an area more than half the size of all the roads, parking lots, and other paved areas of the country. In addition, pollution from mines is carried by water and wind far beyond the mining sites. In the western United States, 16,000 kilometers of streams contain acidic or toxic drainage from mines and mining wastes (Young 1991, p. 42). The smelting of ores discharges into the air sulfur oxides, arsenic, lead, and heavy metals, some of which are carcinogenic and some of which cause acid rain, and they blow across countries and international boundaries. Non-fuel mining produces over 1 billion tons of waste annually, *six to seven times the total amount of municipal waste produced* (Young 1991, p. 42). All of these effects will increase as the world mines lower-grade ores; more land will be devastated, and an ever-greater part of what is mined will be waste materials.

Reducing Demand: The Limits of Recycling and Conservation

If we cannot expect to expand supply very substantially by new discoveries, easy substitutions, or (with the possible exception of several abundant metals) using ore of ever-lower grades, can we reduce demand sufficiently to produce the same net effect? This may indeed be a more promising approach to the problem of resource scarcity, but we must still confront limits on our economies. It is often said that once the price of a metal has climbed high enough, it will pay to recycle the metal scrupulously, so that we shall be able to keep reusing the same stock. Yet first of all, we must remember that in accordance with the second law of thermodynamics, the recycling process can never approach 100% efficiency.* When all the inevitable losses—in production, from friction and wear, through corrosion and other chemical processes, from outright

---

* Every time energy is transformed from one state to another, a certain penalty, called entropy, is exacted. Entropy is unavailable, unusable, or dissipated energy. For example, when water flows over a dam into a lake, it can turn a turbine and generate electricity. In the lake below, the water is no longer in a useful state to do such work. Any time any work occurs in the world, some amount of energy becomes unavailable to do future work. For thermodynamic reasons, products differ radically in the entropy created in their recycling process (the "theoretical efficiency" with which they can be recycled). Thus metals can be recovered with high efficiency, but petrochemicals cannot. For example, four old tires are needed to make the raw materials for one new tire.

loss and other human failure—are added up, we should be miraculously lucky to achieve the 90% level in recycling efficiency. (Actual recycling efficiency for used metals is now on the order of 30% or less.) The only metals for which we could expect a higher ratio of recovery are the chemically nonreactive precious metals, which are hoarded rather than used. Thus, to maintain the stock in use, we shall still need raw materials. Second, many recycling processes are rather dirty, so recycling will continue to add to our pollution problems. Third, recycling runs thermodynamically uphill—that is, scattered materials must be collected, transported, and transformed from a high-entropy to a low-entropy state—which requires energy, typically in rather large amounts (this is one reason why the United States recycles only 11% of its solid waste).* Fourth, the potential efficiency of recycling and the design of products are directly related. For example, even small amounts of contamination may make a particular metal useless for many industrial purposes, and designs that make recovery impossible without cross-contamination will frustrate recycling (this is the major technical impediment to recycling the metals in junked cars) or compel us to use inordinate amounts of energy to purify scrap metal. Thus the effective recycling of many major metals will require a revolutionary change in industrial processes, business practices, and design standards—and perhaps also in the nature of our cities.

A second approach to reducing demand is conservation or source reduction. If we can do more with less, or even the same with less, by making more efficient, more durable and smaller machines or structures, then we can reduce levels of demand accordingly. This is often not so much a matter of technological wizardry—the savings to be made by further miniaturization and other innovations are probably rather small— as of not using materials wastefully. The typical high-rise building, for example, contains far more steel than is really required for structural integrity; the typical residence still uses more wood than is needed. Another way to avoid waste is to reuse materials. Parts of automobiles,

---

* The energy used in reducing ores to a pure metal is higher—sometimes much higher—than the energy used in recycling a product for reuse. (In the most extreme case, producing an aluminum item from recycling reduces energy use between 90% and 97% percent compared to the energy needed to produce that same item from raw materials.) But governments frequently subsidize the mining of raw materials. In the United States, for example, mining companies receive massive tax exemptions, called depletion allowances, which offset the energy costs of producing raw materials. So far the government has not granted equivalent tax breaks to recyclers.

refrigerators, washing machines, plumbing materials, aluminum siding, and small tools and machinery can be cleaned or reshaped and reused. Consumer cans and bottles can be reused as well. The savings in materials and energy, even compared to those possible through recycling, are huge. A 12 ounce "refillable" bottle, for example, can be reused 50 times. To ready it for reuse consumes only 2.4% of the energy needed to create a recycled container for reuse and only 1% of the energy needed to create a bottle or can from virgin materials. Despite this, few "bottle bills" have been adopted to encourage people to abandon their "throwaway" habits.

Another approach to conservation is to make products more durable. If the average car were built to last twice as long as current models do, it would reduce by nearly half the auto industry's gluttonous appetite for materials. This is the kind of technically simple economy measure that would have the greatest immediate impact on the life of reserves. Paradoxically, this is also the kind of measure that is least discussed. Economic factors are probably the chief explanation: Products that were more durable, were easily reusable or recyclable, and fit other resource-conserving criteria would cost more, and business turnover would be less. The "iron law" of American marketing, according to Vance Packard, is to create "maximum sales volume [which] demands the cheapest construction for the briefest interval the buying public will tolerate (Young 1991, p. 47). Until the price of ordinary metals approaches that of gold, therefore, resource-conserving production will probably be at a disadvantage if matters are left to be decided purely by market forces.* In any event, conservation only buys time; if growth continues, and especially if we use some of the resource-intensive technological "fixes" that have been proposed, then any conceivable conservation program will have little more effect than a mere doubling of reserves.

Minerals: An Emerging Crisis

To summarize, there are demonstrable limits on the expansion of mineral resources. After spending most of its history as a mineral-rich nation, the United States now finds itself increasingly dependent on imports for a very large proportion of its essential requirements. Barring a deliberate national policy of maximum autarky, or national self-sufficiency (which would be possible in the short run if we were willing to bear the economic and ecological costs), the dependency ratio is bound to in-

---

* We shall explore these kinds of political, economic, and social impediments to ecological rationality in Part II.

crease in the future. As a result, we will probably face major political, economic, and psychological adjustments, especially because worldwide competition for minerals appears to be increasing. The United States trade deficit in mineral materials, which has varied between $9 billion and $16 billion annually since 1983 (Corson 1990, p. 184), will present us with increasing difficulties, particularly in view of the trend toward overwhelming dependence on foreign supplies of energy. All of the industrial countries will have similar difficulties. All of them (except the Soviet Union) run trade deficits in mineral materials. These countries will have to compete among themselves and with the developing world for diminished supplies. All will have to make adjustments in consumption patterns as supplies become uneconomic to obtain or the environmental costs of obtaining them become unacceptable.

Energy

Can Energy Supplies Surmount Ecological Limits?

Humanity needs energy. To produce more food and to mine the earth's minerals (especially those that are becoming depleted) require energy. To cook food, to heat homes, and to replace human labor in agriculture and industry require energy. To transport ourselves—indeed, to do all work—requires energy. What are our energy resources and how rapidly are we consuming them? What new sources are under development? What are the dangers, if any, attached to current and future energy production and use? These last two questions are the most important. Current methods of energy production and use, as we shall see, are destroying the planet. At some point, the biosphere will be so badly damaged that the costs, in terms of human suffering and death, will force us to abandon these methods—even though, if we wait to be forced, *anything* we do will be too late. On the other hand, humanity theoretically has the capacity both to reduce energy's contribution to environmental havoc and to meet its energy requirements for some time. To do so would require us to use existing energy vastly more efficiently than we do now and rapidly to develop new, nonpolluting energy sources. Depending on the choice we make, energy production will either contribute to our pollution problems or become part of their solution.

Rising Energy Demand and the Supplies of Fossil Fuels

In 1987 the world consumed 20% more energy than it consumed a decade earlier. Most of that energy, 90% of it, came from fossil fuels, biologically stored solar energy from the remains of finite numbers of plants and animals that lived millions of years ago. Most of the increase in consumption took place in developing countries, whereas in Europe and the United States, as energy conservation measures enacted in the 1970s began to take effect in the 1980s, energy consumption went up only slightly. In Africa, however, consumption went up 68%, in South America 30%, and in Asia 54% (WRI 1990-91, p. 316-17). Worldwide, energy consumption is expected to increase by 2% per year through the year 2000. As they try to industrialize, as they expand their electrical supplies, and as they base their transportation systems increasingly on the use of motor vehicles, developing countries will be responsible for the bulk of increased energy consumption; their projected increase is 4 to 5% per year (Davis 1990, p. 58).

World supplies of oil and natural gas, if they can be used, are adequate to meet expected levels of demand until about the middle of the next century. Table 2-5 shows the pertinent figures for coal, petroleum, and natural gas. Coal is very abundant; at 1989 levels of use, coal would last for 218 years. In the United States, domestic petroleum will be effectively used up by about the year 2000 (as is common knowledge, the United States has become dependent on foreign supplies well in advance of this date). Domestic natural gas is projected to last for another decade, until about 2010. But like the world as a whole, the United States does have huge quantities of exploitable coal (a reserve projected to last 286 years at the 1989 use rate).

Unfortunately, however, fossil fuels cannot be consumed at the 1989 rate for very much longer. As with nonfuel resources, we have already skimmed the cream, and much of what we can reasonably anticipate drilling or mining in the future will be increasingly uneconomic. For

Table 2-5   Proved Fossil Fuel Resources at 1988 Production Rate

Fossil Fuel	Years Remaining
Oil	41
Natural gas	58
Coal	218

Source: Adapted from British Petroleum, *Statistical Review of World Energy* (BP London, 1989), p. 2, 20, 24. Cited in World Resources Institute 1990-91, p. 145.

example, Gever and his associates calculated that the amount of energy produced from oil compared to a given amount of energy used for exploration, extraction, and processing—the energy output/input ratio— declined from about 100 in the 1940s to 23 in the 1970s (quoted in Daly and Cobb 1989, p. 406). In 1990 the energy output/input ratio had declined to 8; by 2005 it is expected to drop to less than 1 in the United States, and no new domestic oil production will occur (Daly and Cobb 1989, p. 406). Nor can all the extractable reserves of fossil fuels be drilled or mined. Humanity is not likely to tolerate the burden of increasing environmental damage caused by their production. Oil exploration and drilling are already controversial in the United States. For example, the Trans-Alaska Pipeline was controversial when proposed. When Congress waived environmental standards and allowed it to be built, more controversy developed after the project created an environmental mess, the most visible (but not the most destructive) manifestation of which was the Exxon *Valdez* disaster (see Box 8). Controversy also persists in the United States over the drilling of oil in the Arctic National Wildlife Refuge (which, if permitted, would yield only about a year's supply of oil) and over drilling off the coast of the lower 48 states. Some offshore oil drilling has been prohibited because of environmental concerns. The National Academy of Sciences estimates that drilling a single oil well produces between 1500 and 2000 tons of drilling muds and cuttings, most of which is contaminated with hazardous chemicals (quoted in Holing 1991, p. 15). Drilling companies dispose of millions of pounds of these materials by dumping them on the sea floor, suffocating lobsters and other bottom dwelling creatures. Drilling also produces waste water contaminated by oil, grease, cadmium, benzene, lead, and radioactive and other hazardous materials. The oil industry discharged 1.5 million barrels of such contaminated waste water into the Gulf of Mexico *each day* in 1986. Oil drilling also causes air pollution, including nitrogen oxides, sulfur oxides, and hydrocarbons. These combine to produce ozone, smog, and acid rain, the consequences of which we will discuss in the next chapter. The daily air emissions from one exploratory drilling rig are equivalent to those from 7000 cars driving 50 miles (Holing 1991, p. 16).

Finally, whatever way oil is produced, it almost always must be transported, which creates additional environmental problems. Each year between 3 and 6 million metric tons of oil are spilled—most deliberately—into the oceans. (Oil tankers routinely wash themselves out with clean seawater, dumping the oily ballast water back into the sea). "Accidental" oil spills are also commonplace. For example, 3 weeks before the Exxon *Valdez* incident, the Exxon Houston spilled 117,000 gallons off Hawaii; 2 weeks after the *Valdez* spill, another ship released 10,000

8

Environmental Effects of North Slope Oil Production

Many people are aware only of the 1989 Exxon *Valdez* disaster that spilled more than 10 million gallons of oil into Alaska's Prince William-Sound, killing thousands of birds and other wildlife. But even before that disaster, tankers transporting North Slope oil had been involved in 200 spills of some 13 million gallons of oil. Port Valdez was classified as an impaired waterway under the Clean Water Act because of routine discharges of toxic materials into it. Air pollution from the port was ranked as one of the worst on the west coast. Ground-level concentrations of benzene, a carcinogen, were 4 times the level of Chicago's.

Building the Alaska pipeline resulted in the spillage of 2 million gallons of oil and other materials; keeping the oil moving through the pipeline emits formaldehyde, benzene, toluene, xylene, and 700 tons of sulfur dioxide and 600 tons of nitrogen oxides per year. 100,000 metric tons of methane are emitted each year and 70,000 cubic yards of drilling wastes are produced *each day*. Billions of gallons of toxic wastes have been injected into the ground. What was once 800 square miles of pristine wilderness is now dotted with piles of hazardous solid wastes, 350 miles of roads, and 1000 miles of pipelines (Holing 1991, p. 17-18).

gallons off another Hawaiian coast. On June 23, 1989 there were 3 major spills of oil, totaling over 1 million gallons, into United States waters. From 1970 to 1985, the number of oil spills off the United States coast increased by 196%; the total volume of the spills increased by 57% (Commoner 1990, p. 38).*

---

* Oil spills have been predicted by some environmentalists and industry critics since the 1950s. They demanded legislation requiring new supertankers to be built with double hulls. But industry dismissed their fears of oil spills as exaggerated and successfully persuaded Congress to kill the environmentalists' bills (Bookchin 1990, p.82-83). Repeated efforts to require double hulls on supertankers were beaten back until 1990, when the United States Congress enacted a weak law in the aftermath of the *Valdez* and other incidents (cited in Chasis and Speer 1991, p. 21).

9

Can We Meet Our Energy Demands with Coal?

Many believe that American energy needs can be met from our substantial reserves of coal until technological invention rescues us altogether from dependence on the stored-up energy in fossil fuels. Unfortunately, even if some of the technological improvements now on
the drawing board (for example, magnetohydrodynamic power generation) prove to be practical methods of extracting work from coal more
efficiently and cleanly than we now do, burning more coal will still
magnify our existing pollution and ecological problems. As we will see
in the next chapter, perfect control of effluents is rarely possible (physically or economically), and in any event, technological controls on the
final step in production do nothing to mitigate ecological damage by
all the other links in the chain from extraction to final use. The most
striking and uncontrollable side effect of increased coal mining will be
land devastation, for the most readily accessible and cleanest (that is,
low-sulfur) coal must be strip-mined. The notoriously high human and
ecological costs of strip mining are blatantly evident in Appalachia, and
now the west and southwest are also under siege. Reclamation of strip-
mined lands will always be inadequate. The well-known reluctance of
mining companies to pay for it is the least of the problem. Even under
the most favorable conditions, the land is left sunken and maimed, and

The exploitation of the remaining coal also presents problems. Underground coal mining is a hazardous occupation. Apart from risking injury and
deaths in mining accidents, coal minors suffer from coal induced black lung
disease, emphysema and cancer. Surface mining causes land defacement and
acid mine drainage. Industrial countries have passed laws intended to reduce
these problems, but many laws have not been strictly enforced, and certain
restoration problems have so far been intractable. The United States government, for example, has not enforced the Surface Mining Control and
Reclamation Act of 1977; in issuing mining permits to coal operators,
administrators routinely grant exemptions from soil reconstruction requirements and performance standards. In addition, coal operators do not have the
equipment or technological means to restore farmlands to their pre-mining
capabilities. Finally, as we shall note in Chapter 3, mined land is subject to
erosion, landslides, and floods. Even reclaiming it does not eliminate the acid

water quality as well as other environmental values are inevitably impaired. In the arid west, there is not enough water even to attempt restoration.

If all the coal mines, power stations, and liquefaction or gasification plants now projected were to be built, they would require for their operations, *exclusive* of reclamation, between three and four times the *total* amount of water now used throughout the entire country (McCaull 1974). But because water is generally coming into short supply, and farms and industry as well as ordinary individuals also need that water, it will clearly be impossible to operate all the projected mines and plants. Moreover, making coal the basis of our energy economy will require money, labor power, and a general logistical effort (for example, to build transportation facilities) of staggering dimensions, especially if we strive for a maximum of "energy independence" in the near future (A. L. Hammond 1974a). Maximum coal development would also necessitate the destruction of good arable and forested land—in short, the sacrifice of long-term agricultural productivity for short-term energy production.

Thus the amount of coal we can reasonably expect to obtain and use is far less than the amount theoretically available in the ground. At best, coal offers only a temporary, stopgap satisfaction of our short-term energy requirements, and we shall very soon have to conceive and construct alternative technologies of energy supply that do not depend on such nonrenewable and environmentally damaging fuels.

drainage problem, which is contaminating streams and groundwater beyond the mining areas themselves.

Even if humanity could live with the pollution caused by the extraction of fossil fuels, it is questionable whether humanity can live with the pollution caused by their *burning*. In Chapter 3, we will see the extent to which acid rain is damaging aquatic life in eastern lakes and streams and how it corrodes buildings, weakens some species of trees, and adversely affects human health. We also will observe how smog causes human health problems, damages agriculture, and also kills forests. Technological means exist to reduce the amount of smog and acid rain formed with the burning of fossil fuels, both by producing "cleaner" fuels in the first place and by using new control technologies during burning. Some advanced repowering technologies for producing electricity, such as the integrated coal-gasification combined cycle system (IGCC), can reduce methane and remove almost all of the sulfide by-products of coal before the synthesis gas is burned (Fulkerson et al. 1990, p. 131). At the same time, the efficiency by which power is produced in these plants is raised from 34% (for a coal-burning conventional plant) to 42%. But utilities are slow to adopt advanced technologies. The new technologies may be perceived to be (and indeed may at first be) more costly, or they may be perceived as unnecessary. Utilities do not want to retire today's conventional power plants before their useful life (which may last from 20 to 40 years) is over, and it may take a very long time from the day a new technology is conceived until it is designed, developed, tested, refined, and finally installed.

Moreover, even if utilities and industry were persuaded or forced to quickly adopt state-of-the-art technologies to reduce the emissions of fossil fuels that cause smog and acid rain, carbon dioxide is the inevitable product of combustion. The buildup of carbon dioxide in the atmosphere causes the greenhouse effect. Six billion tons of carbon are added to the atmosphere each year, and the amount has been rising by 400 million tons annually since 1986 (Flavin and Lenssen 1991, p. 24). A United Nations study concluded in 1990 that these levels of carbon emissions will produce rapid and highly disruptive climate changes in the coming years—changes that will harm agriculture, forests, and people directly. (We will discuss this further in Chapter 3.)

Unfortunately the carbon emissions of burning coal, which we have in abundance, are higher than those of burning any other fossil fuel. Coal produces 25% higher carbon emissions than an equivalent amount of oil and 80% higher carbon emissions than an equivalent amount of natural gas. Producing synthetic fuels or synthetic natural gas from coal does not overcome this problem. Although advanced powering technologies, such as IGCC, * can reduce carbon dioxide emissions slightly by improving the efficiency of electrical production, coal gasification by conventional

---

* Integrated coal-gasification combined cycle plants would convert coal to a gaseous mixture, using steam and oxygen. The gases would power a gas turbine to produce electricity, and the heat from the gas would be harnessed to vaporize water to run a steam turbine (Fulkerson et al. 1990, p. 132). Because the process powers two turbines, it gets more electricity for the energy supplied than do today's coal plants. Theoretical models suggest that efficiency may be as high as 42% with advanced power technologies, compared to about 33 to 37% with conventional plants. The latter simply burn coal to vaporize water to run a steam turbine; the gases themselves just go up a smokestack as waste.

means *increases* carbon dioxide emissions by 50% or more per unit of energy produced (Corson 1990, p. 194).*

Scientists believe that in order to stabilize the atmospheric concentration of carbon dioxide, worldwide carbon emissions must be *reduced* by 60 to 80% (Flavin and Lenssen 1991, p. 25). To produce and consume coal at 1989 rates is utterly incompatible with this objective.

In any event, humanity may not stabilize atmospheric concentrations of carbon dioxide in time to avert highly disruptive climate changes. It takes decades to achieve reductions in carbon emissions, even when there is—as in the United States there is not—the political will to achieve this goal.† As population continues to grow worldwide, analysts believe that Western Europe must cut its per capita carbon emissions to 25% of what they are today. The United States, which consumes almost twice as much energy per capita as Western Europe and Japan, would have to reduce *its* per capita carbon emissions to less than 15% of what they are today. To accomplish this would require changes not only in America's methods of producing energy but also in American industry, agriculture, transportation, and housing and consumption patterns.

The Potential and Peril of Nuclear Energy

Many regard the coming changes and foreseeable end of the fossil-fuel era with complacency, for they believe that, just as coal replaced wood when the latter became scarce and expensive, new technologies will take over the burden of energy production, allowing material growth

---

* Oil shale and tar sand deposits were thought at one time to be alternative sources of liquid hydrocarbons. The United States has substantial reserves of the former, Canada of the latter. However, obtaining liquid fuel from these resources required enormous quantities of rock to be mined, processed, and disposed of at high monetary, energetic, and environmental cost. In addition, it required the use of more water than could be supplied. Finally, oil shale and other synthetic fuels have a high carbon content per unit of energy produced. The United States abandoned its government-subsidized oil shale development program in the 1980s as uneconomic.

† Fifteen nations, mostly from the European Economic Community, recently established a goal of reducing their carbon emissions by 20% or more. But the United States spurned this undertaking. At several international conferences beginning in 1989, it has prevented international agreement from being reached on specific goals and timetables for stabilizing and reducing carbon emissions (Meyer 1990, p. 3).

to continue. In the relatively short term, say these optimists, we can turn to nuclear power. But is nuclear power the panacea many of its proponents claim? Indeed, is it even a safe and sensible stopgap source of energy until we develop thermonuclear fusion and explore other long-range possibilities? This is an exceedingly controversial issue. Leaving aside some of the more esoteric technical problems, let us examine the issues of nuclear safety and waste, for these seem to be the main points of controversy.*

As we will see in Chapter 3, very small amounts of radiation can cause severe harm to ecosystems and people, particularly if the release of radioactive compounds continues for a number of years. Thus virtually perfect radionuclide-emission control is required. The boosters of nuclear power believe that we have the engineering and management capacity to achieve this level of control for large-scale generation of nuclear power. The critics contend that this hasn't been achieved so far and that it can't be achieved.[†] What are the major points at issue?

Proponents of nuclear power reiterate tirelessly that nuclear generating plants are designed to keep emission during normal reactor operation low enough that any threat to public health is negligible. Even if one

---

* At one time, some thought that the development of nuclear power would be restricted by finite global supplies of uranium. Light-water reactors (LWRs), the predominant type, burn only uranium 235, which constitutes only a tiny fraction of naturally occurring uranium (composed largely of uranium 238). The uranium-235 must be concentrated by laborious and highly energy-intensive techniques—techniques that release substantial quantities of carbon dioxide—before it can be used for reactor fuel. Added to the inefficiency with which fissionable uranium is obtained, light-water reactors are also relatively inefficient energy converters. If all the world's supply of uranium is consumed in light water reactors, only 70 terawatt-years of thermal energy will be produced, compared to the 154 from already discovered reserves of oil and 130 from already discovered natural gas (Hafele 1990, p. 138). LWRs fission less than 0.6% of available uranium atoms; over 99% are wasted. Nevertheless, uranium is plentiful. The current estimated world supply of uranium ore is between 6 and 7 million tons—enough to power the number of nuclear plants currently in operation and the 96 under construction for about 100 years (Hafele 1990, p. 137).

† One indication of the radiation being released from nuclear power plants is the concentration of krypton-85, which is uniquely associated with their operation. From 1970 to 1983, the average annual concentration of this radioactive gas in the atmosphere increased by 80% (Commoner 1990, p. 34).

grants the validity of this position,* it is hardly decisive, for the very assumption of normalcy begs most of the important questions. Nuclear power generation can be safe *only* if the design and construction of the reactor are flawless; there are no accidents or operating errors; *only* if reactors, fuels, and other nuclear installations can be perfectly protected from acts of God, terrorism and sabotage, criminal acts, and acts of war, civil or foreign; and *only* if the release of radionuclides during all other phases of the fuel cycle (mining, processing, transportation, reprocessing, and disposal) can be rigidly controlled. As critics point out, this is a rather alarming list of "ifs." In fact, the nuclear industry has run into trouble in almost every one of the areas mentioned.

Design and operator competence have been far from perfect. The catastrophe at Chernobyl, which is expected to cause 70,000 cancer deaths and whose ultimate costs will surpass the Soviet government's total prior investment in nuclear power, was initiated by operator errors. But design defects—control systems that proved to be inadequate—compounded the effects of operating errors (Flavin 1987, p. 33). The nuclear industry proclaimed that such an accident could not happen in the United States because western plant design was superior. But in March 1979, a minor problem developed in the plumbing of the Three Mile Island-2 plant near Harrisburg, Pennsylvania. Operator errors and design defects then led to a partial meltdown in the TMI-2 core. Some control systems did not work, meters displaying crucial data were hidden from operators' view; a critical valve got stuck, an important sensor didn't work, and hundreds of warning lights came on and alarms sounded that only confused the operators more. One reactor operator, Ed Frederick, later told investigators that operators had closed the stuck valve (which was allowing cooling water to drain away from the overheated reactor) only because "no one could think of anything else to do" (quoted in Nogee 1986, p. 12). Some experts believe that the United States came within 30 minutes of a catastrophe that would have far exceeded the one at Chernobyl. (Nogee 1986, p. 12) Chernobyl released between 50 and 100 million curies of radioactivity into the biosphere; the containment at TMI-2 held in 18 billion curies of radioactivity (Corson 1990, p. 196).

---

* Many critics do not grant it. Because it has now become clear that even tiny doses of radiation have long-term adverse effects on human and ecological health—that, in other words, no radiation exposure can be considered risk-free (see Chapter 3)—critics contend that even the low-level, "normal" emissions from plants now operating constitute a threat to health.

Some nuclear advocates point out that the ultimate safety device, the containment structure, did hold at TMI-2 and that no one was killed, demonstrating a key design superiority. But the fact that TMI-2's containment held may have been a matter of luck. An official of General Public Utilities, TMI's owner, has recently admitted that "we're surprised the reactor vessel contained the accident" (quoted in Flavin 1990, p. 60). Moreover, the Soviet containment structure at Chernobyl was up to some Western standards; it was, in fact, similar to the containments at plants built in the United States by General Electric (about one-third of U.S. reactors).*

More important, United States nuclear plants have been plagued by serious accidents, shutdowns, and near misses. Between 1979 and 1987, there were 27,000 mishaps at licensed nuclear power plants. In 1985, for example, there were 3000 plant mishaps and 764 emergency shutdowns, 18 of which were serious enough to lead to core damage (Flavin 1987, p. 42).† In 1987 there were 3000 mishaps, 104,000 incidents of worker exposure to radiation, and 430 emergency plant shutdowns. There were more than 150 serious accidents in 14 Western nations from 1971 to 1984. Significant nuclear incidents have been initiated by field mice, a loose shirttail, and an improperly used candle. When accidents reveal design flaws and regulators force utilities to add safety features to existing plants, these additions are unavoidably piecemeal and are poorly integrated into the facility's layout.

---

* Actually, in the early 1970s, the Atomic Energy Commission's top safety officer concluded that GE's containment was inadequate and should be banned. Exhibiting an attitude that is commonplace among nuclear regulators, Joseph Hedrie, who later became the Chairperson of the Nuclear Regulatory Commission, rejected his safety officer's recommendation. A memo obtained by the Union of Concerned Scientists quotes him as writing that the proposed ban "could well be the end of nuclear power. It would throw into question the continued operation of licensed plants, would make unlicensable the GE [and some Westinghouse] plants now in review, and would generally create more turmoil than I can stand thinking about." (quoted in Nogee 1986, p. 13).

† One, at the Davis–Besse nuclear plant near Toledo, Ohio, involved 16 equipment failures, including the same stuck valve and human error involved in the near disaster at TMI-2. Operators averted core damage because they quickly shut that valve, allowing auxiliary pumps to cool the core.

10

Nuclear Safety: The Big Gamble

That nuclear safety may be a bad gamble is indicated by the refusal of commercial insurance companies, bastions of prudence and the careful calculation of risks, to cover the nuclear industry until they were assured of drastically limited liability and government reinsurance through the Price-Anderson Act. This Act, renewed in 1988, limits payment for nuclear accidents to $7 billion, even though a 1982 study done for the Nuclear Regulatory Commission reported that a major nuclear accident could result in more than $100 billion in injuries and property damage. Even this larger figure does not include most of the special ecological and societal costs of radioactivity, such as chronic, long-term ecosystem damage and adverse genetic effects.

The nuclear industry attempts to assure the public that nuclear safety is a good gamble by publicizing the results of their "probabilistic risk analysis" of nuclear reactors. This defines risk as the probability that an event will happen multiplied by its consequences. These analyses predict that core-damaging accidents should occur once every 10,000 to 20,000 years of reactor operation (Hafele 1990, p. 141). But Three Mile Island occurred after only 1500 years of cumulative reactor operations, and Chernobyl occurred after only another 1900 years (Flavin 1987, p. 40). Probabilistic risk analysis models have certain difficulties. For example, they make assumptions that may not reflect reality. One such assumption is that redundant safety systems will not be destroyed simultaneously. But when a technician used a candle to check for air leaks at the Browns Ferry, Alabama, nuclear plant in 1975, a fire destroyed several redundant electrical systems at the same time, actually shutting down the control room! Probabilistic risk analysis also does not take into account unknown dangers of reactors built in developing countries, which have frequently been plagued by mismanagement, substandard construction materials and techniques, and bribes paid to inspectors. As a result, nuclear critics claim that probabilistic risk analysis is not much better than guesswork. No one knows when and where the next nuclear accident will occur, but everyone knows that, somewhere, it will.

In some nuclear incidents, radiation has been released into the environment.* Nuclear regulatory bodies, which are usually protective of the industry, minimize incidents and nuclear hazards. Radiation releases are regularly characterized as presenting no danger to the public; in France, as Chernobyl's radioactive cloud passed overhead, French officials stated repeatedly that it had missed the country. Utilities sometimes have not reported nuclear incidents or have delayed reports of them. GPU delayed reporting the TMI-2 incident and then repeatedly issued misleading statements minimizing its seriousness.

Additional safety hazards loom. Nuclear plants have a relatively short life—at most 40 years. By 1990, 35 nuclear plants were at least 25 years old. Already, many are showing signs of aging. Neutrons bombard steel pressure vessels, causing them to become embrittled. Steam generators corrode. Pipes burst in unexpected places. In 1986 a hot-water pipe burst in the Surry Nuclear plant in Virginia, and four workers were killed (Flavin 1987, p. 45). Nuclear radiation builds up continually over the life of a reactor; dangerous levels of radionuclides will remain for thousands of years.

Therefore, unlike conventional power plants, nuclear reactors cannot be destroyed with a wrecking ball. They must be dismantled and buried (decommissioned), or they must be mothballed for several decades until short-lived radioisotopes decay and must then be decommissioned. (A third option, entombment of the reactor on site, is now regarded as impossible. No matter how it was designed, the tomb would decay before the radioactivity did and would release the radioactivity into the biosphere.) Not only do the plants contain high levels of radioactive elements but everything in the reactor is radioactive—the pipes, the equipment, the concrete. Even the solvents used for cleaning up the reactor

---

* A more insidious safety problem is the accumulation of minor releases of radionuclides from all the other phases of the fuel cycle. Such releases are in fact rather common today. An example is the planned release of long-lived radionuclides during fuel reprocessing. In addition, refining uranium for use in nuclear power plants produces uranium mill tailings; by 1987, the nations' licensed mills had produced 186 million metric tons of tailings. Radioactivity from uranium mine tailings is already a problem in some localities. Tailing piles emit radon gas into the atmosphere and contaminate the groundwater below them with dangerous levels of radioactivity. At some sites, radioactivity levels in the groundwater exceed EPA standards by a factor of a hundred or a thousand (Critical Mass 1989, factsheet 5).

11

Low-Level Nuclear Wastes

Nuclear power plants also produce large amounts of low-level nuclear wastes each year. When they are decommissioned, they will produce 16,000 cubic meters more—almost half as much as they produce cumulatively over their operating life.

No one knows what to do with these wastes. Right now, they are dumped into special commercial landfills. But half of the landfills designated for this purpose have been closed because their radioactivity is contaminating adjacent property. In Illinois, Kentucky, and New York, plutonium and other contaminants from these landfills have leaked into the groundwater.

The Nuclear Regulatory Commission's proposed solution to the problem is simple. It wants to label some of these wastes "below regulatory concern" and allow producers to dispose of them in unregulated municipal landfills (Critical Mass 1989). The NRC apparently believes that the solution to pollution is dilution—that hardy folks everywhere will not mind *just a little* radioactivity in their drinking water.

become contaminated; any solvents that spill contaminate the soil or whatever surface they splatter. All machinery in contact with a contaminated surface becomes radioactive. Furthermore, neutron-activated parts of the reactor, including its pressure vessel, its internal components and structures, and its concrete shield, become 1000 times more radioactive than other contaminated components; because they become composed of radioisotopes, they cannot be washed clean (Pollock 1986, p. 10). A huge amount of materials must be dismantled, chopped up into pieces and removed by remote-control devices, and buried. No machinery has yet been designed for this task. It will be very labor-intensive, and the peoples who operate the machinery and do other decommissioning tasks will have to be rotated frequently to avoid radiation overexposure. Unplanned radiation leaks are inevitable. Decommissioning will be complicated, hazardous, time-consuming, and expensive—as expensive, perhaps, as building each nuclear plant in the first place.*

And there is still nowhere to bury these materials safely. Disposal of wastes, a half-century into the nuclear age, is an unsolved problem. The fundamental reason is that the wastes are dangerous for millennia but planetary changes cannot be predicted for millennia. The metals alone in each of the nation's nuclear plants will have built up an average of 4,600,000 curies of radioactivity. Each reactor produces 30 tons a year of highly radioactive spent fuel. These materials will take 3,000,000 years to decay sufficiently to no longer be a serious risk. Spent fuel continually accumulates and no one knows what to do with it. Temporarily, it is stored in pools of water at each reactor: 21,000 metric tons are now sitting in these pools, and 40,000 metric tons will accumulate by 2000. Some reactors are running out of capacity to store spent nuclear fuels. This material is very dangerous. If it were unshielded, a person nearby would receive a lethal dose of radiation in seconds.

The nuclear industry contends that radioactive waste disposal is a political problem, not a technical one. It is the victim, they say, of the NIMBY phenomenon—the "not in my back yard" reaction of an ignorant public. But is it? No one disputes that nuclear wastes must be perfectly contained (and guarded) for millions of years. Nor is there any dispute that the pools at power plants, some of which have already leaked or are reaching capacity, are inadequate for this purpose. The authorities have been casting about for years for a viable alternative. Unfortunately, all the alternatives explored so far seem to have serious drawbacks. One of the most attractive schemes, solidifying the wastes and placing them in unworked salt mines in Kansas had to be abandoned when the dangers of water percolation became apparent. The Department of Energy has poured over $700,000,000 into a "test" facility in another salt deposit under the desert in New Mexico. But it, too, has water leaking into it in at least one location at the rate of a gallon a minute. This water will create

---

* These expenses, moreover, have not been figured into the costs of producing nuclear power or into the electric bills of customers. (If they were, nuclear power would be ruled out as an energy source). Nor do regulators (with few exceptions) force utilities to save for decommissioning costs. It is highly unlikely that many will have the money available for decommissioning when it is needed. Furthermore, customers of utilities at the time of decommissioning will undoubtedly claim that it is unfair and unethical for *them* to pay for it, because they did not use the power the plant produced.

a brine solution that will, within decades, corrode the steel drums in which the radioactivity is contained. When they corrode, the drums will leak. The Department of Energy, which has enshrouded the project in secrecy, has released no details on how or whether it can solve this problem, but it is proceeding with the project. In 1987 Congress designated Yucca Mountain, Nevada, as a nuclear waste repository without scientific investigation of whether the site was geologically or hydrologically suitable. (The site was, however, politically suitable. It is underpopulated and remote, and Nevada had only one unhappy congressional representative and two unhappy senators.) It turns out that the site is on volcanic rock that no one can be certain will be stable for 3,000,000 years; furthermore, no one also knows whether nearby underground nuclear tests have cracked the volcanic tuff. Like the New Mexico project, the Nevada project is proceeding as though there were no scientific problems, while the industry describes local opposition as merely an example of the NIMBY phenomenon.

It may be that government and industry intransigence reflects a desperation that engineers will not admit publicly. Radioactive wastes are accumulating everywhere; as we will see in Chapter 3, the power industry is only part of the problem. Military sites, and the groundwater and streams near them, are already seriously contaminated. Nuclear authorities may know that they have their backs up against the wall and that they simply must override normal democratic processes in order to get a site. Certainly some of their other proposals for dealing with nuclear wastes have an air of desperation about them: using Antarctic glaciers as repositories (extreme transportation hazards; serious risk of upsetting the delicate heat balance of the glaciers, with potentially momentous climatological consequences) or rocketing the wastes into space (staggering expense; potentially grave consequences of rocket failure). Their latest proposals also sound desperate: constructing "temporary retrievable storage facilities . . . on islands or peninsulas" (Hafele 1990, p. 144). Nuclear authorities believe that this will give the industry (which has already had 50 years) "the time it needs to develop scientific, technologic and institutional final waste disposal methods" (Hafele 1990, p. 144). Another "advantage" of this proposal is that it would "encourage the development of new global institutions [to do the siting] immune to national politics" (Hafele 1990, p. 144).

In sum, there are many difficult and for the most part unsolved safety, security, health, and pollution problems connected with the use of nuclear technology. The resolution of these problems is not assured, no matter how much money and effort is expended. As Christopher Flavin has written,

> Nuclear power is not the mature industry that proponents claim, but rather a sick one sustained by government subsidies.... The noble visions of the fifties did not include shoddy construction practices, billion-dollar cost overruns, disinformation campaigns by government officials, thousands of tons of accumulating nuclear wastes, or exploding reactors that contaminate foodstuffs a thousand kilometers away (1987, p. 64-5).

Already the nuclear problem is imposing very heavy management burdens and other social costs, important matters that will be explored in Chapter 3 and later in Part II. Above all, nuclear power does not appear to be the panacea that some of its proponents claim. Indeed, it does not even seem very attractive as a short-term stopgap. The earth does not seem large or stable enough to accommodate its wastes; human beings will never be infallible and thus will never be able to control all its hazards. Even if we factor in the possible adverse effects of global warming, it is foolhardy to choose a source of energy that is devilishly unforgiving of the slightest human failure and that produces lethal toxins, some of which have half-lives longer than the span of recorded history.

Fusion Power: Infinite Potential Fraught with Problems and Limitations

In theory, controlled thermonuclear fusion constitutes a potentially infinite source of energy. Many therefore regard it as *the* long-range answer to all problems of energy supply. However, although fusion is undeniably attractive on many grounds, it is by no means free of problems and limitations.

Above all, no one has yet demonstrated its practical feasibility, even in the laboratory, despite over 40 years of sustained international effort. Thermonuclear reactions take place in plasmas of ionized gases at temperatures and pressures comparable to those found in the sun and other stars. These plasmas cannot be physically contained, so confinement by magnetic field and other difficult and esoteric techniques have been employed in an effort to attain the levels of temperature and pressure necessary for a continuous reaction. Very little progress toward this objective has been made. The best achievements still fall considerably short of what is needed. Most scientists working in this field are confident that the laboratory breakthrough will eventually come, but the history of research in the field suggests that the solution to each particular problem either reveals a worse one behind it or proves incompatible with the solutions to other problems. Even optimists concede that the

12

Safe Nuclear Power?

Nuclear advocates are pursuing research that they claim will lead to the production of safe nuclear reactors. One type of new reactor would incorporate passive safety systems into new light-water reactors; the other type would be gas-cooled, the reactor being theoretically unable to reach temperatures that would melt the fuel particles. But nuclear scientists have not yet reached a consensus about which design to pursue. Once that was done, building and completing tests on a prototype would take at least a decade. And even if the new reactor passed all safety tests, it would probably not be until 2010 that these "safe" reactors could come on line.

There is no guarantee that the newly designed reactors would pass all safety tests. "Nucleonics Week," an industry publication, concludes that "experts are flatly unconvinced that safety has been achieved—or even substantially advanced by the new designs" (quoted in Flavin, 1990, p. 24).

In addition, how much these new reactors will cost, compared to other energy sources then available, has not been addressed. The nuclear industry has a history of substantially underestimating the costs of nuclear projects. The industry has also not addressed the problem of nuclear wastes, which the new reactors will still produce.

breakthrough is unlikely before 2050, at the earliest (Hafele 1990, p. 142). In any event, laboratory feasibility is only the first of a very long and expensive series of steps in research and development that will be required to make fusion practical. The engineering problems to be solved are enormous; temperatures and pressures of stellar intensity are far beyond anything technologists and engineers have hitherto tried to tame. Even research costs are skyrocketing. Governments seem unable to afford present and prospective costs, and fusion budgets are declining.

If fusion power ever does become a possibility, it will not be without problems. Although a virtually infinite supply of fuel is claimed by fusion enthusiasts, in fact the reactors now being invented will fuse deuterium and tritium, the latter of which must be produced from lithium. Although deuterium is so abundant in seawater that it will be for all practical purposes infinitely available, lithium is relatively scarce, and lithium-6, the isotope needed for the fusion reaction, is scarcer still. Thus, once readily exploitable lithium ores have been used up, it will have to be obtained by "burning the rocks." In short, the problem of fuel supply will not necessarily be abolished by the generation of fusion power.

Producing tritium from lithium involves two other problems. First, production must be done in a breeder reactor. Breeder reactor technology is complicated and extremely dangerous;* in 1984 the United States abandoned the Clinch River Breeder Reactor after spending $1.5 billion for its development. Scientists believe a lithium-blanketed breeder reactor would be especially difficult to design, because lithium is highly reactive and even explosive. Second, even if the tritium is successfully produced, tritium is radioactive. Some tritium releases are inevitable.† Neutrons from the deuterium–tritium reaction will make the containment vessel radioactive. Fusion technology, as presently conceived, is not clean.

---

* A breeder reactor uses plutonium (instead of relatively scarce uranium-235) for fuel; because it simultaneously converts a surrounding blanket of comparatively abundant uranium-238 or thorium into more plutonium, it in effect creates more fuel than it burns. This essentially eliminates any concern that uranium supplies might run out, except in the very long term, and it markedly diminishes the possible side effects of uranium mining. But this advantage comes at a price. Breeders require reprocessing. The substances they produce are extremely toxic: Plutonium is the most toxic substance known. Some of this waste must be discarded into nuclear waste disposal sites—and a safe one has not been developed. The plutonium created during the fission of uranium, along with the unused uranium, must be extracted from the spent fuel. Plutonium's extreme toxicity greatly magnifies the health hazards of a reactor accident or a release of radioactive materials at any other stage in the fuel cycle. Finally, unlike the fuels used in light-water reactors, plutonium is a material that can be used to manufacture weapons. Thus many thorny security issues are raised by the plutonium fuel cycle.

† Although tritium, one of the radioactive isotopes of hydrogen, is less hazardous than some radionuclides, it is still extremely dangerous both because it cannot be fully contained—at least 0.03 percent of the total inventory in reactors would escape each year (Metz 1972)—and because it is especially apt to be taken up and concentrated in living systems (where its half-life of 12 years makes it dangerous for over a century).

It is true that other scientists envision alternative fusion technologies that, as far as can be determined, at this stage of their conceptualizing, would be clean. But some of these may never pan out. One, for example, envisions substituting helium-3 for tritium. Helium-3 is not radioactive. Instead of producing it in problematic breeder reactors (which is possible), we could mine it from the moon, where it is abundant. How this would provide us with a net energy gain, however, has not been explained.

In fact, the potential net energy yield of a fusion reactor economy remains a question mark. First—just as with ordinary nuclear technology, only more so—starting up a fusion power system will require an enormous expenditure of energy. Containment and heating of the plasma require great quantities of energy. Especially if the efficiency of the fusion reactors proves to be less than the hoped-for 60%, a very large proportion of the power output of each reactor will have to be used just to keep the reactor in operation. Producing deuterium fuel from seawater will also require the expenditure of energy. Finally, any conceivable form of thermonuclear reactor technology will demand, in addition to lithium, large quantities of scarce minerals such as helium, vanadium, and niobium, many of which are not produced domestically. Eventually, these too would have to be extracted from rock at high cost in energy. Thus the net yield of energy after all costs are counted may be very low.

In sum, the Promethean attempt to provide humans with inexhaustible stellar fire is a bold enterprise that may bring great benefits if it succeeds. But given the extraordinary challenges, the development of fusion power may take several generations, if indeed it proves possible at all. And the promise of thermonuclear fusion should not be allowed to obscure the many problems and hard choices that are evident even today.

The "Energy Options"

With fusion energy not in sight, fossil fuel supplies finite and their burning a source of serious pollution, and nuclear energy a fool's alternative, what are the options for humanity? One option, of course, is to go all out in producing ever more supplies of energy from fossil fuels and nuclear power, allowing carbon dioxide, radiation, and other pollutants to build up in the biosphere. Unfortunately, political leaders in many countries support this option, and none more strongly than

Table 2-6 R & D Expenditures, Fiscal Year 1991, United States Government

Renewable	$158
Nuclear fission	$305
Nuclear fusion	$274
Fossil fuels	$459

Source: *Nucleus,* Spring 1991, p. 5. Amounts are in millions of dollars.

the United States government (see Table 2-6). The probable consequences of agricultural disruption and increasing starvation, cancers, and death will be inflicted primarily not on people who are the decision makers today but on future individuals, some not yet born. Another possibility is to choose the opposite path: to quickly and sharply reduce the production of energy from fossil fuels and nuclear power, develop energy sources that have few or no harmful consequences, and use the energy that is produced far more efficiently than it is used today. This option would have beneficial environmental impacts, but it would involve disruptive changes in the way people work and live, especially in the industrialized world. No one takes this option seriously. A less drastic choice would be to (1) discontinue use of nuclear energy as a power source, (2) phase down the use of fossil fuels, (3) rely more heavily on natural gas—the least polluting and most efficient of the fossil fuels—among the fossil fuels that are used, (4) bring into widespread use already-available technology that increases energy efficiency, and (5) support research into renewable energy alternatives that are not yet competitive, and develop and immediately use those that are. This option would also have beneficial environmental effects, the most important of which would be to reduce the amount of additional radiation and carbon dioxide emitted into the biosphere. It would require some changes in our habits and involve some decentralization of energy production, but the changes themselves would have relatively minor effects on the advantages we obtain from energy or overall economic activity. What follows is a review of theoretical possibilities for using energy more efficiently and for developing nonpolluting sources of energy.

Conservation and Improved Efficiency

By becoming efficient in the use of the energy we have, humanity can reduce the ever-escalating damage it is doing to the environment while giving itself more time to phase in the use of nonpolluting energy sources. Although cutting per capita energy use by one-half will not in itself be adequate to avoid global warming, one study shows that this level of efficiency in industrial countries can be achieved without reducing present standards of living (WRI 1990-91, p. 146).

Countries already vary significantly in how much energy they consume in order to produce the same amount of economic output. The United States, for example, uses about twice as much energy as West Germany and Japan to produce the same output. The Rocky Mountain Institute estimates that the United States can cut its electricity use by 75% at an average cost of 6 cents per kilowatt-hour. Accordingly, it is cheaper (on average) to improve efficiency than to produce more electricity by burning fossil fuels. And is far cheaper than producing more electricity by nuclear energy (Fickett et al. 1990, p. 66). Some utilities in the United States are already convinced. Sixty of them promote the use of electrically efficient lamps, appliances, windows, insulation, motors, and other devices among their customers rather than building new power plants. Regulators in each case allow the utility to keep part of the money saved via the efficiency programs, so both utility and customer gain.

The potential and benefits of energy efficiency in some sectors of the economy are huge. For example, if everyone in the United States did nothing else but substitute currently available efficient lights for incandescent bulbs, utilities could avoid building power plants costing from $85 to $200 billion to construct and $18 to $30 billion a year to operate (Fickett et al. 1990, p. 67). Comparable savings can be obtained with efficient motors and superefficient appliances. (Switching to efficient motors can save 80 to 190 billion watts; the changes pay for themselves in an average of 16 months (Fickett et al. 1990, p. 68). It is possible for builders to construct super-insulated homes, offices, stores, schools, and hospitals that consume as little as 10% of the energy needed for heating their conventional equivalents (see Box 13). In Sweden, new office buildings need no manufactured energy at all to take them through their cold winters. Sunlight, plus the heat generated by the people, lights, and office equipment, are sufficient to keep buildings warm (Bevington and Rosenfeld 1990, p. 79.) The Swedes install super insulation and building materials that store solar heat; the extra cost of doing so is partially offset by not building central heating and ventilation systems. Air exchangers ventilate interior space.

13

Home Energy Savings

The benefits an individual can achieve by using electricity efficiently are astonishing. According to *Scientific American,* "if a consumer replaces a single 75-watt bulb with an 18-watt compact fluorescent lamp that lasts 10,000 hours, the consumer can save the electricity that a typical U.S. power plant would make from 770 pounds of coal. As a result, about 1600 pounds of carbon dioxide and 18 pounds of sulfur dioxide would not be released into the atmosphere.... Alternatively, if that electricity were produced by an oil-fired electric plant, the compact fluorescent lamp would save 62 gallons of oil—enough to fuel an American car for a 1500-mile journey. Yet far from costing extra, the lamp generates net wealth and saves as much as $100 of the cost of generating electricity" (Fickett et al. 1990, p.74). The consumer also benefits personally. The bulb will cost him or her about $18.00. But over its lifetime, it will save that customer $57.00 (at $0.10 per kilowatt-hour) in electricity and replacement bulbs.

Individuals can also achieve astonishing savings in providing energy to heat their homes. Superinsulated homes are standard in Sweden, for example. Using materials that store solar heat and superinsulation, these homes get all the winter heat they need by diverting hot water from their hot-water heaters to small heating units. In the United States, builders can save $1000 to $2500 on central furnaces and ventilating systems to pay for superinsulation; one superinsulated home in Chicago had a winter heating bill of $24.00 (Bevington and Rosenfeld 1990, p. 82). But superinsulated homes are rarely built in the United States. The United States has been content so far to begin to attain only modest residential energy savings, as, for example, window manufacturers supply more efficient low-emissivity (low-E) windows. (Americans lose as much heat through their windows each year as the energy equivalent of entire annual flow through the Alaska pipeline).

Builders can also retrofit existing buildings. Retrofits usually obtain 30 to 70% energy savings. Customers typically spend nothing for these efficiency measures. Instead, energy-services companies finance the efficiencies and are paid back with 50 to 70% of the savings accrued from the efficiency investment until their costs plus profits are achieved (Bevington and Rosenfeld 1990, p.78).*

All in all, then, industrial countries can gain major efficiencies in the use of energy in many sectors of modern life without changes in the way their people live or the way they do business. It is possible to adopt them cost-effectively, with either a relatively quick payback of up-front costs or, as in the case of innovative utility programs and energy services companies, the consumer's incurring of no up-front costs at all. (Southern California Edison, among other things, gives away compact fluorescent light bulbs. Its program "creates" energy at an average cost of 2 cents per kilowatt-hour, far less than the cost of any power plant (Fickett et al. 1990, p. 71). These methods of boosting efficiency cost less than any other method of producing power (Table 2-7). Achieving improvements in efficiency can also displace the buildup of seven times as much carbon dioxide in the atmosphere, per dollar invested, as expanding nuclear power.

People in industrial countries can make greater gains in energy efficiency if they are willing to modify some of their living patterns. Consider, for example, the transportation sector—the only one in which highly efficient products are not yet available. Great gains in transportation efficiency are nevertheless possible if people travel by public transit rather than private automobile. In the United States, a light rail vehicle carrying 55 passengers uses an average of 640 BTUs of energy per passenger per kilometer, and a bus 690. This compares to 4580 BTUs per passenger-km to travel by automobile (Lowe 1991, p. 59). Despite this fact, and despite the pollution and traffic congestion that result from automobile commuting, over 80% of the population commutes to work by private vehicle. By contrast, in Europe, where efficient rail and bus systems are maintained, the figure is half that. In Tokyo, only 16% commute by private automobile.

---

* Comparable efficiencies in the transportation sector are not yet available. This is regrettable because, as we will see in Chapter 3, automobiles are major polluters and the size of the global fleet is increasing rapidly. But significant improvements in automobile efficiencies are possible, despite manufacturers' claims to the contrary. Volvo, for example, has produced a prototype compact car that is designed to cost no more than and accelerates as rapidly as a typical compact car, meets all countries' crashworthiness requirements, and achieves 63 mpg city and 81 mpg highway (WRI 1990-91, p. 151).

Table 2-7 Costs of Avoiding Carbon Emissions Associated with Alternatives to Fossil Fuels, 1989

Fossil Fuel Alternative	Generating Cost[1] (cents/kwh)	Carbon Reduction (percent)	Estimated Pollution Cost (cents/kwh)	Carbon Avoidance Cost[2] (dollars/ton)
Improving efficiency	2.0-4.0	100	0	<0[3]
Wind power	6.4	100	0	95
Geothermal	5.8	99	1	110
Wood power	6.3	100	1	125
Solar Thermal (with gas)	7.9	84	0.2	180
Nuclear	12.5	86	5	535
Photovoltaics	28.4	100	0	819
Combined cycle coal	5.4	10	1	954

[1] Levelized cost over the life of the plant, assuming 1989 construction costs and a range of natural gas prices.

[2] Compared with existing coal-fired power plant.

[3] Some energy-efficiency improvements cost less than operating a coal plant, so avoiding carbon emissions is actually free.

Source: Flavin 1990a, p. 27.

Developing countries present another story. By taking advantage of new, highly efficient materials and technologies, it is possible for developing nations to increase their energy consumption more slowly than their growth rate. However, with their rapidly increasing populations, their energy use will increase no matter what they do. Moreover, most developing countries are investing little or nothing in efficiency. Despite the fact that they have no hope of financing their energy-expansion proposals, and despite the fact that efficiency measures would be a much cheaper way for them to "create" power,* developing countries have shown little interest. To the extent that efficiency measures have been adopted, they have been adopted in the industrial world.

Why isn't humanity fully exploiting the possibilities for efficiency when doing so is clearly in not only its long-term interest but its short-term interest as well? The obstacles, which are institutional not technological, will be discussed in Part II. But a few preliminary observations are in order here. First, there is the matter of inertia. In industrial countries, for example, heavy investments have already been made in modern infrastructures; industry and government, therefore, are reluctant to invest anew in energy efficiency.[†] Second, people and industry tend to discount future savings in favor of lower initial costs; most consumers will buy a cheaper item even if making a higher initial investment has a payback time as short as two years (Gladwell 1990, p. A3). Third, people and industry resist government policies to encourage conservation whenever those policies burden them. In the United States, for example, voters frequently reject increased gasoline taxes, mass-transit expenditures, land-use controls to prevent suburban sprawl, and even mandatory recycling measures. Industry lobbies heavily against mandatory conservation measures.

However, these generalizations are not always true. France, Belgium, the Netherlands, Ireland, Portugal, and the Scandinavian countries have already decoupled economic growth from increased energy consumption. Germany is committed to reducing energy consumption by 25% over the next 15 years (Flavin and Lenssen 1991, p. 25). Japan has an energy manager in every industry. In Europe, public opinion seems to take the greenhouse effect seriously. By investing in energy efficiency, these countries not only slow the greenhouse effect but improve their

---

* For example, it is cheaper for developing countries to replace inefficient motors and lighting with efficient devices, and to substitute solar-powered water heaters and stoves for electric ones, than to build new coal, nuclear, or hydroelectric central power plants. Efficiencies are even possible in the use of energy by the world's poor. New, inexpensive solar cookers have been developed that can be produced at the village level and used in place of fuelwood to cook food and pasteurize water (Corman 1990, p. 206).

† This is undoubtedly part of the reason why the United States continues to pour its transportation investments overwhelmingly into highways instead of building and maintaining transit systems and encouraging its citizens to use them.

economic competitiveness. Second, not all of industry or all consumers make purchases on the basis of the short-term bottom line. For example, we have already seen that some utilities are learning to create both market push and market pull to influence the purchase of efficient devices. The majority of them pay some form of rebate to purchasers to create market pull; some pay suppliers rebates, creating market push (Fickett et al. 1990, p. 71). These methods seem to be effective in changing purchasing habits. Finally, the European experience shows that people and industry will accept some government policies to encourage conservation even when it is burdensome; high gasoline taxes, mass-transit facilities, controls on land development, energy standards for buildings, bicycle pathways, and recycling are commonplace on the continent.

However, these encouraging developments notwithstanding, institutional obstacles are preventing efficiency measures from being adopted as fast or as extensively as is necessary (even in Europe). Thus, if improvements in energy efficiency are confined to the industrialized world (as they are, for the most part, today)* global changes in climatic and other environmental problems will probably be severe. Population growth and economic development in developing countries will probably result in a doubling of their carbon emissions per person by 2030. So even if the industrialized countries (including the United States) were to halve their per capita carbon emissions by that time, annual emissions of carbon dioxide would be 2.5 times what they are now (Gibbons et al. 1989, p. 141). At present, global carbon dioxide emissions are *increasing* by 400 million tons each year. They must be *reduced* by 60 to 80%—to the level of the 1950s—to stabilize the amount of carbon dioxide in the atmosphere (Flavin and Lenssen 1991, p. 25). Scientists have calculated that to achieve a sustainable energy system, the world as a whole will have to produce goods and services with one-third to one-half as much energy as today, and renewable energy sources must quadruple (Flavin and Lenssen 1991, p. 26).

Finally, whatever the pace or extent of efficiency, it is not a panacea for eliminating the pollution caused by humanity's current methods of energy production. For example, even if we were to succeed in cutting energy consumption in half, it would merely have the same effect as doubling the supply. Rapid population increases and economic development will eventually cause more energy to be used and overrun the effect of increased efficiency. Consequently, energy conservation alone can

---

* Only China among the developing countries has implemented significant energy efficiencies, cutting energy required per unit of economic output by 4.7% a year for the past decade (Chandler et al. 1990, p. 125).

never be more than a short-term palliative. To the extent it is adopted, conservation must be seen as an interim measure to avoid some global pollution and give humanity more time to develop nonpolluting energy sources.

Geothermal Power: Tapping the Heat of the Earth

The technology to use the earth's interior heat both to heat buildings directly and to generate large amounts of electricity is essentially in hand. Moreover, given that due care is taken, the environmental consequences of exploitation are comparatively benign. Global geothermal capacity increased by 16% per year from 1978 to 1985 (Corson 1990, p. 197). Today it produces 6 million kilowatts of electrical power. A few countries get a substantial portion of their electricity from geothermal sources: El Salvador gets 40%, Nicaragua, almost 30% (Flavin and Lenssen 1991, p. 30). Some estimate that the United States can get as much as 10% of its total energy supply in the year 2000 from geothermal power.

But geothermal power has limits. Although some countries have geysers, natural steam, and hot-water fields that can be exploited for power production, others do not. (For a variety of reasons, not all natural geothermal resources are available as power sources. Consider as an example the controversy over whether large geothermal fields should be developed in the United States' only tropical rainforest in Hawaii (Wysham 1991, p. 12).) To produce substantial amounts of power in countries without these fields, artificial geothermal reservoirs must be created by drilling down to areas of high heat flux in basal rocks, fracturing the rocks, pumping in water to be heated, and then extracting the heat from the resulting steam. The technology for this is becoming available, and early experiments with the technique are promising, but much more research, development, and exploration must be done before a clear picture of the extent of geothermal resources emerges.

The exploitation of geothermal power is also not pollution-free. The steam or hot water used to produce power almost always contains noxious gases and corrosive compounds; many wells emit significant quantities of radionuclides. Well blowouts sometimes occur, releasing the steam into the atmosphere. Environmentally compatible ways of both capturing and disposing of used steam and water are improving, but the problem is not yet solved. Another problem is that geothermal reservoirs must be located in geologically suitable areas, which requires that power be produced at some distance from markets. This means that the monetary and environmental costs of power transmission (as well as the attendant energy losses) may be considerable. (On the other hand, this problem can be surmounted if scientists develop materials that are super-conductive at temperatures that are practical to work with. "Practical" superconductivity would achieve more than making geothermal and other location-specific power sources more available. It would, in effect, "create" more electricity. Presently, 50% of the electricity produced is lost in transmission.)

All in all, geothermal power has significant potential as an energy source; optimists foresee this form of energy as ultimately contributing more than 20% of supply. * For it to do so, more effort and money than are currently being invested will be needed to develop the technology for geothermal's use in more areas and to overcome its minor but stubborn pollution problems.

Biomass

Plants are the oldest source of energy known to humans. Green plant matter is created in photosynthesis. Humans eat the plants containing this stored solar energy; they burn other plants, especially wood, to do their work. Traditional fuels, primarily wood, provide 40% of the energy supply of developing nations. The obvious problems, however, are air pollution from wood burning, conflict with other uses for wood, and demand for space (leading to potential conflict with needs for food and fiber or, at least, for living room). Unfortunately, as we have seen, many nations are destroying their forests faster than they are replacing them. Out of hunger for energy, people are continuing the ancient pattern of reducing forested mountains to bare rock skeletons. In fact, this pattern is prominent today in many poorer countries, where the populace ravages the land for wood fuel.

Deriving energy from biomass, however, is usually distinguished from the burning of fuelwood. It refers to converting a variety of plant materials into efficient fuels without producing pollution. First, the source plant material must be grown sustainably, so that the amount of carbon dioxide produced when the plant is processed and burned is the same as that which was consumed as the plant grew. Second, the fuel must burn without creating other kinds of pollution. The production of ethanol from sugar cane residues satisfies these requirements and is a major industry in Brazil. Ethanol can power existing motor vehicles with minor modifications of their engines. In 1986, ethanol supplied half of

---

* Geothermal's potential could be still higher if, in addition to achieving practical superconductivity, it were possible to develop the technology to extract energy from underground masses of hot rock.

Brazil's automotive fuel (Corson 1990, p. 211). Another possibility is to gasify plant materials. The gas can power turbines, and its use would be considerably less polluting than that of fossil fuels. Gasifier–gas–turbine plants are only at the theoretical stage. But scientists have calculated that 75% of Africa's current electrical capacity can be generated by using plant wastes alone (Flavin and Lenssen 1991, p. 29).

Unfortunately, biomass energy has a significant disadvantage as an energy source: the scarcity of its resource base. With other solar technologies, as we shall see, the fuel (sun, wind, or water) is plentiful and renewable. What is required of technology is the ability to harness the energy in the fuel. But although plants are also plentiful and renewable, biomass fuel is not. One possibility is to grow energy plantations. But in a world where there will be insufficient food and fiber for rapidly growing populations, diverting agricultural land for this purpose seems unlikely to win approval. Even some developed countries, such as those in Western Europe, where an adequate food supply is not a problem, do not have enough land available to grow large amounts of biomass. Alternatively, biomass will rely on plant residues. However, many of these residues are presently plowed back into the soil to enhance soil fertility. Diverting it to other purposes will only reduce agricultural yields. In Nepal, the diversion of biomass from the fields has reduced grain yields by 15% (Corson 1990, p. 151).

Biomass, then, has some limited potential as a future energy source. Some of the support for biomass is based on its capacity to produce alcohol, which can be used in (modified) gasoline engines. But there isn't enough free land or plant residues in the world to power today's vehicle fleet, to say nothing of one expected to double in 30 years. A more realistic possibility is that biomass (mostly wood, cane sugar, and beverage industry wastes) will supply the alcohol for an alcohol–gasoline mixture (gasohol). This gasohol could be used as a transitional vehicle fuel, reducing pollution until some other means of powering motor vehicles is developed.

Hydropower

An established technology that relies on a theoretically renewable, inexhaustible source of energy is hydropower. About 20% of the electricity of the world is generated from falling water. Norway obtains 50% of its electric power from hydro.

Hydroelectric power's theoretical capacity has also hardly been tapped. Whereas North America had developed 60% of its large-scale hydroelectric potential in 1980, Europe had developed only 36%, Asia

9%, Latin America 8%, and Africa 5%. Developing countries have made it a priority to expand hydroelectric capacity since that time. From 1980 to 1985, 31 developing countries doubled their capacity. But even with this doubling, the hydroelectric potential in developing countries is 10 times what is currently being generated.

Hydropower, however, is not without adverse environmental effects. As we will discuss in Chapter 3, the large dams many countries have been building spread diseases. The reservoirs, which are usually too deep and sterile to support fish, often expand the breeding grounds for the carriers of malaria, schistosomiasis, and river blindness. The large dams also inundate forests, farms, and wildlife habitats; entire species of plants and animals have been wiped out in submerged areas. Many people upstream of the dam are physically displaced, and their way of life may be destroyed. Silting is a serious problem, reducing the storage capacity and power potential of the dams over time, and depriving the soils downstream of the fertilization that comes from silt-bearing floodwater. As a result, the people who have farmed or fished downstream of the dam in these floodwaters for millennia are dispossessed of their food supply. Their way of life may also be destroyed.

Hydropower, therefore, cannot be developed to its full theoretical potential. But with due care to limit its environmental effects, it still has much room for growth. Projects using small dams and reservoirs have fewer adverse ecological effects, and their disruptive effects on human populations are more manageable. China may provide something of a model of what can be done; it has built over 86,000 small hydroelectric projects that provide some power to almost every province in the country (Corson 1990, p. 197). By itself, however, small-scale hydroelectric power generates relatively small amounts of electricity, and the potential varies greatly with locality. Therefore, small-scale power production from water can be only a minor part of the answer to the needs of an energy-intensive industrial civilization.

Solar Power: The Ultimate "Fuel"

In the final analysis, almost all the energy available to people is solar. Fossil fuels are simply the stored legacy of past photosynthesis; the fissionable elements were formed in a solar furnace; and a thermonuclear fusion reactor is essentially a miniature sun. However, the term *solar power* ordinarily refers to the use of the direct energy of the sun's rays via solar heat collectors or photovoltaic conversion cells and to the exploitation of the indirect results of solar heating—falling or moving water, wind, natural heat traps, and photosynthesis.

Hydroelectric power is a form of solar energy and is renewable: The sun's heat evaporates water which falls as rain, which flows into rivers. The dams hold back and channel the energy of the falling water to rotate turbines. But hydroelectric power differs in a critical way from the new solar technologies. Whereas hydroelectric projects can have serious environmental effects; the new technologies have inconsequential ones. Biomass also differs critically from other new solar technologies; it alone does not rely on an ample resource base. Thus only the new solar technologies would appear to be appropriate sources of energy for humanity's long-term future.

The problem with solar energy is that it is difficult to harness. It is also diffuse, unequally distributed around the globe, and variable with season and weather, and some forms of it are available only in limited quantities during any given period. Nevertheless, the total amount of solar energy is so huge that harnessing even a fraction of it could satisfy the energy requirements of humanity.*

Wind Power

Wind power already appears to be an almost perfect energy source. Uniquely among the new solar technologies, it is economically competitive with energy produced from fossil fuels. It can produce electricity more cheaply than the most efficient nuclear power plant. 1660 megawatts of wind-generating capacity was installed worldwide in the 1980s, 85% of it in California. The California windmills in 1991 produced electricity at 5 cents per kilowatt hour, the same cost as electricity produced from a new coal plant. Moreover, in contrast to a coal plant, which takes years from the design stage to the production of electricity, today's windmills can be mass-produced and can be properly sited and "on line" in a matter of months. According to the U.S. Department of Energy, advances in windmill technology are expected to bring the cost of wind power down to 3.5 cents per kilowatt-hour at good sites within the next 20 years (Weinberg and Williams 1990, p. 148).

Plenty of good sites are available for wind-generated electricity. In the United States, 90% of these are in 12 contiguous mountain and plain states, where windmills can be placed on ranches and farms without disturbing present agricultural activities. In California, cattle graze under

---

* The amount of solar energy *theoretically* available is staggering. Each year the earth receives about 5000 Q (5×10^{18} BTU) from the sun. By contrast, humans have thus far consumed only about 15 Q of fossil fuel!

existing wind machines while the ranchers reap royalties for the use of their land. Scientists have calculated that enough electricity to meet the entire electrical needs of the United States could theoretically be generated by means of wind energy alone (Weinberg and Williams 1990, p. 148). To do so would require only 10% of U.S. land area and would be compatible with existing (mostly agricultural) uses.

Scientific obstacles to this development still exist, however. (There are political obstacles as well. These are discussed in Part II.) Wind-generated energy is intermittent. Therefore, using wind power can save the fossil fuels that would otherwise be needed to generate the same power in conventional power plants and can make it unnecessary to build the conventional power plants needed to supply electricity in periods of peak demand. But wind power cannot serve as an exclusive power source. Base-line power must be generated by some other method. One way around this limitation, which affects most solar technologies,* is to develop much better methods for storing electricity. Batteries, at least batteries of the types currently in use, clearly will not do. One promising method is to use electricity to compress air into underground storage. When energy is needed, the air is released from the cavern, heated, and funneled through turbines to produce power. This method is 70% efficient and is already in use in Germany.

Another difficulty with wind-generated electricity is the same one we observed with geothermal power and affects most of the other new solar technologies: the cost (in dollars and energy losses) of transmitting the power produced from rural locations in relatively unpopulated areas to urban areas in distant states where local demand for electricity is high. Apart from research on superconductivity, another approach to solving this problem is to develop a "hydrogen economy," the technology for which is unfolding.

In sum, scientists are surmounting the technical obstacles to full exploitation of wind-generated electricity. As they do, wind power can become a major provider of electricity in many countries. Production of wind power causes no pollution. It takes up land, but most of the land can be used for other purposes simultaneously. Noise used to be a problem,

---

* Hydroelectric power and biomass use stored energy as needed; thus their power is not intermittent except in extreme situations. In the case of hydropower, when seasonal variations in water availability can be anticipated, electrical power can be used to pump water up into the reservoirs in good times, to be used in periods of droughts. These pumped hydroelectric storage systems achieve about 70% efficiency (Flavin and Lenssen 1991, p. 30).

but modern wind turbines are quiet. About the only remaining environmental damage done is that birds sometimes fly into the windmills and are killed.

Solar Thermal Power

Solar thermal power is also a promising new technology. It uses collectors to track the sun and collect its heat. That heat warms a fluid up to $3000°C$, and that fluid is then used in a power-generation cycle. The Luz Corporation has built commercial solar thermal facilities in California's desert northeast of Los Angeles. These facilities produce 350 megawatts of power at 8 cents per kilowatt-hour with 95% reliability. This far exceeds the capabilities of new nuclear plants, which produce power at 13 cents per kilowatt hour with 60% reliability. It only slightly exceeds the cost of producing electricity at coal-fired plants, which costs 6 to 8 cents per kilowatt-hour on average—a figure that does not however, include the costs of lost productivity, health-care expenses, and cleanup of coal's environmental pollution.

Like wind power, solar thermal components can be mass-produced. Their production costs are likely to come down further over time. Also like a wind facility, a solar thermal plant takes only months to construct. Luz constructed one plant in nine months. A coal plant takes 6 years to come on line; a nuclear plant can take up to 15 years.

Solar thermal plants are practical only in sunny and mostly sunny areas. The full exploitation of solar thermal power, therefore, can occur only as the technology for efficient electrical transmission and storage capabilities continues to develop. Solar thermal power, however, itself provides a means for storing electricity. Sun-tracking mirrors can concentrate tremendous heat in water, oil, and bedrock during hot summer months; 85% of the stored heat can be recovered later for use as a heat or power source.

Solar thermal plants render the land they occupy unavailable for most other uses. But they do not take up more land than conventional power sources, if one includes the land used for mining or drilling for the fuels that conventional facilities require. Solar thermal plants also don't pollute. The only "pollution" associated with them occurs in the manufacture of some of their high-tech components.

Ocean Energy

Another emerging solar technology is ocean thermal energy conversion (OTEC). The principle of OTEC has been understood for many years; it can best be described as an air conditioner in reverse. An air conditioner uses electricity to create a difference in temperature; OTEC uses the difference in ocean temperatures between surface and deep water to create electricity. Several countries have built small prototype OTEC plants. The technology, however, will work most efficiently with large central power stations wherever there is at least a 20°C difference between surface and deep ocean waters. These plants are expected to produce power at a cost of 7 cents per kilowatt-hour (Corson 1990, p. 198).

When it is developed, island countries with population centers close to the sea can profit most from OTEC technology. There are 300 population centers close to ocean waters that have sufficient differences in temperature for OTEC, theoretically, to work. Many of these countries, such as Japan, are short on natural resources; they now import almost all of their fossil fuels. The "fuel" of OTEC, seawater, is plentiful. Governments and utilities have drawn up plans to build or are already constructing commercial OTEC plants off the coasts of Japan, Tahiti, and Bali (Penney and Bharathan 1987, p. 87).

OTEC is unique among the new solar technologies in that it is not intermittent; its fuel, the difference between seawater temperatures, varies only slightly from day to day. (Seasonally, its greatest potential is in the summer, when power demands are highest.) Hence it does not depend on the development of new electrical storage technologies. OTEC is also pollution-free. In fact, OTEC's by-products are one of the main arguments *for* developing the technology (The Economist 1987, p. 94). One by-product is fresh water, formed when steam is condensed by the cold seawater. Another is aquaculture, which can be sustained by the rich nutrients in the cold seawater that is pumped into an OTEC plant and then discharged at the ocean surface.

OTEC does have some remaining technical problems. It is inefficient, converting only about 3% of the heat energy to power, compared with 35% in conventional power plants (The Economist 1987, p. 94). In one sense, this doesn't matter; OTEC's fuel is practically infinite. But to take advantage of this bounty of seawater requires enormous heat exchangers that will resist corrosion and fouling, and it requires the means to pump and pipe huge amounts of cold water from the ocean's depths (Penney and Bharathan 1987, pp. 89-90, 92). Keeping a plant seaworthy under all storm conditions and sufficiently immobile to attach an electrical cable to it are two other problems. These problems are being solved, but it is clear that more research is needed before OTEC can be used widely.

Other ways of harnessing energy via seawater are also under development. Norway has built two small wave-power plants. At least one is financially viable, and a Norwegian company is selling the machine commercially. British engineers have designed a large wave power plant weighing 23,000 tons and intended to be located underwater, in the sea bed. The plant will operate on the same principle as the Norwegian plants: Oscillating water that rises and falls inside a cylinder will generate air pressure, which in turn will power a turbine to produce large amounts of electricity. France has built a tidal power plant that can generate 240 megawatts of electricity (Corson 1990, p. 198). These technologies, like OTEC and other renewables, produce no acid rain, no carbon dioxide buildup, and no radioactivity. Thus they would seem to have good long-term potential.

Photovoltaics

The ultimate energy source is photovoltaic energy. Photons (individual particles of light) from the sun are absorbed in a semiconductor to displace electrons and produce a current. Once installed, they need no maintenance and there are no operational costs. They require no water or wind, cause no noise, and, unlike other new solar technologies, can be placed close to all users of electricity, even those who are in partly cloudy locations. They can be constructed in small units; using current technology, a 40-square-meter array on a south-facing roof in a locality receiving an average amount of sunlight in the United States can supply all the electricity needed by that household (Weinberg and Williams 1990, p. 149). Or they can be built in large arrays to provide a central source of power for utility companies. PV arrays can be constructed relatively quickly (though not so fast as windpower or solar thermal units). A small array can be built in less than a year, a large array in less than two.

Their disadvantage, so far, is price. The price of PV power has declined dramatically in the past 20 years, from $60 per kilowatt hour in 1970 to 20 to 30 cents per kwh in 1990 (Weinberg and Williams 1990, p. 149). These declines were achieved despite minuscule government research support. But at these prices, PV power is still about five times more expensive than conventional power. Its use is justified predominantly in remote commercial or home applications and military and communications facilities where power lines are not available. Some developing countries find PV power economically justified in rural electrification projects.

Still, PV power is expected to continue to achieve further dramatic declines in price in the near future. Further technological breakthroughs may be possible. For example, scientists have developed solar cells in the

Table 2-8   Land Use of Electricity-Producing Technologies in the United States

Technology	Land Occupied (Square meters per gigawatt-hr. over 30 years)	Location	Compatible
Coal	3642	Various; mining usually despoils local environment	No
Solar thermal	3561	Sunny areas	No
Photovoltaics	3237	Deserts, rooftops, wasted building space	Depends on site
Wind	1335	Barren grazing sites, farms	Yes
Geothermal	404	Natural "hot" areas	No

Coal figure includes coal mining sites. Wind includes turbines and service roads.
Source: Adapted from Flavin and Lenssen 1991, p. 36.

laboratory that cost about one-tenth as much as cells currently on the market (Weinberg and Williams 1990, p. 150). Being one-fiftieth the thickness of a human hair, these cells use only tiny amounts of raw materials, and they can potentially be mass-produced. Other laboratories have built solar cells that are 35% efficient, comparable to the efficiency of conventional power plants. It seems only a matter of time before at least some of this (and other) laboratory research materializes in marketable products.

PV power is not a competitive source of electric power—yet. But because it can be installed in a variety of applications and arrays, it is the most flexible solar technology with the greatest long-term potential. In addition, once in operation it causes no pollution. PV power generation does require space for its solar arrays. Land used for PV is not generally available for other purposes. On the other hand, the land required for PV is less than that required for conventional power, when mining, transportation, and waste-disposal are considered. All of the electricity the United

States currently consumes could be supplied by PV arrays taking up 0.37% of the land area of the country (Weinberg and Williams 1990, p. 149).* Furthermore, a substantial amount of space needed for PV power will not be on land at all; it will be on rooftops and other such structures. In Spain and Italy, the government is funding rooftop installations on homes.

PV energy is intermittent. It produces no power at night and little on overcast days. Thus, for humanity to fully exploit PV power requires improved energy-storage technology. In this respect, PV energy is like wind and solar thermal energy.

The Hydrogen Economy

The technology to store large amounts of electricity is being developed in parallel with the development of renewable energy sources. We have previously noted the possibilities of compressed-air and high-temperature (solar) thermal storage. An even more promising technology is solar hydrogen. Energy from the sun can generate electricity, which can be passed from one electrode to another in water. The process splits water into two parts hydrogen and one part oxygen. The hydrogen can be burned in place of oil, coal, and natural gas. It can also be transported to distant locations where it is needed, much as natural gas is today.

Hydrogen is an extraordinarily clean-burning fuel. Unlike fossil fuels, it emits no carbon monoxide, carbon dioxide, particulates, volatile organic compounds, or sulfur dioxide. To burn hydrogen is simply to recombine it back with oxygen to produce water. Combustion also produces small amounts of nitrogen oxides, which catalytic converters can almost completely eliminate.

Because hydrogen can be stored in tanks and transported through pipelines to where it is needed, humans can substitute a hydrogen economy for one based on fossil fuels. Hydrogen will allow solar-generated power to be used in places and at times dissociated from when the sun is shining or the wind blowing. Scientists have projected that PV cells can be located in desert locations to manufacture the hydrogen, because the annual rainfall that most deserts get supplies more water than is needed in electrolysis (Weinberg and Williams 1990, p. 154). Moreover, scientists have calculated that it will cost less to transport hydrogen by pipeline than it now costs to transport electricity by wire (Weinberg and

---

* PV arrays used to produce hydrogen can be located in unpopulated deserts.

14

The Multiplex Energy Economy of the Future

It seems likely that in the future we shall make eclectic use of many different energy sources, from the age-old to the ultra-modern, in a "multiplex energy economy." The centralized power production characteristic of today's advanced industrial civilization is encountering various types of limits. Old fuels will become too noxious to burn and eventually run out; some new sources of energy, such as nuclear fission, are ecologically dangerous and socially problematic; and other promising new forms of energy supply, such as geophysical or solar power, need not fit exclusively into the current system. Energy from all sources will no longer be so cheap and abundant as it has been. The result will be a tendency toward (1) greater decentralization and energy self-sufficiency and (2) intensive exploitation of every nonpolluting source of energy (combined with scrupulous conservation).

The possibilities for greatly increased energy self-sufficiency are virtually endless. Roofs can hold solar collectors and small windmills, small water turbines in local streams can produce electricity, local underground heat can be tapped, and so on. In India today, village children collect cow dung and other organic wastes to produce methane gas, which yields the villages' electricity supply. The do-it-yourself home techniques that are now being pioneered by "soft" tech-

Williams 1990, p. 153), so desert production of electricity should be practical. (In addition, wind-generated electricity can be used to produce hydrogen. As we have seen, windmills can be sited in agricultural areas without adverse effects.) In some cases, existing natural gas pipelines can be used as part of the (wind or) PV-hydrogen's distribution network.

Hydrogen does pose problems, however. Most often mentioned is its explosiveness. But in fact, hydrogen does not pose a greater danger of exploding than does natural gas; although accidents occur, users have learned how to handle it. For example, hydrogen has been used safely since the 19th century as "town gas" to heat homes. Industries have also developed processes for handling hydrogen harmlessly in a variety of commercial applications, such as making ammonia, fertilizer, dyes, and rocket fuels. An important limitation of hydrogen is that it does not readily substitute for gasoline as a motor vehicle fuel. Compared to gasoline, hydrogen has both a low density of energy and a low density of

nologists should become at least one aspect of our future energy economy even if, as will probably be the case, most electricity and other forms of energy will still be centrally generated to run cities, factories, mass transit, and the like.

In addition, we can probably expect energy sources to be much more varied than they now are. For example, we will see windmills on farms, grasslands, and mountain passes, solar collectors in deserts, small hydroelectric projects on rivers and streams, and OTEC and wave projects offshore. Small biomass facilities will be located in the villages of developing countries and near forests and sugar plantations. No safe source of energy will go unexploited. It is likely that wastes of all kinds—industrial, commercial, residential, agricultural, silvacultural—will be much more widely reused to achieve energy efficiency, burned to produce heat, or distilled to produce gaseous and liquid fuels, making a modest but locally significant contribution to the energy supply. In some areas, draft animals seem likely to be retained, even though their upkeep often requires heavy energy expenditure, for in many cases the alternative to draft animals is human labor.

We seem, then, to be headed for a kind of multiplex, two-tier energy economy in which centralized, industrial power production will support certain key sectors but in which individuals and localities, employing a curious combination of pre-modern and post-industrial means, will be important players in meeting humanity's energy needs.

power (Bleviss and Walzer 1990, p. 106). The former limits the range and load of a hydrogen-powered vehicle, the latter its ability to accelerate (Bleviss and Walzer 1990, p. 106). These problems do not seem insurmountable, but they have not yet been conquered. * A few automobile manufacturers are working to develop hydrogen-powered cars, but unlike

---

* One possible solution is to generate electricity by combining hydrogen and oxygen in a fuel cell, a chemical reaction that produces no nitrogen oxides. The generated electricity would fuel still-to-be-developed electric motor vehicles. Another possibility is to produce methanol from hydrogen and carbon dioxide. Methanol, like ethanol, can fuel motor vehicle engines directly or can be combined with gasoline to produce another type of gasohol. This "solution," however, creates a serious pollution problem: The combustion of methanol produces formaldehyde, a carcinogen (Schmidt-Perkins 1989, p. 21).

the case with fuel-efficient and electric models, they have not yet produced prototypes.

This drawback, however, should not detract from hydrogen's practicality in other energy applications. Hydrogen both stores and trans-
ports solar power and thereby makes solar power available when and
where it is needed.* It is a means of converting solar electricity into
alternating current, a more usable form of electricity. It is also a means
of converting electrical energy into gaseous or liquid fuel, which is
important because only about 25 percent of our current energy needs
are supplied in the form of electricity. A solar-hydrogen economy, if
combined with the efficient use of energy and perhaps with less ex-
travagant transportation patterns, offers the hope of meeting humanity's
energy needs. The manufacture of some solar collectors may involve a
small amount of pollution. But compared to the processes, methods, and
effects of energy production today, solar-hydrogen is virtually inexhaustible and pollution-free.

Energy, Heat, and Climate

Although progress toward a solar-hydrogen economy has been made, and
such an arrangement will undoubtedly be in humanity's long-term
future, the critical question is whether its present limitations will be
surmounted in time to avoid extensive environmental degradation and
human suffering. The answer to that question is still uncertain. Solar
technology requires more research and development support than it is
getting; as Table 2-6 shows, research is grossly biased in favor of fossil fuels
and the nuclear option. Other than sheer inertia, this probably reflects the
fact that solar energy, unlike fossil fuels or uranium, is not a resource that
can be owned. Moreover, some forms of it (such as rooftop collectors)
initially appeared to lend themselves primarily to local and perhaps even
individual exploitation, which is of no interest to gigantic regionally
oriented utilities. As we have seen, this perception is no longer accurate;
with hydrogen as a storage and transportation mechanism, solar collectors
can be located in deserts, wind farms can be located in agricultural areas,

---

* Hydrogen actually complements the weaknesses of solar-generated electrical
power. For example, PV power is in the form of low-voltage direct current, but
today's machinery and appliances run on high-voltage alternating current. One
solution is to redesign applications, where possible, to work on low-voltage
direct current. But the alternative is to use the low-voltage power to make
hydrogen, for which it is ideally suited, and later to burn the hydrogen to
produce high-voltage alternating current.

and ocean energy can be tapped in offshore locations. All solar technologies are adaptable to either local or central production facilities.

How much time humanity has before it *must* develop solar energy is also uncertain. The impact on climate is the ultimate limit on human energy use, a limit that no amount of technological ingenuity can remove. Ever since humans became technological beings by inventing fire, they have significantly altered the climate of the earth, creating semiarid savannahs where once there were grasslands. Nevertheless, the impact of industrial humanity on global climate is potentially far greater, for the continuation of certain current trends could render the earth quite literally uninhabitable by the human race and most other species as well. But we do not really know what we are doing to climatic mechanisms. In effect, with only minimal theoretical knowledge, we are running an enormous experiment on the global climate. By discharging carbon dioxide, particulates, chlorofluorocarbons, and other substances into the atmosphere, we are tampering with the watch-like perfection of the global climate system, not understanding clearly how it works or what the consequences of our actions will be. Worse still, as with the destruction of the ozone layer, some of our actions *are* producing visible consequences just as we come to understand what we have wrought. Now, any remedy we try is too little and too late to halt the inevitable and worsening effects. In the United States alone, 12,000,000 skin cancers will result and 200,000 people will die from just this intrusion on climate mechanisms. All life on earth will suffer harm from intense ultraviolet radiation, and some sensitive organisms crucial to global ecology will probably succumb.*

Most scientists are now predicting that our emissions of greenhouse gases will result in global warming. But again, humanity does not know what it is doing. We do not know how much or how fast this warming will occur; we do not know how much it will alter our capacity to grow

---

* It was once thought that the release of human-made heat might also disturb the mechanisms of climate balance. The second law of thermodynamics ordains that all forms of energy must inevitably decay into low-grade heat, so energy use and heat release are ultimately synonymous. However, the extra heat due to human use of energy now appears to be pitifully small compared to the total energy flow involved in the global heat balance (the flow of human-made heat is only about 1/15,000 or 1/20,000 the absorbed solar flux). It appears that it would take 200 years at a 5% growth rate in fossil, fission, and fusion fuel production for humanity to produce enough heat to cause noticeable global warming. This is extremely unlikely, for all the reasons we have cited, and so the heat we are now putting into the atmosphere is not likely to be a problem.

15

The Thermodynamic Economy

Although the currency of nature's economy is energy, the current human economy takes energy into account only indirectly, via the monetary cost of energy production and use. It is therefore in conflict with basic laws governing our physical existence. The laws of thermodynamics tell us that we cannot get something for nothing. The matter and energy (which are thermodynamically interchangeable) from which we derive economic benefit have to come from somewhere, and the inevitable residuals remaining after we have obtained the benefits have to go somewhere. Unless the thermodynamic cost of obtaining the energy and disposing of the residuals is less than the benefits of the use to which the energy is put, the system as a whole loses. Thus a "thermodynamic economy" based directly on an accounting of energy or entropy has become essential, for otherwise social decisions based on traditional economic criteria will continue to compromise the system through so-called externalities, or side effects, that create more entropy—increased disorder or reduced energetic potential—in the system as a whole.

The basic insight of thermodynamic economics is that entropy is the real basis of economic scarcity. There is a fundamental difference between classical scarcity and thermodynamic scarcity—that is, between a scarcity of land and other reusable or flow resources such as solar radiation and a scarcity of coal or other nonrenewable resources that, once used, are gone forever (or can be recycled only with limited efficiency and at a high cost in energy). Thus, says the thermodynamic economist, our nonrenewable resources are exceedingly precious capital stocks that can never be recreated, so to waste them or even to expend them primarily on current consumption is absurd, no matter how rational it may be in terms of dollars and cents.

In practical terms, what this means is that we are eroding 26 billion tons of topsoil every year; we have destroyed half of our tropical rain forests; we have polluted most of our rivers and lakes and some of our groundwater; and the planet has become one-third desert. Yet none of this counts when a country calculates its gross national product. Perversely, it also means, for example, that the Exxon *Valdez* oil spill *improved* the United States' economic performance. After all, Exxon spent billions of dollars on the cleanup; it created jobs and consumed resources.

What to do about this is being debated. Economists have focused on schemes of natural resource accounting, a century-old idea that originated with Alfred Marshall (1842–1924). Germany is attempting to calculate a green gross national product. The task is difficult. For example, the Germans can easily calculate the revenue loggers have lost because pollution has killed much of the Black Forest. It is more difficult to calculate the value of lost plants, lost animals, lost whole species, lost ecosystems, and lost habitats resulting from forest destruction. Forests also clean the air, conserve soil, purify water, and provide recreational opportunities. Some American economists have devised a way to deal with some of these problems: an "index of sustainable economic welfare" that incorporates goods and services provided by the environment into economic accounting (Daly and Cobb 1989, p. 401). Thus we know how much a sewage plant costs to build and operate. If the forest performs the same service "free," Daly and Cobb's suggestion is that the forest provides the equivalent value of the sewage plant. Other economists propose panel interviews to determine what forests or species (and so on), are worth to samples of people. However these matters are ultimately resolved, certain preliminary conclusions have already emerged. First, it will never pay humanity to run a completely technological world, as some extreme technological visionaries urge, for life support is cheap if we let nature do it and fantastically expensive if we take on this burden ourselves. Second, even though collection and conversion percentages may be quite low compared to some of humanity's engineering creations, and despite the apparently greater economic costs, when pollution and all the other thermodynamic costs are considered, solar energy and other forms of decentralized energy production are actually more efficient than methods based on using up nonrenewable resources and are therefore a thermodynamic bargain. Third, human labor may become thermodynamically cheaper than capital or other factors of production; industry may therefore become increasingly labor-intensive. Similarly, less energy-intensive materials such as wood and steel will displace such materials as aluminum, which are highly energy- and capital-intensive. In short, a thermodynamic (or, to put this concept in its proper context, sustainable) economy would aim at careful husbandry of resources, dependence on natural flows and processes, decentralization, more labor-intensive production, and a combination of ultra-sophisticated technology with some of the energy-saving methods that sustained our forebears.

food, how much it will shift areas of endemic disease, and the like. We don't know how able we will be in coping with its effects. We do know a few things, however. First, if climate change does cause death, disease, and human suffering, anything we try to stop it then will be woefully inadequate. Second, if climate change produces unacceptable consequences and forces humanity to drastically reduce carbon dioxide emissions, the costs of doing so will be higher than what it would have cost to develop renewable energy sources before that time. Unfortunately, the standard wisdom of the economic and political elite in the United States is that, as far as energy is concerned, we must have more of the same: more fossil fuel production and more development of nuclear power. Government funds pour overwhelmingly in the direction of supporting these technologies—by research, by direct subsidies, even by military means where necessary.

3

Deforestation, the Loss of Biodiversity, Pollution, the Management of Technology, and an Overview of Ecological Scarcity

So far, we have discussed the components of ecological scarcity that became apparent in the 1960s and 1970s: the exploding growth in human population, the difficulty of fulfilling expanding demand for food when the supply of agricultural land cannot keep up (and is being lost), the depletion of essential minerals, and the limits to satisfying the growing demand for energy by relying on sources that are both finite and polluting. But in the 1980s, observers recognized that these are not our only difficulties. Human demands on the environment are causing the problems shown in Table 3-1. Thus by the year 2000, according to projections, the growth in human population and activities will eliminate not only one-sixth of our grain land per person and one-tenth of our irrigated land, but nearly *one-fifth* of our forest and grazing land per person. These losses will occur in just one decade. In addition, human activities are causing increased pollution of the air, water, and land—and even of the stratosphere—despite expensive pollution-control efforts over the past two decades.

Table 3-1 Basic Natural Resources Per Person with Projected Population Growth

Resource	1990 (hectares)	2000	Percent Decline in One Decade
Grain land	0.13	0.11	15
Irrigated land	0.045	0.04	11
Forest land	0.79	0.64	19
Grazing land	0.61	0.50	18

Source: Adapted from Brown 1991, p. 17.

In Chapter 1, we mentioned the extent and futility of the destruction of tropical forests. What follows is an expanded discussion of tropical deforestation and the resulting loss of habitats and biological diversity. Thereafter, we turn our attention to the dangers of pollution. What these topics bring into sharper focus is the futility of expecting technology to solve the problem of ecological scarcity, at least in a timely way. As we shall see, our problem is that we are putting stresses on the environment in a multitude of ways. Technology may be able to offer solutions, at acceptable costs and with acceptable side effects, to one or another of these problems. (For example, a technological solution to our energy problem is within reach, and technologists may develop substitutes for some scarce minerals in time, such as fiber optics in place of copper.) But all-encompassing solutions do not seem to be within reach. And some problems seem insoluble. No one has yet suggested technological creation of an ecosystem or complete habitat, and we shall see that technological pollution control is at best a temporary tactic that will only make the growth in human activities possible a while longer.

Tropical Deforestation

Moist tropical forests may once have covered 1.5 billion hectares, an area twice the size of the continental United States (Corson 1990, p. 118). At present, fewer than 900 million hectares remain. The rate at which tropical forests are being destroyed is increasing dramatically. The Food and Agricultural Organization estimated in 1980 that the annual rate of tropical deforestation was 28 million acres. By 1990 that rate had increased by 40%—to 42 million acres per year (Booth Sept. 9, 1991, p. A18). Some estimates of tropical forest destruction are still higher—as much as 50 million acres per year (WRI 1990-91, p. 102).

The loss of tropical forests is a mark of how fast human beings are exterminating whole species of plants and animals, destroying watersheds, and affecting both regional and global climates. Tropical forests sustainably support people and millions of plants and animals. They provide fruits, vegetables, spices, nuts, medicines, timber, oils, waxes, and rubber consumed by people and industry around the world. Staple grains have relatives of somewhat different genetic makeup in the forests; these wild strains have been interbred with staple crops to provide resistance to diseases and pests. In fact, 40% of all medicines are derived originally from chemicals found in tropical plants. Forests moderate air temperature, recycle wastes, control soil erosion, and regulate stream and river flows, which moderate the floods and droughts that lower agricultural yields. Forests take carbon dioxide from the atmosphere and produce the oxygen all animals need.

When a tropical forest is destroyed, a habitat for living things vanishes. Although some species can adapt to other habitats or can persist in diminished numbers in remaining forests, thousands of species of plants and animals are being wiped out each year. Human communities are also being destroyed. Brazil's Indian population has decreased from 6 million people in the sixteenth century to about 200,000 today (Corson 1990, p. 123). Valuable products are lost. In less than 10 years, fewer than 10 tropical forest nations will be net exporters of forest products—down from 33 such nations today (Corson 1990, p. 124). Deforestation also degrades forest soils. Unlike what occurs in forests in temperate climates, when leaves or trees fall or animals die in tropical forests, they do not nourish the forest soil. Rather, other forest organisms quickly decompose them, and then these creatures support other tropical life. So deforested soils are nutrient-poor to start off with, and they support agriculture or cattle ranching for only a few years. At the same time, deforestation is the second largest human source of atmospheric carbon dioxide, right behind the burning of fossil fuels. One-third of the carbon dioxide emissions that humans produce is caused by deforestation. In 1987 carbon dioxide emissions from deforestation in Brazil alone are believed to have equaled the total carbon dioxide emissions from all sources in the United States (WRI 1990-91, p. 110).

If tropical deforestation has but temporary benefits and is self-destructive even over the middle term, why do people continue the practice? The first reason is to increase cropland. For thousands of years, people slashed and burned small areas of forest, farmed it for a few years,

and moved on to a new plot. Old plots, small in size, would lie fallow for
up to 30 years and revert to their wild state. This practice did the forest no
harm and fed small communities. Today slash-and-burn agriculture still
feeds people, but the way it is practiced is destroying the forests. More
people are clearing forests more frequently and re-cultivating old fields
too fast for them to regain their fertility. Because in many tropical
countries a few wealthy people own most of the land (in Brazil, 7% of the
population owns 93% of the fertile land), thousands of landless peasants,
suffering from hunger, take to the forests to farm. Moreover, people are
increasing cropland to produce commercial crops for export. Such crops
as bananas, coffee, and coca bring in much-needed foreign cash to finance
development projects or to pay off foreign debts. The government of
Indonesia explained in a United States magazine advertisement that its
people have the same aspirations as Americans; it must, therefore, convert
20% of its forests to plantations to produce teak, rubber, rice, coffee, and
other agricultural crops (*U.S. News & World Report,* December 18, 1989,
pp. 80–81).

Second, government policies encourage large development projects,
sometimes to resettle people from teeming cities or to build hydroelectric
dams or roads. The people who obtain short-term benefits from these
developments are not usually the same people who suffer from the
harmful effects a large project may have on the environment. We have
already seen some of the adverse environmental effects of large
hydropower projects.

A third reason to clear forests is to "harvest" timber. At least 5 million
hectares of tropical forests are logged each year, mostly for timber exported to Japan and the United States, to bring in foreign currency to
finance development or pay foreign debts. Unfortunately, most logging
operations are unsustainable. Even in those places where logging operations do not destroy ecosystems, land-hungry peasants often use logging
roads to penetrate once-inaccessible forests and burn down most of the
remaining trees to grow crops. According to a 1988 survey of the
International Tropical Timber Organization, timber is being produced
sustainably on less than 1% of the exploitable tropical forests (cited in
WRI 1990-91, p. 106).

A fourth reason for deforestation is to create pasture for cattle.
Two-thirds of Central America's tropical forests have been cleared, mostly
for cattle ranching. However, Central Americans do not cut down their
forests because they eat a lot of beef; rather, most of their beef is exported
to the United States, to fast food chains and for pet food (Corson 1990, p.
121). Every American hamburger for which meat was imported from
Central America is estimated to have destroyed five square meters of

forest (Corson 1990, p. 121). Finally, local people destroy forests because they need fuelwood. Fuelwood is the primary source of energy for cooking and heating by 80% of the people in the world; it is scarce for over 1 billion people.

Unfortunately, for reasons we have explained, once a large area is deforested in the tropics, it supports a few years of productivity but then becomes degraded. A 3-million-hectare area of the Brazilian Amazon that was converted into cropland in the 1880s is today a vast, uninviting, nonproductive expanse of scrub (Corson 1990, p. 121). Cattle pastures created in the 1960s have already become unproductive. In Brazil, government authorities have seen the futility of clearing forest for pasture and have stopped subsidizing it. Even so, some people still seek pasture in order to get title to land, to secure an inflation-free investment, and to gain status. Ranching merely brings in extra income.

. . . And the Loss of Biodiversity

Wildlife habitats are areas where nondomesticated species find the food, water, and other resources they need to survive. Humans have been converting them to their purposes since pre-agricultural times. But it is only since the Industrial Revolution and the colonial expansion of the nineteenth century that we have been destroying forests, grasslands, wetlands, riparian areas, mangroves, sea grasses, and coral reefs both rapidly and extensively (WRI 1990-91, p. 123). The area of temperate forests has been reduced by one-third, and most of those remaining are new-growth forests, which do not restore the native habitats that the old-growth forests provided (WRI 1990-91, p. 107). Woody savannas and deciduous forests have been reduced by one-fourth. Tropical forests are only 60% of their original size. The data for the loss of dry tropical forests, grasslands, and wetlands are fragmented, but what we have indicates broad trends. For example, in Africa and Asia, human activities have reduced natural grasslands to 41% of their original size (WRI 1990-91, p. 125). In California, 69 percent of the Central Valley's grasslands have been destroyed (WRI 1991, p. 125). 60% of the wetlands have been converted to agriculture or development in Asia, 56% in the United States (WRI 1990-91, pp. 127–128). 26% of the mangroves in the United States and Puerto Rico have been destroyed. In the European heartland, few if any original wildlife habits remain (WRI 1990-91 p. 126). Everywhere, as we have seen, coral reefs are dying.

The extinction of plant and animal species has been rising in rough proportion to the loss of habitats (see Figure 3-1). We may be losing up to 17,500 species a year (Corson 1990, p. 101). Some scientists believe we

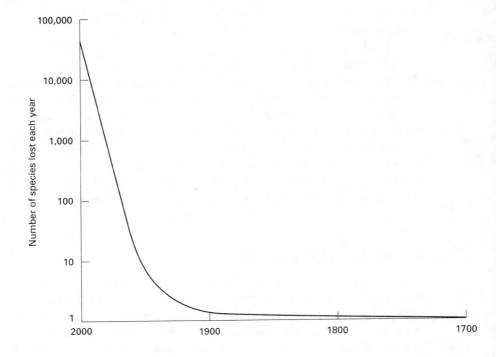

FIGURE 3-1   Estimated annual rate of species loss, 1700–2000.

could lose 25% of the species on the planet in the next few decades, which would be a larger loss than any of the extinctions in geological history. The most threatened habitats today are those richest in species diversity and biological productivity, the tropical forests and the coral reefs. A square mile in a Peruvian lowland has twice as many butterflies as the entire United States and Canada; one river in Brazil has more species of fish than all the rivers in the United States (Corson 1990, p. 100).

Species are becoming extinct for three reasons. First, as human population explodes, more habitats are converted into croplands. Hungry people will convert any place where crops can be grown into food production. Second, people introduce non-native species into new environments; native species that do not have the appropriate defenses against the invader die out. Third, people overharvest desired products and areas. For example, tropical agroforestry, based on small plots, on the intermixing of tree and soil crops, on the inclusion of domestic animals, and on the recycling of plant and animal wastes, is technically possible and would be sustainable over the long term—but it is scarcely practiced. Likewise, people are killing off 80,000 elephants a year, mostly to harvest their ivory; only 15 white African rhinos are believed to be left in the wild, largely because certain humans want their horns as a supposed aphrodisiac; and whales have been reduced below the level of commercial viability because some people want their oils (Corson 1990, p. 103).

Several practical reasons exist for preserving healthy ecosystems and the biodiversity within them. We discussed some of these in the last section. (1) Wildlife performs free environmental services. Plants take carbon dioxide from the air and produce oxygen. They counteract greenhouse emissions. They regulate stream flows and groundwater levels, recycle soil nutrients, and cleanse pollutants from surface water. Both plants and animals degrade wastes, help to control floods and soil productivity, and contribute to pest control. Even insects, usually regarded as pests, are not always so; many are important pollinators and decomposers of waste. (2) Wild species are the original source of over 40% of the chemicals used in prescription drugs (Corson 1990, p. 103). Lurking in one or more of the thousands of species we do not know about may be the cures for such diseases as AIDS and bilharzia. (3) Wild species provide humans with needed products and services. We have already noted how wild plants have made it possible to bestow immunity on cultivated crops. True, technologists may come up with synthetic substitutes for some of these products or services. Synthetic rubber, for example, has replaced natural rubber in two-thirds of its uses (Corson 1990, p. 104). But natural rubber is of better quality and is still preferred for heavy-duty applications such as airplane tires. And technologists seek wild plants for the genes needed to accomplish their goals.

When a species is lost, it is not renewable. Its particular function or potential—its unique combination of genes—is gone, forever. A given loss may or may not have direct consequences for people. But with the possible loss of 1 million species by the year 2000, it would be foolish to assume that the whole phenomenon is inconsequential. This is another area where we are running a large experiment without knowing what we are doing. Along with this uncertainty is the effect of our greenhouse emissions on biodiversity. If climate changes are of the magnitude that a majority of scientists are predicting, global warming will affect the location, size, and character of all wildlife habitats on the planet (WRI 1990-91, p. 130). Many species will not be able to adapt quickly enough, and extinctions will dwarf any that have been predicted, because habitats will shift, shrink, or disappear (WRI 1990-91, p. 130).

Pollution

The Inevitability of Pollution

Nature does not produce "pollution." Plants grow, taking up carbon dioxide from the air and nitrates from the soil. They "excrete" oxygen into the air. Animals eat the plants, excreting carbon dioxide to the air and organic compounds that the plants use as fertilizer. This and other natural cycles repeat themselves.

Humans, on the other hand, do produce pollution. They produce chemicals and products that are not found in nature and cannot be used (eaten) by living things, or they produce them in such great quantity as to overwhelm nature's ability to use them. Many dangerous forms of pollution already abound; the intensification of agriculture and basic-resource production will only aggravate the problem. In fact, because virtually all modern techniques of production and consumption entail pollution, any increase in production and consumption necessarily produces a proportionate increase in pollution. The laws of physics tell us that matter and energy can be neither created nor destroyed, but only transformed. These transformation processes are never completely efficient; by-products or residuals are an inevitable result, and these are almost always noxious, or at least unwelcome, to some degree. In practice, the efficiency of industrial practices is low. For example, the most modern fossil-fuel plant converts less than 40% of the energy in coal or oil into electricity. The rest of the energy escapes up the stack and into cooling water as heat, and large quantities of residuals in the form of ash and gases are produced. Thus pollution is the result of the operation of basic physical laws.

Moreover, this description understates the problem. The petrochemical industry, for example, each year produces about 600 billion pounds of products and 500 billion pounds of toxic chemicals, only 1% of which are destroyed (Commoner 1990, p. 51). But many of the "products" of this industry are themselves pollutants—solid or hazardous wastes, such as pesticides and plastics—that also cannot be used by nature. Thus pollution results not only from the inefficiencies of physical transformation processes but also from some of the very products humans contrive these processes to produce.

Why not simply use technological devices to control pollution? Unfortunately, for thermodynamic reasons, there is no way that we can control many forms of pollution technologically. Waste heat is the prime example, but many others are also important, including fertilizer run-off and carbon dioxide from combustion. Their damage to ecosystems may be mitigated in some ways, but short of forgoing production altogether or adopting radically different technologies, there is simply no way to contain these high-entropy pollutants.

Other forms of pollution are, of course, susceptible to technological control, but even these pollutants cannot be controlled indefinitely, for continued growth in industrial output must inevitably overwhelm any pollution-control technology that is less than 100% effective. For example, if overall industrial output grows at 5% (doubling period 14 years), reducing by 90% the quantity of pollution emitted and maintaining this level of efficiency in the future would buy only about 45 years of grace, for after that amount of time, the absolute level of pollution would be the same as it is today. Even a 99% level of efficiency would extend this period of grace by only 46 years. Thus, with continued industrial growth, at some point pollution will reach dangerous levels under any pollution-control regime. This point may not be very far away, for even the 90% level of overall control effectiveness would be difficult, if not impossible, to achieve in practice, and a considerable price in money and energy would have to be paid for it.

The Dilemmas and Costs of Technological Pollution Control

The same physical laws that make production and pollution two sides of the same coin also tell us that all technology can do is exchange one form of pollution for another. Worse, because energy is generally used to "solve" the original problem, there is bound to be an overall increase in both gross pollution and entropy. For instance, some of the CFC substitutes now coming into use to avert the destruction of the ozone layer are themselves toxic, exacerbate the greenhouse effect, or contribute to photochemical smog. They are also less energy efficient to produce and less effective than the CFCs they replace. As is undoubtedly true in this case, a particular technological "fix" may nevertheless be desirable on economic, ecological, or aesthetic grounds despite the overall thermodynamic loss, because the burden will presumably be shifted away from a sector that is harder pressed ecologically or whose degradation is more obnoxious to our senses and health. However, once the absorptive capacity of the environment has been used up, and all sectors are about equally hard pressed, then simply converting pollution into a different form will no longer be a workable strategy. It is clear, therefore, that no matter how much money and energy are at our disposal, there must be some ultimate limit on our ability to control pollution without also controlling production.

However, we are unlikely to reach this ultimate ecological limit. Our supplies of money, energy, and other resources are not without limit, and at some point the increased costs of pollution control seem likely to render further growth in production fruitless. For example, imagine a lake with a "waste-absorption capacity" of y units of residuals. (In fact, as we shall see later, waste-absorption capacity is an economist's concept that has no ecological validity.) A factory on the shore of this lake produces x units of a certain commodity and at the same time emits exactly y units of residuals. Although the factory now exercises no control over its pollution, the residuals it emits have not yet begun to degrade the quality of the lake water. However, suppose that demand for the commodity doubles, requiring either a second factory of like capacity or an equivalent addition to the old one. Without pollution control the lake would begin to die biologically, because the $2y$ units of residuals resulting from $2x$ units of production would be twice what the lake can safely absorb. Thus pollution-control equipment that is 50% effective must be installed to reduce the residuals to a safe level. If demand doubles again, production of $4x$ units of the commodity produces $4y$ units of residuals to dispose of, and pollution control must now be 75% effective. It is clear that if doubling of production continues, more than 99% effectiveness in pollution control eventually becomes necessary.

These points are supported by reviewing the recent history of pollution-control efforts in the United States. Since the 1970s, the United States has spent about a *trillion* dollars on pollution-control efforts. What does the country have to show for it? Unfortunately, not much. Despite the fact that we have reduced pollution per process or pollution per event, our growth in numbers and productivity has so eroded our accomplishments that our net gain is nil. Both air and water remain seriously polluted, with increasingly serious consequences both to humans and to the planet. Hazardous substances are entering the environment in record (and growing) amounts. These substances, including toxic and nuclear waste, pesticides, and industrial chemicals are causing serious human health consequences, notably cancer, birth defects, and heart and lung diseases. Solid wastes are overrunning our ability to manage them and are contaminating the drinking water of millions of Americans.

Air Pollution

In 1971 the Environmental Protection Agency set standards for reducing the levels in the air of substances known to produce serious health effects on human beings. These included sulfur and nitrogen oxides, which damage the respiratory tract and lungs, carbon monoxide, which deprives the blood of oxygen, and organic hydrocarbons, some of which cause cancer and birth defects.(These substances also adversely affect human health by contributing to the formation of ground level ozone and acid rain. And they also harm the environment.) The EPA thought its standards would result in an 80 to 90% reduction in the emissions of these substances between 1975 and 1985. In fact, the average reduction in the emissions of these substances between 1975 and 1987 was 18% (Commoner 1990, p. 22). Nitrogen oxide emissions actually *increased* by 2% during those years. Why? The answer is not merely the regulatory retreat by the administration that came to power in 1981. For example, new automobiles in the United States emit 90% fewer hydrocarbons and 75% less carbon monoxide than did those of the early 1970s. But the number of vehicles on the road has nearly doubled in the last 20 years (McCloughlin 1989, p. 48). The increased number of motor vehicles, new industrial plants, and new power plants (even though many are equipped with such pollution-control devices as scrubbers), have nullified much of the pollution controls. The same outcome has occurred in other industrialized countries attempting to control pollution—except that in some countries the record is worse.* In West Germany, for example, the average reduction of these emissions was 15% (its nitrogen oxide emissions increased by 29%); in Great Britain, the average reduction was 1% (Commoner 1990, p. 36). In general, countries that have tried to reduce the emission of a pollutant by controlling it—as opposed to preventing the pollutant from being produced at all—have seen their controls negated by the economic "growth" taking place in those countries.†

As a result, and because many countries have no or few pollution controls, huge quantities of pollutants enter the atmosphere from human activities. According to the Organization for Economic Cooperation and Development, 110 million tons of sulfur oxides, 69 million tons of

---

* Sweden is a notable exception. Its reductions of pollution have exceeded its increases in growth; sulfur dioxide emissions, for example, dropped by more than two-thirds between 1970 and 1985.

† Lead is the only notable success story in diminishing an air pollutant. Lead is a toxic metal. It causes neurological disorders (especially mental retardation in children) and kidney failure. It inhibits both respiration and photosynthesis in plants. In the United States, it has almost been eliminated as an air pollutant; total annual emissions of lead were reduced by 94% between 1975 and 1987 (Commoner 1990, p. 22). The reason for this unique success was that the government did not try to *control* lead pollution but rather removed its *source* from the environment. That is, it forced technological change. Because lead clogged catalytic converters, oil companies were forced to use lead substitutes to raise the octane levels of gasoline.

nitrogen oxides, 193 million tons of carbon monoxide, 57 million tons of organic hydrocarbons, and 59 million tons of particulates were emitted into the atmosphere in 1980 (Corson 1990, p. 221). (This listing does not include substances classified as toxic emissions, which we will discuss shortly.) The Office of Technology Assessment estimates that current levels of particulates and sulfates in the air may cause the premature death of 50,000 Americans a year (Corson 1990, p. 223). The American Lung Association concluded in 1990 that pollution from motor vehicles alone costs Americans $40 billion to $50 billion in annual health-care expenditures and causes as many as 120,000 unnecessary or premature deaths (*The Washington Post* Jan. 21, 1990, p. A12).

Moreover, these five pollutants are not the only ones that have worrisome health effects. Ozone is another serious air pollutant. It is created when organic hydrocarbons and nitrogen oxides react with oxygen in the presence of sunlight. Ozone is produced in great quantities around the world, primarily by cars and trucks. In some countries, no controls exist on the emissions of motor vehicles. Where controls do exist, the increasing numbers of vehicles are nullifying the reductions, especially that of nitrogen oxides. In 1988, 96 cities and counties failed to meet the standards the Environmental Protection Agency set for ozone in the United States. Ozone contributes to smog, reduces resistance to infection, interferes with lung functions, contributes to asthma and nasal congestion, and irritates the eyes. It also damages crops and contributes to the greenhouse effect. At current levels, ozone causes crop damage estimated at between $2 billion and $4.5 billion per year.

Acid rain is another air pollutant. Acid rain is weak sulfuric and nitric acid. It forms when nitrogen and sulfur oxides combine with moisture in the air. Acid rain can kill fish and other aquatic life, destroy forests, and damage structures such as metals, (such as railroad tracks), building facades, and paints. Acid rain damage varies in different parts of the world. Half the forest area in 11 European countries shows acid rain damage; over 35% of the forests in Europe as a whole are damaged (Brown 1990, p. 106). 43% of the conifers in the Alps in Switzerland are dead or seriously damaged (Corson 1990, p. 225). 80% of the lakes in Norway are biologically dead; in Sweden 14,000 lakes cannot support sensitive aquatic life; in Canada 150,000—1 out of 7—eastern lakes are biologically damaged. On the other hand, although eastern China has acid rain damage, Japan has found no evidence of large-scale damage from acid rain (OECD 1991, p. 47). Acid rain has caused only moderate environmental damage in the United States. A recent federal government study, conducted over 10 years, concluded that aquatic life is being damaged in about 10% of eastern lakes and streams, that visibility is being reduced in the eastern United States and in some metropolitan areas of the west, that acid rain is contributing to the erosion and corrosion of stone and metal structures, and that it is reducing the ability of red spruce trees at high altitudes to withstand the stress caused by cold temperatures (*The New York Times*, Sept. 6, 1990, p. A24.) It also found evidence that acid rain is contributing to the decline of sugar maple trees in eastern Canada and causing respiratory and other diseases in humans. It predicted that acid rain, if unchecked, would lead to forest decline in the decades ahead.

Acid rain may combine with ozone pollution to have more damaging effects than either substance does alone. Some scientists believe the combination of these two pollutants have weakened trees to the point where 70% of the standing trees above an elevation of 900 meters (2950 feet) in Virginia, Tennessee, and North Carolina are dead (Corson 1990, p. 228). (Ozone and acid rain are more concentrated at higher elevations in a given area.)

Chlorofluorocarbons and halons pose another threat to human health. They are used as propellants, as refrigerants, as solvents, as stabilizers, in fire extinguishers, and in blowing foam (Corson 1990, p. 228). The military is a major user.* But these chemicals, along with nitrogen oxides, cause depletion of the earth's ozone layer. Ozone in the stratosphere is not a pollutant but a filter of harmful ultraviolet radiation from the sun. UV exposure damages plants, causing reduced crop yields. It kills aquatic organisms, including, ominously, the phytoplankton that produce much of the oxygen on which all animal life depends. Increased exposure of humans to ultraviolet radiation increases the incidence of a fatal form of skin cancer, eye damage, and immune system disorders. The incidence of melanoma has increased 83% over the past 10 years in the United States (Shea 1989, p. 82); skin cancer cases are expected to increase from 500,000 to 800,000 a year (Weisskopf 1991, p. A1). The average ozone concentration in the stratosphere declined 2% from 1969 to 1986, but a 1991 report issued by the National Aeronautics and Space Administration concludes that ozone depletion is accelerating twice as fast as previous projections. Despite this, the production of chloroflurocarbons, the most important cause of this decline, is increasing by 5% a year.

---

* The U.S. military, for example, uses 76% of all halon–1211 consumed in the United States and 50% of the CFC–113. Each space shuttle launch deposits 56 tons of chlorine into the upper atmosphere (quoted in Renner 1991, p. 140).

The industrial countries of the world have agreed to phase out the use of CFCs by the year 2000 if possible. But millions of additional tons of CFCs will be released into the atmosphere in the meantime. They will persist there for over 100 years. Furthermore, the chemical companies now producing CFCs will be introducing substitutes that will still deplete the ozone layer, though at a significantly lower rate.* A worse problem is that developing nations have refused to join in phasing out their use of CFCs; they believe that the substitute chemicals will increase their costs and slow their growth rates. The industrialized countries, for their part, agreed not to restrict CFC exports to the developing world. In sum, although in this case the industrialized world has concurred on joint action to combat a perilous environmental pollutant, the endeavor is too little and too late; it will not stop the destruction of the ozone layer. Millions of additional cases of skin cancer, reduced yields of crops, and reduced phytoplankton activity are the predictable consequences.

Finally, carbon dioxide—the main by-product of combustion— along with nitrous oxides, methane, and chloroflurocarbons, produces a greenhouse effect. These gases trap the heat absorbed by the earth from the sun and prevent its escape back into space. Scientists have proved conclusively that the earth is about 33°C warmer than it would be if it did not have carbon dioxide in its atmosphere. Also, scientists have observed that for the past 160,000 years, whenever atmospheric carbon dioxide levels increased, the earth's temperature rose shortly thereafter; CO_2 levels and the earth's temperature have been closely correlated through the centuries (quoted in *Wind Energy Weekly 1990*, #403). Today, atmospheric CO_2 concentrations are 20 to 25% higher than at any time in the pre-industrial period, going back 160,000 years. In addition, the atmospheric concentration of methane, a molecule of which traps 20 to 30 times the heat of a CO_2 molecule, has reached more than double pre-industrial levels. The atmospheric concentration of CFCs, a molecule of which traps 20,000 times the heat of a CO_2 molecule, obviously is infinitely higher than in pre-industrial times, when it did not exist. For these reasons, most scientists predict the earth will warm 2°C to 5°C in the next century (WRI

---

* Substitutes for CFCs exist that (unlike the HCFC's now planned) have no ozone-depleting effects but are significantly more expensive. DuPont, Allied Chemical, and other CFC producers have thus far shown no interest in these chemicals. If enforced, however, the 1990 Clean Air Act will require phasing out of the use of HCFCs by 2030 in the United States.

1990-91, p. 13).* (The earth's temperature during the last ice age was only 3°C to 5°C cooler. The predicted increase in temperature is of like magnitude but would occur 10 to 40 times as fast).†

The implications of such rapid warming for life on earth are not well understood. Oceans may rise. That will contaminate nearby ground-waters with salt, cause coastal flooding, and drive millions of people from their homes. The cost of protecting cities from a 1-meter rise in ocean levels will be in the hundreds of billions of dollars. Forests may suffer. Experiencing a 1°C rise in temperature is equivalent to moving 200 kilometers toward the equator; with global warming, tree species may not be able to "migrate" toward the poles that fast. At the warmer limits of a species's range, trees would be more susceptible to disease and insects and less able to adapt to other human-made environmental stresses, such as

---

* Scientists cannot be sure of the extent or timing of global warming because of uncertainty about several variables, including the absorptive capacity of the oceans and probable changes in cloud cover. A small minority of scientists doubt that greenhouse gases will actually produce global warming; some of the doubters believe that increased evaporation and phytoplankton activity resulting from the warming effect of greenhouse gases will lead to increased cloud cover, *reducing* the amount of solar warmth reaching the earth. In addition, two Danish scientists have linked the global warming we have already experienced to sunspot intensity (Stevens 1991, p. C4). However, another minority of scientists believe that the warmer it gets, the hotter it *will* get. Some cite evidence that methane stored in permafrost will be released as it is warmed, thus causing more warming; others cite evidence that since 1976 plants have dramatically increased their *production* of carbon dioxide and methane in a process called "dark respiration." Dark respiration occurs because plant respiration rates increase with rising temperature—unlike photosynthesis, the rate of which rises to a point and then declines (Moore 1990, p. C3).

† In January 1991, scientists at the Goddard Institute for Space Studies, measuring data from 2000 sites worldwide, and the British Meteorological Office announced that 1990 was the warmest year of the 140 years since weather records have been collected (Booth Jan. 1991, p. A3). This record followed on the heels of several broken in the 1980s; six of the seven warmest years on record occurred during that decade. It is possible, of course, that these records have nothing to do with global warming; temperatures fluctuate from year to year, or in groups of years, for reasons having nothing to do with a greenhouse effect. It is also possible that thermometers were not calibrated exactly the same 50 or 125 years ago, producing false comparisons today. The question is whether it is prudent for humanity to act on evidence that global warming may be under way or to use such pretexts to deny it.

increased UV radiation, ground-level ozone, and acid rain. Agriculture could also suffer. Although some plants' rate of photosynthesis increases in the presence of high CO_2 levels, this characteristic has not been identified in the crops humans have cultivated. As we noted earlier, changing rainfall patterns may make many agricultural lands unsuitable for farming, render obsolete many irrigation facilities, and require new ones to be built elsewhere at great cost.

The sources of carbon dioxide are everywhere fossil fuels are burned (see Table 3-2). The production of most of the energy to power industry, commerce, and homes produces carbon dioxide. The average car spews out prodigious amounts of carbon dioxide: 16,000 pounds a year. The CFCs in a car's air conditioner have the greenhouse impact of another 4800 pounds of CO_2. To the extent that these emissions have been controlled, it has been by increasing gas milage. The average fuel efficiency of American cars, for example, has doubled since 1970, from 13 to 26.5 miles per gallon. This success, however, is nullified by the fact that Americans drive nearly twice as many motor vehicles than they did in 1970. Vehicle miles are increasing by 25 billion miles a year. Worldwide, the problem is worse. Despite the fact that fuel milage in other industrial countries is now slightly better than in the United States, the global average fuel economy is 20 miles per gallon.[*] Most countries have not adopted gas milage standards, but the increase in their numbers of cars has been comparable to that of the United States. In 1950 there were 50 million cars in the world; by 1960 that number had doubled. By 1970 there were double that number again, and by 1990 the number had redoubled again to 400 million cars. They spew 550 million tons of CO_2 into the atmosphere per year. By 2010, there are expected to be 700 million cars! It is true that if all governments agreed to require automobile gas efficiency to be 50 miles per gallon,[†] success would effect

---

[*] Cars in Organization for Economic Cooperation and Development (OECD) countries—that is, the industrialized West and Japan—average 30 miles per gallon. (Bleviss and Walzer 1990, p. 103).

[†] The technology to do this exists. Even in 1991, several manufacturers have built prototypes of cars that exceed 50 miles/gallon. Volvo has a prototype compact car that gets 63 miles/gallon in the city and 81 miles/gallon on highways; the car is designed to meet all U.S. emissions and safety requirements and to exceed the U.S. crash standard of 30 miles/hour. The car would cost little more than current cars to manufacture (Bleviss and Walzer 1990, p. 106; WRI 1990-91, p. 151). Already in the 1992 model year, Honda is selling a version of its popular "Civic" that gets nearly 60 miles/gallon on highways.

Table 3-2   Major Sources of CO$_2$ Emissions in the United States

Source	Category	CO$_2$ Emissions (metric tons/year)
Steam for power	Industrial	649.1
Automobiles	Transportation	613.7
Trucks	Transportation	587.3
Motor drives	Industrial	476.5
Space heating	Residential	469.3
Direct heating	Industrial	380.8
Appliances	Residential	289.2
Lighting	Commercial	278.0
Space heating	Commercial	262.9
Airplanes	Transportation	248.6
Cooling	Commercial	210.0
Water heating	Residential	205.4
Coal	Industrial	130.7
All others	—————	680.3
Total		5,587.8

Source: Adapted from Booth 1990, p. A1.

a slight net reduction of carbon dioxide emissions. If governments combined this efficiency with other measures, including the manufacture of different types of automobile engines, the use of alternative fuels, and the imposition of a carbon tax to reduce emissions further—and if governments required the provision of mass transit, the building of bikeways, the clustering of housing, and so on to keep the number of cars down to 500 million, carbon emissions would be reduced to half of what they are today. As has been true of other pollution-control efforts, however, governments are unlikely to adopt a sufficient number of these measures in time for humanity even to stabilize total automobile carbon emissions, much less reduce them. (Why this is true is the subject of Part II of this book.)

Moreover, cars constitute only 15% of the global carbon dioxide problem (though they contribute 25% of all U.S. carbon emissions). Growth in the industrial, commercial and residential sectors is also increasing carbon emissions faster than increases in efficiencies are reducing them. Increases in energy productivity in the developed world have been modest at best. Most governments have made only token efforts to develop energy from renewable carbon-free sources.[*] Finally, as we have seen, people are destroying 17 million hectares of forests each year. This increases atmospheric carbon dioxide because destroyed forests release trapped carbon dioxide and no longer "consume" carbon dioxide in photosynthesis. The liquidation of the earth's forests is thought to be responsible for 10% of the excess carbon dioxide in the atmosphere; in the tropics, where the forests are wiped out mostly by burning, the fires themselves produce another 1 to 2 billion tons of CO_2 per year.

About 15 nations, mostly in Western Europe, have plans to limit their production of carbon dioxide over the next 15 years. But the United States, which, with 5% of the world's population, is responsible for more than 25% of the world's carbon emissions, has refused to join in CO_2-reduction efforts. And it has repeatedly prevented the Europeans from converting their plans into binding targets (Meyer 1990, p. 3; Stevens 1991, p. 61). (Ironically, the United States is now only half as efficient in using energy as Western Europe and Japan are, so increasing its energy productivity to reduce carbon emissions would improve the "competitiveness" of the United States economy.) Underdeveloped countries also have no plans to reduce carbon emissions. There remains, therefore, a vast gap between projected growth rates in carbon emissions and what scientists believe is necessary to control global warming.

---

* The United States government in particular has invested huge amounts of research and development funds in nuclear energy (the hazards of which were discussed in Chapter 2) while investing trivial amounts in, and eliminating once-existing tax breaks for research and development in renewables. President Reagan even removed a fully functioning solar hot-water system from the White House. Despite this, non-nuclear carbon-free sources of energy exist; they include wind, geothermal, photovoltaic, solar thermal, biomass, and ocean thermal energy conversion. A few are already cost-competitive with fossil fuels. With further research, more could be.

Water Pollution

Water pollution is also worsening in industrial countries. In the United States alone, the national government has spent over $100 million to clean up surface waters, and we have little to show for it. More than 17,000 of the nation's rivers, streams, and bays are polluted. From 1974 to 1982, levels of fecal coliform bacteria decreased at only a few river reporting stations; levels of dissolved oxygen, suspended sediments, and phosphates increased at about the same number of sites as those at which they decreased (Commoner 1990, p. 25). Nitrate levels increased at four times as many as those sites at which they decreased. Arsenic and cadmium levels are also increasing. Overall, water quality has deteriorated at more than three-quarters of the measuring stations. Various states prohibit eating, or warn their citizens against eating, the fish caught in their rivers and lakes. For the hundreds of millions of dollars (including state and local spending for sewage-treatment plants, private spending for septic tanks, and the like) Americans have spent to clean up their waters, relatively few waterways that had undrinkable water or were unsafe for swimming in the 1970s have drinkable or "swimmable" water today. At the same time, many sites that had good water in the 1970s are unsafe today. In addition, sewage, plastic litter, discarded fishnets, tar balls, and toxic and radioactive substances are contaminating coastal waters.* Some coastal areas are so polluted that they are closed to oyster and shellfish fishing. Despite pollution-control efforts, the percentage of coastal waters closed to such fishing has been increasing; about half are now closed. A quarter of the usable groundwater in the United States is contaminated—more than three-quarters in some areas (Corson 1990, p. 164). Sixty pesticides, many of them carcinogens, have been found in the groundwater of 30 states (Corson 1990, p. 164). Other toxic chemicals, as well as saltwater and microbiological substances, are also polluting groundwater. Groundwater pollution is particularly insidious because there is no practical way to clean it up: Once the groundwater is polluted, it remains so.

---

* Oil from oil spills generates more publicity than other kinds of water pollution; witness the *Exxon Valdez* spill. However harmful, the effects of oil spills on coastal waters are less grave than routine unheralded municipal and industrial practices. For example, *every day* the oil industry dumps into the Gulf of Mexico over 1.5 million barrels of waste water containing oil, grease, cadmium, benzene, lead, and other toxic organics and metals!

Many causes for water pollution exist; they are summarized in Table 3-3. We have also discussed additional problems caused by acid mine drainage and acid rain. Agricultural practices are the source of some of the organic chemicals in water. For example, the groundwater contamination from pesticides results from repeated (and increasingly poisonous) sprayings, and only 1% of pesticides actually reach the target pests. But fertilizers used in agriculture also contribute to water pollution; fertilizer not used by the plants either runs off into surface water or leaches down into the groundwater. The runoff distributes large amounts of nitrogen and phosphorous to the water. Finally, water used for irrigation picks up salts on the land and carries them back to rivers and streams.

Industry causes another large share of toxic water pollution. The largest producers of toxic water pollution (and, as we shall see, of hazard-

Table 3-3   Causes and Effects of Water Pollution

Substance	Cause	Health Risk
Pesticides	Agriculture	Cancer, birth defects
Nitrates	Agriculture, airborne nitrates from cars, power plants	Globinemia
Chlorinated solvents	Chemical degreasing, machinery maintenance	Cancer
Trihalomethanes	Chemical reaction between organic chemicals and water treated with chlorine	Liver, kidney damage; possible cancer
Pathogenic bacteria, viruses	Inadequately treated sewage, leaking septic tanks	Gastrointestinal illnesses, diseases
Metals	Industrial, mining processes; oil production	Cancer, neurological disorders

Source: Adapted from Corson 1990, p. 166 (adapted from *Time,* March 27, 1989, p. 38) and Commoner 1990, p. 28.

ous solid wastes) are the chemical and plastics industries. Metal finishers, steel makers, and the pulp and paper industries also generate toxic water pollution. At least 627 industrial firms, along with 250 city sewage facilities, routinely discharge toxic wastes into American surface waters. Two-thirds of industry's toxic products are dumped into landfills or injected into injection wells or pits. The chemicals in all of these disposal sites eventually seep or leach down into the groundwater. 77,000 such disposal sites are known to exist in the United States (Corson 1990, p. 163). Nearly 1000 of them have been identified by the Environmental Protection Agency as urgently requiring attention because they are already leaking into the groundwater or are threatening to do so.

These problems are repeated in other industrial countries. 90% of the rivers monitored in Europe have nitrate pollution; 5% have nitrate concentrations over 200 times the unpolluted level. Many rivers have high levels of such metals as zinc, lead, chromium, copper, arsenic, nickel, cadmium, and magnesium, as well as organic chemicals. Some European nations, such as the United Kingdom, Finland, Belgium, and Spain, have higher levels of chlorinated hydrocarbons (DDT-type insecticides and PCBs) in their waters than does the United States. Worse yet, many pesticide-using countries in the developing world have higher levels of chemical residues in their waters than either Europe or the United States. Examples include Thailand, China, Colombia, Tanzania, Malaysia, and Indonesia.

Sewage causes water pollution in all parts of the world. In this instance, the developing world has the severest problems by far, because sewage is often not treated at all. Fecal coliform counts in Latin American rivers are as high as 100,000 per 100 milliliters (WRI 1990-91, p. 162). This compares to a World Health Organization recommended level for drinking water of 0 per 100 ml. 80% of all human disease in the world is linked to unsafe water, poor sanitation, and lack of basic knowledge of hygiene and disease mechanisms (Corson 1990, p. 162). 25 million people die each year from waterborne diseases (Corson 1990, p. 162). *Every hour* over 1000 children die from diarrheal diseases (Corson, p. 162). Developing countries are increasing their expenditures on building adequate wastewater treatment facilities. But to eradicate these conditions while their populations are growing is simply beyond their resources.

By comparison, the United States and Europe have minor sewage treatment problems. 80% of the people in West Germany, Switzerland, Denmark, and Sweden are connected to sewage-treatment plants. Only 106,000 people in the United States picked up waterborne diseases between 1971 and 1983, although that figure excludes cancers caused by toxic pollutants (Corson 1990, p. 165). Still, the expense of building adequate wastewater treatment facilities is huge, even for industrial countries. In the United States, the Clean Water Act of 1972 forbade municipalities to discharge sewage after 1977 until 85% of the bacteria and pollutants in it had been removed. Then the deadline for compliance was extended until 1988. By 1989, after that deadline had passed, an EPA study reported that two-thirds of the 15,600 wastewater treatment plants in this country still did not meet federal standards and that it would cost 83.5 billion dollars to bring them into compliance. That figure is 17 times more than the whole EPA budget for all antipollution controls.

To preserve fresh water that human beings can safely drink is thus a daunting task. (Making fresh water from the sea is not an easy way out of our difficulties. It takes 3 kilowatt-hours of energy to make 1 gallon of freshwater. Not surprisingly, two-thirds of the world's desalinization plants are on the Arabian Peninsula (Postel 1990, p. 150).) Water pollution controls, like air pollution controls, have been very expensive and, except for sewage, have produced little in the way of results. As with the case of the near elimination of airborne lead pollution, humans may have to eliminate the sources of pollutants rather than trying to control them after they are produced. Rather than building expensive facilities to control nitrogen levels in water, for example, it may be that agribusinesses will have to stop using chemical fertilizers to grow crops. The production of human sewage cannot be eliminated, but using such devices as composting toilets may be more practical in many areas than expensive centralized sewage collection and wastewater control facilities. Where centralized waste-water control facilities *are* built, they must be designed to produce toxic-free compost, which in turn can be used as a substitute source of fertilizer for agriculture. Similarly, no control mechanisms for pesticides sprayed on the land exist; they will end up and accumulate in groundwater until agribusinesses are forced not to use them. As we shall see, no economic means of controlling the toxic chemicals produced by the petrochemical industry exist; that industry endures only because it does not pay its environmental costs.

Still, in only a few instances (PCBs from industrial and electrical products, DDT and related chemicals from pesticides, and, in a few states, phosphorous from detergents) has an isolated harmful product or process been eliminated from production. Where used, such elimination has dramatically reduced particular pollutants. The reductions cannot be nullified by increases in economic growth. But the elimination of most industrial and agricultural products and practices that cause pollution is discussed very little.

Hazardous Wastes

Air and water pollution controls, which have achieved only modest reductions of a few pollutants, look like a roaring success compared to efforts to control hazardous wastes. A hazardous substance is one that harms human health or the environment; it includes "toxics," which are directly poisonous to humans. Humanity is engulfed by hazardous substances. In the United States alone, 260 million metric tons of hazardous substances are produced each year—more than 1 ton for every person in the country. The three most common types of hazardous substances are chemicals, (70,000 of them, mostly synthetic organics such as vinyl chloride or dioxin), pesticides (1 billion pounds of them used each year), and heavy metals. 70% of all hazardous substances are produced by the petrochemical industry. Except for pesticides, which are sprayed widely over agricultural land, they are dumped in a variety of sites in and out of the country.

We have discussed the pesticide problem before. The National Academy of Sciences has estimated that in the United States, 20,000 new cases of cancer are caused each year by pesticide residues in the food supply. In addition, EPA tests show that pesticides contaminate the groundwater in 34 states; 1 in 9 wells tested was contaminated (Allen 1990, p. 129). 95% of rural Americans rely on groundwater as their drinking source. In the San Joaquin valley of California, investigators have found pesticides in 2000 wells, including 125 public water systems (Corson 1990, p. 252). There, and in some other places of heavy pesticide use, cancer rates are up, especially among children.

The World Health Organization estimates that 1,000,000 cases of pesticide poisoning occur worldwide each year, 5000 to 20,000 resulting in death (French 1990, p. 14). People in developing countries suffer most of these deaths, partly because pesticide companies do not label their products, the people cannot read pesticide labels, or they are not trained in proper pesticide handling. Developing countries also use extremely toxic pesticides that industrial countries ban in their own countries. 25% of the 400 to 600 million pounds of pesticides that the United States exports are either banned or severely restricted in this country (French 1990, p. 14). Three-quarters of all pesticides used in India, for example, are chemicals banned in the United States (Corson 1990, p. 252). United States consumers reap some of the poisons their government permits chemical companies to sow: A Natural Resources Defense Council sampling of coffee beans in 1983 revealed that all samples had residues of DDT, BHC, and other banned pesticides. More than one-third of all fruit sold in the United States is grown in countries where few or no controls on pesticide

Table 3-4   Toxic Substances Discharged by U.S. Industry, 1987

Destination	Millions of Pounds
Air	2,700
Lakes, rivers, and streams	550
Landfills, earthen pits	3,900
Treatment and disposal facilities	3,300
Total	10,450

Source: EPA, reported in *The Washington Post*, April 13, 1989, p. A33.

use exist, and less than 1% of food imports are inspected for pesticide residues (Corson 1990, p. 253).* Some believe that 50% of all imported fruit is pesticide-contaminated (Weir and Matthiessen 1990, p. 119).

Other than pesticides, two-thirds of the hazardous and toxic wastes produced in the United States are disposed of in ways that eventually contaminate groundwater. The rest of the wastes are either spewed into the air or discharged into streams and rivers (see Table 3-4). Fifteen thousand uncontrolled hazardous-waste landfills and 80,000 contaminated surface lagoons have been identified in the United States (Corson 1990, p. 248). Moreover, American industry exports 3 million tons of hazardous wastes to underdeveloped countries that, for the most part, are even less aware of their hazards than Americans are.

The Environmental Protection Agency has identified 1200 hazardous-waste sites as the most dangerous in the country. These sites qualify for cleanup paid for out of federal funds under a "superfund" law passed in 1980. From 1980 to 1986, the agency spent *1.6 billion dollars* to clean up *13 sites* (Corson 1990, p. 249). At that rate, the cost of cleaning up the most dangerous sites would be 148 billion dollars. In fact, the cost of cleaning up toxic wastes is starkly prohibitive. Barry Commoner has noted that in 1986, the annual output of the chemical industry, as represented by its top 50 products, was 539 billion pounds

---

* For example, the United States imported 17,620,000,000 pounds of bananas from 1983 to 1985. Of these, the Food and Drug Administration examined 160 for pesticide residues. Moreover, it tested for fewer than half of the pesticides used on bananas (Weir and Matthiessen 1990, p. 119).

(Commoner 1990, p. 89). The industry that same year discharged 400 billion pounds of toxic chemicals into the environment. Assuming that the industry were forced to incinerate these chemicals—a process that emits dioxin and other toxic chemicals but is the only "control" available—at an average charge of $100 per ton, it would cost industry $20 billion. That same year, Commoner reports, the chemical industry's total after-tax profit was $2.6 billion (Commoner 1990, p. 90). In short, industry cannot do it. As long as petrochemical products are produced in anything like current quantities, there is no realistic prospect that the accumulation of toxics in the environment will even be stabilized, much less reduced.

The problem is worse than that. The *products* of the petrochemical industry themselves become "wastes." Plastics, for example, are often used just once—as is the case with household grocery bags, packaging, and bottles—and then thrown away. These plastics do not, like paper, leather, and the other products they replace, decompose in the landfills where they wind up. In addition, the sheer volume of solid wastes is filling up existing landfills and making it most unlikely that we can find enough new landfill sites.* The petrochemical industry does not pay the costs of managing the solid wastes that its products become any more than it pays the costs of detoxifying the hazardous wastes generated when it produces those products. Finally, some petrochemical products release harmful substances even when used as intended; examples include carpets and automobile interiors that emit formaldehyde, gasolines that emit benzene, and solvents that emit carbon tetrachloride. Some of these chemicals are carcinogenic, are mutagenic, or damage the nervous system. The industry does not pay the medical expenses of those persons who contract disease as a result.

The EPA estimates that the 2.7 billion pounds of toxic substances discharged into the air alone cause 2000 new cases of cancer each year in the general population. But that figure underestimates the harmful consequences of these emissions. First, it does not account for the possibly carcinogenic effects of hundreds of suspect substances that are vented into

---

* Some European countries reduce the production of trash via public policies that encourage the use of returnables and recycling. Norway taxes non-returnable containers. Denmark prohibits their use (Young 1991, p. 49). Germany has ordered the packaging industry and retailers to recover 80% of all packaging materials by 1995 (Environmental Action March/April 1991, p. 29). In the United States only a few places, such as Seattle, Washington, have enacted comprehensive recycling programs. A 1987 survey found that the states as a whole were spending 39 times as much money on incineration as on recycling (Young 1991, p. 45).

the atmosphere but the EPA has not studied. From 1970 to 1989, for example, that agency listed only 8 hazardous substances and set national standards for only 7 of these (Maillet 1989). Second, the EPA estimate does not examine the health effects of chemicals that have been identified with birth defects, sterility, central nervous system damage, and other serious ailments. Third, EPA estimates are based on risks imposed on the average person from one chemical at a time. It does not consider the fact that people are exposed to multiple chemicals and that such exposure has additive or synergistic consequences. It also does not consider effects on vulnerable and hypersensitive segments of the population, such as children and the elderly. Moreover, it is known that more than 200 industrial plants around the country emit toxics into the air at levels over 1000 times the level considered safe by the Agency; 7 million Americans who work at these sites are thus subject to added health risks.*

In a study of the fat tissue of 900 people representative of the United States population, two-thirds of the subjects were found to exhibit 33 of the 37 toxic compounds for which the tests were conducted (Commoner 1990, p. 32). Among the carcinogens found were benzene, chloroform, and dioxin.†

Toxic wastes are a problem elsewhere in the world. A few industrial countries (one is Denmark) detoxify some hazardous wastes before disposing of them, thus lowering the speed at which toxics are accumulating and the costs of cleaning up hazardous waste sites. In other

---

* Workers at chemical plants suffer risks not only from normal toxic emissions but also from industrial accidents. From 1981 to 1985, 7000 such accidents were reported, and thousands more were unreported. The reported accidents resulted in 138 deaths, 5000 injuries, and 200,000 people forced to evacuate their homes. The EPA estimated that 17 of these accidents had the potential to produce worse casualties than the accident at Bhopal, India, where over 3000 died and 200,000 persons were injured (Waxman 1989).

† Dioxin exposure of the general population results in part from trash incineration—in particular, the combustion together of chlorinated plastics and wood products. Because it creates mountains of trash, the public incinerates it when it runs out of landfills, but incineration produces its own toxic wastes. In addition to dioxin, incineration discharges nitrogen and sulfur oxides, carbon monoxide, acid gases, and furans into the air, along with such heavy metals as cadmium, mercury, and lead (Young 1991, p. 46). On new "clean" incinerators, scrubbers and filters remove most of these pollutants. But most of the "removed" pollutants then accumulate in the ash. This ash is disposed of by being injected into pits or dumped into landfills. Eventually, therefore, the toxics removed from the air end up in the groundwater.

countries petrochemical wastes and heavy metals are dumped, just as they are in the United States. In the developing world, they are sometimes dumped into the water supply or on agricultural lands. Heavy-metal and organic-chemical contamination appear in vegetables and other foods. Millions of tons of hazardous wastes are imported into these countries from the United States and Europe, usually illegally.* These too are dumped in uncontrolled ways.

Thus it is evident that the world has not come to grips with the hazardous substances it is producing. Human efforts to reduce this pollution have so far had minimal effects, but it hasn't mattered much, because effective "control" of these substances seems to have been provided "free" by the environment. But worsening pollution and ineffective pollution control cannot go on indefinitely. Reports about rising cancer rates in industrial countries are already appearing in scientific journals. These rates cannot be explained by the aging of the population (Okie 1990, p. A1). When the incidence of environmentally caused disease becomes higher and more obvious, humanity will recognize that this is the bill from past environmental neglect. The environment in effect will force stern action on us. (Unfortunately, by that time, there will be little we can do to lower immediately the rates of cancer, birth defects, and so on that result from long-term exposure). We may then eliminate from production any products that are themselves hazardous or the production of which creates hazardous wastes. The substitute products may be somewhat more expensive, less efficient, or less convenient. In that event, the net effect will be to reduce either our purchasing power or our "standard of living" as that term is conventionally understood today. But that loss may seem less painful (or costly, in terms of medical costs) then the rising incidence of disease and death.

If we don't eliminate hazardous products altogether, we will incur substantially higher costs to "control" pollution. However, because pollution control makes us pay for something that used to cost us nothing and often makes no contribution to productivity or product improvement, the net effect of increased commodity prices due to pollution control is a reduction of our purchasing power. Such price rises foretell the coming of the day when marginal costs of growth equal the gains and when growth will therefore cease.

---

* The European Economic Community agreed in 1990 not to export toxic and radioactive waste to their 68 former European colonies. These ex-colonies, in turn, agreed not to import toxic wastes from anywhere else. The United States and 50 developing countries are not parties to this treaty (French 1990, p. 13).

A problem more serious than financial costs is that even where technological control is theoretically possible, the technical problems may be extremely demanding. For example, we still have no really workable technological means of detoxifying hazardous wastes. Of course, future inventions may improve conditions greatly in many areas, but certain pollutants appear to be so intractable that effective technological control may never be achieved.

Radiation: The Insidious Pollutant

Radiation is the primary case in point. It is especially important to examine it in detail because many theorists seem to rely heavily on the generation of nuclear power both to circumvent the unacceptable pollution that would result from the expansion of conventional fossil-fuel power production and to compensate for the eventual disappearance of fossil fuels altogether. Moreover, radioactive compounds are only the most vicious of the wide array of dangerous chemicals we now discharge into the environment without any real knowledge of their ultimate potential for harm, so the case of radiation can serve as a model of the general long-term dangers of pollution and of the dilemmas that confound technological pollution control.

Radioactive isotopes, or radionuclides, are dangerous in extremely small doses. It has become clear that even tiny doses have long-term adverse effects on human and ecological health—that no radiation exposure can be considered risk-free.* Thus experts agree that virtually any increase in radiation exposure is to be avoided. Why are radionuclides so

---

* That was the conclusion of the National Research Council in 1989. Its study also concluded that the incidence of fatal cancers increases in proportion to increases in radiation exposure (Smith 1989, p. A3). A recent study published in *The Journal of the American Medical Association* reported that workers at the Oak Ridge National Laboratory in Tennessee who were exposed to very low levels of radiation—well below permissible levels and well below exposure levels at commercial nuclear power plants—had a leukemia death rate 63% higher than the general population (quoted in Lippman 1991, p. A3). In the past, some scientists had believed that, below a certain threshold, radiation either had no harmful effects or the harmful effects would be unmeasurable because any cancer that showed up later could have been the consequence of inducers other than radiation exposure. But the Oak Ridge study controlled for other cancer-causing factors, and it revealed that the longer workers were exposed to tiny levels of radiation, the higher the incidence of cancer. Of course, these studies do not end all controversy about the health effects of low-level radiation; the U.S. Council for Energy Awareness, a nuclear industry trade group, criticized both studies.

dangerous? First, physical and biological concentration of radionuclides is a pervasive phenomenon. It can take the form of geographical concentration in lakes, estuaries, airsheds, or any other places where the circulation of air and water is restricted. Or it can take the form of physiological concentration in the body, for example, inhaled plutonium oxide particles tend to lodge permanently in the lung and therefore to irradiate surrounding tissues intensely over a long period.*

Concentration can also take the form of biological magnification, which was described in Chapter 1 in connection with pesticides. Once a biologically active radionuclide enters a food chain, it is concentrated approximately tenfold at each higher trophic level. For example, the modest quantity of radioactive strontium 90 that falls on a pasture is concentrated in grass, again in the cows that eat the grass, and last in the child who drinks the cow's milk. Finally, for metabolic reasons radionuclides are concentrated selectively in particular tissues or organs. For example, strontium-90 mimics calcium in the body and is selectively concentrated in bones and bone marrow, iodine-131 is trapped by thyroid glands, and cesium-137 concentrates in muscles and soft organs, such as the liver and gonads. Thus radiation standards set in terms of averages or so-called whole-body doses may not be very meaningful in two ways: (1) *any* radiation exposure causes cancer. (2) No matter how low the levels are set, because radioactive substances are certain to be concentrated in particular locations, a dose of radiation that is well within the putative limits of tolerance on the average or over the whole body may nevertheless prove lethal.

In addition, there is no such thing as an "average" person for whom certain levels of radioactivity can be judged safe. An organism's vulnerability to damage by radioactivity (or any other pollutant, for that matter) is directly related to the stage in its life cycle as well as to other accidents of personal history. The fetus is particularly vulnerable to poisonous compounds of any kind. Growing children, with their high metabolism, are also at higher risk, and the affinity of a radioactive compound such as strontium-90 for bone and marrow can make it particularly devastating at certain developmental stages. Thus standards must take into account the basic ecological principle that the reproductive period is critical (E. P.

---

* Plutonium's peculiar properties and extreme toxicity make it by far the most virulent of all radioactive substances. This fact has generated a major controversy over the dangers of plutonium to public health. The more extreme critics of the radiation standards-setting process claim that inhalation of as little as one particle may be sufficient to produce a significantly increased risk of contracting cancer.

Odum 1971, p. 108): What to adults or the general public may be an acceptable risk may be very much more damaging to the young and to others who are peculiarly vulnerable.

Furthermore, the effects of radiation combine synergistically with the effects of other pollutants or environmental stresses to produce disproportionate damage to bodies and ecosystems. Synergism occurs when two or more causes combine to produce a net effect that is greater than the mere sum of their separate effects. Thus, to use a well-known medical example, exposure to a modest quantity of either carbon tetrachloride or alcohol has no serious consequences, but simultaneous exposure to both causes serious illness or death. Many environmental problems are either produced by synergism (for example, photochemical smog) or aggravated by it (for example, poisoning by heavy metals). The theoretical basis for this was discussed in Chapter 1: Any environmental stress tends to simplify an ecosystem and therefore to reduce its stability. Few studies have been done on the synergistic effects of pollutants. So many biologically active compounds have been released in such large quantities that neither the money nor the labor power is available for studying even a tiny fraction of the more important interactions. However, the deleterious effect of low-level radioactivity on the structure and functioning of ecosystems and on the human body is amply documented (Smith 1989, p. A3; Wallace 1974; Conney and Bums 1972; Woodwell 1969). In effect, radioactivity "softens up" a biological system, making it more vulnerable to disruption by other pollutants and stresses (and vice versa, of course). The addition to our environment of such a potent stress as chronic radioactivity is therefore a matter for deep concern, even if no immediate effects can be observed.

It is, in fact, the long-term effects of radiation exposure that are the most worrisome. The general epidemiological evidence for increased mortality due to chronic low-level pollution over the last 50 to 100 years is incontrovertible. Epidemiologists agree that the sharp rise in death rates from emphysema, various forms of malignancy, and a number of other prominent modern ills is primarily due to pollution. Researchers estimate, for example, that between 60 and 90% of cancer cases (1,040,000 discovered each year in the United States) are caused by environmental factors, mostly chemicals (Maugh 1974). Moreover, as Figure 3-2 reveals, the overall damage to public health is likely to be far greater than mortality statistics alone indicate. In addition, the effect of stress on natural systems is nonlinear (so that doubling the dose more than doubles the resulting illness), and we can confidently expect future increments of low-level radiation (and of other pollutants) to produce disproportionately more damage to human health than past levels of pollution.

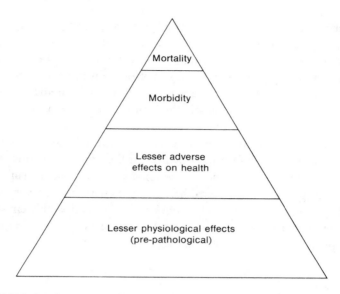

FIGURE 3-2 Spectrum of biological responses to exposure to pollutants (after Newill 1973).

Even more important is the fact that radiation is by far the most powerful mutagen known. The radionuclides released by human activities are therefore doubly dangerous: By exposing us to a level of radioactivity significantly higher than the so-called background radiation of the earth (which our evolutionary history has prepared us to withstand), they threaten the genetic integrity of unborn generations. It is entirely conceivable, for example, that today's adults and even today's children could escape serious harm but that our grandchildren would be grievously injured by current levels of radioactivity. In sum, despite the remaining areas of controversy, the long-term dangers from low levels of environmental radioactivity are essentially indisputable.

Finally, the threat from radionuclides is intensified by their persistence. For example, the half-lives of tritium, strontium-90, and cesium-137 are 12.4, 28, and 33 years, respectively. Because the period of biological danger is roughly 10 times the half-life, the tritium we release today will be a problem for 124 years, and the other two isotopes will remain dangerous for about 3 centuries. A few other radionuclides have such long half-lives that they will be with us in virtual perpetuity. The most dangerous of these is plutonium-239 (it has a half-life of 24,000 years), which takes a quarter of a million years to decay to the point of harmlessness. (Plutonium is used in weapons and is part of the nuclear fuel cycle, either as a by-product or as a fuel.) Consequently, tiny amounts of radionuclides released year after year can, over the not-so-very-long run, build up a large inventory of dangerous radioactivity in ecosystems and human bodies. Thus, even if we and our immediate progeny escape harm, we could still saddle our more remote posterity with a lethal burden of radioactivity.

The conclusion is inescapable that the release of radioactive compounds to the environment in any but the most trivial amounts, made up preferably only of those compounds with very short half-lives and the least capacity for biological harm, carries with it some very serious risks. Anything less than virtually 100% control of radioactive emissions is truly dangerous to the future biological health of the planet and its inhabitants, especially its human inhabitants.

Toxic Nuclear Wastes

Given the dangers of radiation, the wastes generated by the military and civilian uses of nuclear power pose a formidable long-term threat to humanity. The military facilities of the United States and the Soviet Union have generated huge quantities of nuclear wastes. The Soviet military has openly dumped its nuclear wastes into lakes and rivers, forcing the evacuation of people living along their shores and contaminating the Arctic Ocean, thousands of kilometers away. A Soviet high-level nuclear storage facility exploded in 1957, contaminating 15,000 square kilometers of land on which 250,000 people lived. In the worst civilian nuclear accident in history, a reactor at the Chernobyl power station exploded in 1986, releasing between 50 and 100 million curies of radioactive material into the environment. Up to 50 different radioactive isotopes were included in the release, with half-lives ranging from 2 hours to 24,000 years. A huge quantity of cesium-137 (half-life, 33 years) was released and was detected in high amounts in the milk of countries as far away as Switzerland, Germany, and the United Kingdom. One hundred million people in Europe were put under food restrictions, as the radiation contaminated fruits, vegetables, and the grass on which livestock feed. Soon after the explosion occurred, 116,000 people living within 18 miles of the reactor were moved out; the land in the vicinity of the reactor may remain uninhabitable for 15,000 years. In 1990 the Soviet government allocated $26 billion to move 200,000 more people (Shogren 1990, p. A1). But 4 million people still live on contaminated land. By 1990 doctors in the area were reporting dramatically higher levels of skin cancer and other cancers, miscarriages, genetic mutations, birth defects, gynecological problems, anemia, heart attacks, and enlarged thyroid glands. Cancer *deaths* attributable to the accident are projected by the National Research Council to be as high as 70,000 people. Over 30,000 of these fatalities will occur outside the Soviet Union. The radiation release from Chernobyl is expected to cause cancers to develop in tens of thousands of additional people who will survive the disease.[*]

The United States military has also generated huge amounts of radioactive waste—an estimated 1.4 billion curies. (This compares to the 50 to 100 million curies of radioactive material released at Chernobyl (Renner 1991, p. 147).) United States military waste has contaminated the soil and groundwater at over 3200 sites owned by the United States Department of Energy. More than 50 Nagasaki-sized nuclear bombs could be built just from the wastes that have already leaked from storage tanks at the Hanford, Washington, nuclear reservation (Renner 1991, p. 147); 4.5 million liters of high-level radioactive wastes have leaked from these tanks so far (Renner 1991, p. 148). Radiation from the Rocky Flats nuclear facility in Colorado, including plutonium, has spewed in unknown quantities throughout the Denver region; strontium and cesium have leaked into the ground water (Renner 1991, p.148). Rocky Flats employees suffer elevated levels of brain tumors, malignant melanoma, respiratory cancer, and chromosome aberrations (Renner 1991, p. 147). The Oak Ridge, Tennessee, nuclear facility has emitted thousands of pounds of uranium into the atmosphere; the facility at Fernald, Ohio, has emitted at least 250 tons of uranium oxide. The facility's wastes have contaminated nearby surface water and groundwater with cesium, uranium, and thorium. The aquifer under the Savannah River, South Carolina, nuclear facility contains radioactive substances and chemicals at 400 times the concentrations the government considers "safe" (Renner 1991, p. 148). All of this radiation (and more, at thousands of other facilities) is hazardous to biological systems and is accumulating. No one knows what to do with the portion of these nuclear wastes (and civilian nuclear wastes) that can be "cleaned up," because no satisfactory facility for storing nuclear wastes has been found. Worse still, some of these nuclear wastes cannot *be* "cleaned up"; either we don't know how or the wastes are already dispersed into the air, the water, and particularly the groundwater where they will remain harmful, depending on the contaminant, for dozens, hundreds, or thousands of years.

---

[*] Apart from the possibility of catastrophic accidents, nuclear power is beset with other difficulties, which were discussed in Chapter 2.

The Fallacies of Technological Pollution Control

The difference between radioactive pollutants and more ordinary pollutants is one of degree, not of kind. Radionuclides are particularly insidious and especially toxic, but many other pollutants behave in a similar way. In fact, the vast majority of the thousands of synthetic chemical compounds released in any quantity are physically and biologically concentrated or magnified so as to attack certain body tissues selectively; have critical effects on reproduction or early development; react synergistically with other compounds or "soften up" natural systems; have potential or demonstrated delayed effects on ecosystems and human populations; are mutagenic; and are sufficiently persistent to allow long-term buildup of toxic material. Consequently, we already confront a public health and ecological dilemma of unknown but clearly large dimensions. Adding additional quantities of pollution of whatever form can only worsen this situation.

For these reasons, ecologists and environmental health specialists find completely unacceptable the usual economist's position that we should rely on the "natural assimilative capacity" of air and water, controlling pollution only when recreational or other economic "use values" will be damaged. This economic criterion can have ecological validity, if at all, only for nonsynthetic pollutants (such as sewage and other organic materials) that ecosystems are naturally adapted to handle in reasonable amounts. But it has no validity at all for synthetics like petrochemicals, to say nothing of radionuclides. Thus it may make ecological sense to speak of using the environment to dispose of natural organic compounds (providing, contrary to current fact, that they are not contaminated with synthetics), but these are the least dangerous and least troublesome pollutants. By contrast, we must achieve virtually 100% control of the much more dangerous and troublesome synthetics; economics notwithstanding, there is little or no safe assimilative capacity for such unnatural compounds.

Furthermore, even if we were to allow that the environment could absorb with impunity a minimal quantity of these compounds, we could not determine whether there existed a "safe" level of usage except by trial and error. Thus it has taken the observation of increased incidence of cancers in human beings for us to recognize that there is no safe exposure to radionuclides. Our ignorance about safe levels of toxic substances is paralleled, and in most cases exceeded, by our ignorance of what ecosystems can tolerate over the long term and of numerous other factors essential to administering an "economic" pollution-control strategy. Human beings will continue to be the guinea pigs, while administrators of pollution-control laws face the task of setting pollution controls despite this ignorance of what the environment can absorb.

Nor are the technical means available to the administrator at all prepossessing. In fact, because so much pollution derives, like agricultural runoff, from "non-point" sources that can be controlled, if at all, only by truly heroic technological measures, even strict controls on readily accessible "point" sources of pollution may be almost fruitless. General urban runoff, for example, makes a contribution to water pollution about equal to that of sewage, yet there is no way to control it short of a massive hydrological reengineering of our cities (at astronomical cost). This is part of the reason why the $100 billion spent so far to meet the standards of the Clean Water Act has been an "almost meaningless enterprise" (after Abelson 1974).

In this light, our hypothetical pollution-control model now appears rather conservative. Not only are the original x units of residuals already an ecological and public-health threat (though probably not yet a large one), but because pollution is due as much to the general side effects of development as to the more easily controlled discharges of factory effluent, each doubling of production from the factory will more than double the incidence of pollution. Thus both the necessity for stringent pollution control and its expense will increase more rapidly than a simple proportional model would predict.

In sum, therefore, as a strategy of coping with the ever-increasing load of pollution that inevitably accompanies increased production, technological pollution control has some very serious limitations. *Even granting, for the sake of argument, that engineers will come up with systems of pollution control that are, especially for radionuclides, 90% or more effective, growth of production cannot continue forever.* In fact, if we try to double and redouble current levels of production in the United States, it seems very likely that we shall soon be restrained by rising costs of pollution control (at least for some commodities) and by rising levels of pollutants that we cannot control. Pollution control (as distinct from eliminating the production of pollutants in the first place) is therefore only a temporary tactic that will allow growth of production to continue for just a while longer. It is not a genuine solution to the problem of pollution, even under the most optimistic assumptions about our technological capacities and energy supply.

In addition, the basic problems of pollution control are at least as much political, social, and economic as technological. For example, it may well be that we shall come to accept levels of pollution and damage to public health that we would regard as intolerable today. If so, then growth could continue for somewhat longer, until mortality and morbidity grow

16

Ecological Pollution Control

From almost every point of view except short-term profitability, an ecological strategy of pollution control and waste recycling, involving as much as possible the planned use of natural recycling mechanisms, would be preferable to relying on expensive, energy-devouring, and failure-prone technological devices to perform these functions. Specifically, we should create waste-management parks—that is, portions of the environment deliberately set aside as natural recycling "plants" (E. P. Odum 1971, Chap. 16). In these parks, sewage and other controllable wastes capable of being naturally recycled (and not used directly as fertilizer on farms) would be sprayed on forests and grasslands, which would remove and reuse the nutrients that cause water pollution and return the purified water to aquifers. Forestry, fish ponds, grazing, and other means of exploiting the potentially productive energy contained in the recycled wastes would help pay the costs. This concept of pollution control is the most efficient and economical overall, for nature would do most of the work free, and productive use would be made of wastes. Also, because such a park would be in a quasi-natural state, it would serve ecologically as a protective zone, balancing more intensive development elsewhere.

to catastrophic proportions. Also, technological fixes are not ecologically optimal solutions to many important pollution problems. For example, attempts to dispose of urban organic wastes technologically are misdirected, for these wastes are really an unused resource that could be recycled as fertilizer. Unfortunately, even though it is thermodynamically and ecologically rational and would probably save money and energy in the long run, recycling wastes is seriously impeded not only by the initial financial and energy costs of transition, but also by the numerous changes in our general way of doing things that would be required to make this kind of ecological pollution control feasible (see Box 16). Finally, as we shall see later, it is "rational" to pollute and to avoid paying for pollution control. Thus even where control measures—technological or otherwise—are readily available, they usually cannot be implemented except within a general framework of political, social, and economic reforms.

Unfortunately, such a strategy has limits. In the first place, the concept of the waste-management park assumes that resources are not so intensively exploited that we cannot afford to reserve large potentially productive areas for protective purposes. Moreover, as we have seen, many forms of uncontrollable pollution will remain despite any conceivable changes in pollution-control strategy. Also, materials such as radionuclides and many synthetic chemicals, which are dangerous to introduce into ecosystems, will still have to be treated and controlled industrially. Thus the problems of technological pollution control discussed in the text may be insuperable. (Scientists are studying the possibility of transmuting nuclear wastes, that is, bombarding them with neutrons to break them down into substances that would, in about 300 years, be no more dangerous than natural uranium. (Browne 1991, p. C1).) But transmutation has not yet been proven practical; moreover, most transmutation methods would produce plutonium. Scientists envision that the plutonium would be processed in a breeder reactor, but the United States previously abandoned an earlier breeder reactor program as too dangerous). Therefore, it would seem that genuine control of pollution will oblige us not to produce pollutants in the first place; this will require significant industrial changes as well as changes in our social habits and values.

Technology and Its Management

Is There a Limit to Technological Growth?

One of the main components of the argument against the limits-to-growth thesis is technological optimism (see Box 17 for a description of the type of technology at issue). The optimists believe that exponential technological growth will allow us to expand resources and keep ahead of exponentially increasing demands. The eminent British elder statesman of science Lord Zuckerman (quoted in Anon., 1972a) complained about *The Limits to Growth* that "the only kind of exponential growth with which the book does not deal, and which I for one believe to be a fact, is that of the growth of human knowledge"; Zuckerman went on to assert categorically that "the tree of knowledge will go on growing endlessly."

17

Bulldozer Technology

Unless they return to a life of hunting and gathering without either tools or fire, humans are incurably technological in the sense that they will always have to transform nature for utilitarian ends by some kind of applied science. However, radically different modes of technological existence are possible. What we are concerned with here is the peculiar kind of technology that grew out of Baconian experimental science, which first had a social impact during the Industrial Revolution in England, and which has as its explicit purpose giving people power over nature in order to promote "the effecting of all things possible," to use Francis Bacon's arresting slogan (Medawar 1969).

For our purposes, the most important characteristics of this kind of technology are its dependence on fossil fuel and other nonrenewable or man-made resources, its linearity and lack of integration with natural processes, its dominating scale, and its narrow concept of rationality or efficiency. Because all attempts at an exact yet reasonably succinct definition fall short, it might be best to resort to symbolism and call it "bulldozer" technology. The bulldozer and other earth-moving machinery make possible the airports, dams, highways, skyscrapers, and most of the other vaunted achievements of modern technological civilization. Moreover, its violent power, the single-minded way in which it reshapes nature to human design, and its dependence on human-made energy and a complex industrial infrastructure make the bulldozer a paradigm of modern technology. It is on such a technology, rather than on some of its conceivable alternatives, that proponents of exponential technological growth appear to rely.

Zuckerman's boast is not merely a personal opinion but a widely held article of modern faith. Since the age of the Enlightenment *philosophes,* the ideology of progress through science and technology has been our social religion. Indeed, according to its more utopian proponents, such as Karl Marx, eventually scarcity itself (and therefore the age-old evils of poverty and injustice that are rooted in it) will be abolished. Thus to challenge endless scientific and technological progress amounts to a kind of secular heresy.

Yet, as we have seen in the preceding discussions of the particular technological solutions proposed to deal with the problems of growth in the production of food, minerals, and pollution, there are demonstrable limits to the technological manipulation of ecological limits. The time has come to generalize about some of these limits and to discuss certain practical problems of technological management that are only partly connected with the physical limitations of the earth. It will be seen that neither in theory nor in practice can technological growth be as endless as Lord Zuckerman asserts. Already, in fact, limits to knowledge and to the human capacity to plan for and manage technological solutions to environmental problems have begun to emerge.

Limits to Knowledge and to Its Application

Most scientists and technologists believe, with Zuckerman, that necessity unfailingly brings forth invention. However, although we cannot specify the exact limits and must always be aware of potential "failures of imagination and nerve" that would tend to make us overly pessimistic about future possibilities (Clarke 1962), there is at least reasonable doubt that "the tree of [relevant] knowledge will go on growing endlessly." This does not mean that the enterprise of science will end, and it is clear that such an assertion in any event is less true of some fields than others. Nevertheless, it appears that the process of *relevant* scientific discovery must eventually cease. That is, just as we have turned mechanics and classical optics into engineers' tools and, therefore, into played-out fields of scientific investigation, so too shall we come to the end of scientific discovery in other fields *relevant to the problem of surmounting the limits to growth.*

Indeed, the history of science clearly illustrates the law of diminishing returns, for the more scientific work that is done, the more likely it is that new theories will be corrections or refinements of previous ones, leaving most of the old structure of knowledge intact. Thus new knowledge may not be translatable into new technology. In physics the clockwork celestial-mechanical theories of Isaac Newton have been superseded by the relativistic and quantum-mechanical theories associated with Albert Einstein and Werner Heisenberg, but neither relativity nor the uncertainty principle have a significant practical impact on the ordinary physical reality of our species's biological and social existence. Thus even very great future discoveries— ones that totally change our scientific world view or our view of ourselves, may contribute little to removing the ecological limits now confronting the human species.

Moreover, a greater scientific and technological research effort does not seem possible, for the scientific enterprise itself is now struggling with numerous limits to its own growth. For example, the costs of basic research in many areas have risen inordinately in recent years, a clear symptom of diminishing returns. And even when theory clearly favors real-world technological advance, acceptable engineering solutions may not be achievable because the technical difficulties are too great. As previously suggested, fusion could be just such an area.

Finally, as we have had occasion to note in connection with pollution control and energy production, one cannot improve a technology indefinitely without encountering either thermodynamic limits or limits of scale beyond which further improvement is of no practical interest. Many technologies are already near this point, and the rest soon will be, for the substitution of one ever-more-efficient form of technology for another simply cannot continue forever. In effect, the better our current technology, the harder it is likely to be to improve upon it. (In the real world, moreover, there is frequently a trade-off between efficiency and reliability, such that maximizing efficiency can be self-defeating.)

In sum, there may be limits to relevant scientific and technological knowledge or to the human capacity to discover such knowledge. If so, basing our strategy of response to ecological limits on the assumption that scientific and technological knowledge will grow endlessly or even at the rate typical of the recent past appears to be imprudent.

The Overwhelming Burden of Planning and Management

Even if lack of scientific and technological knowledge proves not to be an obstacle, implementation of technological solutions to the full array of problems we have discussed will place a staggering burden of planning and management on our decision makers and institutional machinery.

The 1987 report of the World Commission on Environment and Development viewed the problem as follows:

> When the century began, neither human numbers nor technology had the power radically to alter planetary systems. As the century closes, not only do vastly increased human numbers and their activities have that power, but major, unintended changes are occurring in the atmosphere, in soils, in waters, among plants and animals, and in the relationships among all of these. The rate of change is outstripping the ability of scientific disciplines and our current capabilities to assess and advise. It is frustrating the attempts of political and economic institutions, which evolved in a different, more fragmented world, to adapt and cope (Corson 1990, p. 3).

Furthermore, the environmental crisis is not a series of discrete problems; it is a set of interacting problems that exacerbate each other through various kinds of threshold, multiplicative, and synergistic effects.* Thus the difficulty and complexity of managing the ensemble of problems grow faster than any particular problem. Moreover, all the work of innovation, construction, and environmental management needed to cope with this ensemble must be orchestrated into a reasonably integrated, harmonious whole; the accumulation of the side effects of piecemeal solutions would almost certainly be intolerable. Because delays, planning failures, and general incapacity to deal effectively with even the current range of problems are all too visible today, we must further assume that our ability to cope with large-scale complexity will improve substantially in the next few decades. In brief, technology cannot be implemented in an organizational vacuum. Something like the ecological "law of the minimum," which states that the factor in least supply governs the rate of growth of a system as a whole, applies to social systems as well as ecosystems. Thus technological fixes cannot run ahead of the human capacity to plan, construct, fund, and staff them—a fact that many technological optimists (for example, Starr and Rudman 1973) either overlook or assume away.

Foresight, Time, and Money as Factors in Least Supply

Our ability to achieve the requisite level of effectiveness in planning is especially doubtful. Already the complex systems that sustain industrial civilization are seen by some as perpetually hovering on the brink of breakdown, and current management styles—linear, hierarchical, economic—appear to be grossly ill adapted to the nature of the problems.

One very troublesome problem for social planners is that the consequences of our technological acts cannot be foreseen with certainty. There exist no scientific answers to such "trans-scientific" questions as what risks are attached to nuclear energy or to the use of certain

---

* Some examples: (1) Even if per capita consumption and waste remain constant, a small increase in population can change a healthy river into a sewer once the river's capacity to digest wastes and pollution has been exceeded (the threshold effect). (2) Even if population and per capita consumption grow separately at quite modest rates, the total environmental demand multiplies more rapidly, such that a doubled population that uses twice as much has four times the impact on the environment (the multiplicative effect). (3) Various forms of pollution, from noise to radiation, interact to produce more ill health than would result simply from the addition of the particular effects (the synergistic effect).

chemicals (Weinberg 1972); the only way to determine these risks empirically is to run a real-life experiment on the population at large.* The potential social consequences of technological innovation are even more obscure. Thus there are no technical solutions to the dilemmas of environmental management, and policy decisions about environmental problems must be made politically by prudent citizens, not by scientific administrators. This being the case, technology assessment, the remedy proposed for the general political problem of technological side effects, can never be the purely technical exercise many of its proponents seem to envision; instead, the planning process will come to resemble a power struggle between partisans of differing economic, social, and political values. The difficulties and delays such an adversary planning process entails are foreshadowed by current conflicts over nuclear power-plant siting and safety and over other environmental issues.

Of course, such drawn-out political battles may well be essential for the creation of social consensus and commitment on these difficult issues. Yet it is becoming increasingly apparent that we can ill afford the associated delays, for time will be one of our scarcest resources. Difficult as it seems, dealing with very large increments of growth is really the lesser part of the problem. Exponential growth is dangerous primarily because it is so insidious: As the example of the lily pad and pond illustrates, until a limit is very close in time, it seems very far away physically and psychologically. Thus the all too human tendency to let things slide until they are pressing is potentially fatal, for by then even heroic action may be too little and too late.

For example, we Americans have allowed the private automobile to become so central to our economy and our private lives that we cannot live

---

* Perhaps the worst local example of this phenomenon in the United States is the 85-mile corridor from Baton Rouge to New Orleans, Louisiana, which is home to 135 chemical plants and 7 oil refineries. *The Washington Post* reported that "the air, ground, and water along this corridor are so full of carcinogens, mutagens and embryotoxins that an environmental health specialist defined living [there] as 'a massive human experiment....'" (Mariniss and Weisskopf 1988, p. A1). Although several towns in the corridor report high rates of cancer and miscarriage among their populations, nothing special (beyond weak federal anti-pollution laws) is being done to stop the emission of pollutants from these plants or to relocate the people away from them. The reason is that the effects of chemicals in promoting these diseases is long-term and "creeping"; furthermore, as federal officials tirelessly repeat in discussing everything from pesticides to radiation, no one can prove what role what particular chemical played in causing what particular disease in a particular person, compared to other possible sources of that person's cancer or miscarriage.

without it in the short term, yet because of air pollution, we can no longer live with it in its current form. We are forced to alleviate the worst of its side effects with stopgap technological responses, but this strategy will not even enable us to meet necessary clean-air standards without additional social and institutional changes. At this point, however, we are almost helpless to do better, for we ignored the problem until it became too big to handle by any means that are politically, economically, and technically feasible now or in the immediate future. Similarly, warnings of the destruction of the ozone layer were ignored, and the world is finding itself in a predicament in which nothing we can do will avert millions of additional cancer cases.

Nor is it enough merely to foresee an emerging problem. Planners must also anticipate the lead time necessary to take delivery of even readily available technological solutions, such as using hydrochlorofluorocarbons instead of chloroflurocarbons (10 to 20 years), or to replace a harmful technology with one that does not use chlorine (20 to 40 years), as is apparently going to be necessary to bring ozone depletion under control. Often, however, the replacement technology does not yet exist, so even more lead time must be allotted for its invention. Worse, it may take a very long time for any significant results to appear once a technological fix has actually been applied in the real world. CFCs, for example, remain intact in the upper atmosphere for 100 years. Even if we stop producing them immediately, each chlorine atom already created has the capacity to destroy around 100,000 ozone molecules. For another example, the sheer quantity of the toxic wastes dumped in this country is now so overwhelming that cleaning it up has already proved to be all but impossible.* Even if we were to stop producing toxic wastes quickly, with scientific breakthroughs, a crash program of development, and the political will to implement radical change (none of which exists), much more groundwater contamination is inevitable from the wastes already dumped. So too are cancers among the people who use groundwater for their drinking supply. In sum, coping with exponential growth at our advanced stage of development requires the exercise of foresight, and a planning horizon of 30 to 50 years is the minimum consistent with the

---

* The Superfund Law, which provided for the cleanup of toxic waste sites in the United States, has thus far been a farce. Although 1177 sites were identified as "priority" waste sites by 1988, from 1980 to 1986 the EPA cleaned up only 13—and that at a cost of $1.6 billion (Corson 1990, pp. 249-250). And this may be only the tip of the iceberg: In 1987, the General Accounting office estimated that over 425,000 hazardous waste sites may exist in this country.

existence of innumerable natural and social lags. Moreover, our past failure to exercise foresight means that we have already fallen far behind.

In spite of the fact that money is also a significant practical limitation on technological growth, there are abundant examples of failure to count the financial cost of technological schemes. One is found in the assertion, which has unfortunately begun to achieve some currency, that the way out of our ecological bind here on earth lies in space (Chedd 1974). It is abundantly clear that, whatever the ultimate potential for founding extraterrestrial colonies or whatever the ultimate cosmic destiny of the human race, space offers no escape from the limits to growth *on this planet*. To rocket into space a number of individuals equivalent to just one day's growth in the population of the world (approximately 250,000 people) would be a major undertaking (assuming 100 persons per shuttle flight, 2500 flights would be necessary). This alone would generate colossal environmental problems—enormous quantities of energy for fuel, pollution of the atmosphere (especially the vulnerable stratosphere) by toxic exhaust gases (not to mention chlorine), and so on—and trying to keep pace with population growth would be totally out of the question. Moreover, the expense would be staggering. Not only would it cost billions to lift into orbit these 250,000 persons, but that would be just the beginning. There would also be the costs for space colonization and other life-support costs in space. Unquestionably, keeping pace with the world's population growth for just one year would require a sum exceeding the U.S. gross national product. It is apparent that even highly developed and routinized space travel is not likely to involve large-scale movement of people and materials to and from the earth, at least not in any foreseeable future.

Even less grandiose technological schemes may cost too much. One of the reasons why there is no American supersonic transport (SST) program (and why the AngloFrench Concorde SST project is continually embroiled in political controversy as well as red ink) is that neither government nor private industry was willing to undertake the financial burdens. Indeed, the mere expansion of currently feasible technology will strain our capital resources in the coming decades. To stop carbon emissions and meet our electrical needs by increasing nuclear power production would cost upwards of two trillion dollars over the next 25 years, not including the costs of decommissioning plants or the disposal of radioactive wastes. If we were to replace all coal-fired power plants with nuclear plants by 2025, we would have to build one plant every two and one-half days every year! The staggering expense is one of the major reasons why utilities, despite government support, are not likely to choose the nuclear route. Moreover, given the general shortage of capital, when dollars are spent for expensive technologies to solve a problem in one area, investment must be foregone in other kinds of new plant and equipment needed to cope with another environmental pollutant, or in housing, or in social welfare, and so forth. Insufficient investment capital is therefore likely to be a very serious limitation on continued technological expansion.

Vulnerability to Accident and Error

Because major and irrevocable commitments of money, materials, and effort are necessary to stay ahead of population growth and because major risks are inherent in certain technological choices if all does not go well, it has become supremely important to make the right decisions the first time, for we may have no second chance to solve the problems being thrust upon us so rapidly. Yet even supremely foresightful, intelligent, and timely decision making may do little to reduce the growing vulnerability of a highly technological society to accident and error.

The main cause for concern is that some especially dangerous technologies are beginning to be deployed. We have seen, for example, that there are inherent in nuclear power production (especially with the breeder reactor) certain risks that make virtually perfect containment mandatory, and the evidence does not suggest that such perfection is achievable (see Box 10 and the related discussion). In addition, many other modern technologies—such as the chemical industry, the transport and storage of natural gas, and the supertanker—are capable of inflicting catastrophic ecological or human damage. Experience with these technologies also shows clearly the near impossibility of preventing all accidents. Especially in the developed world, people depend so heavily on a basic technological infrastructure that even less intrinsically dangerous accidents—for example, a sustained electric-power failure—can have devastating consequences.

As population grows and civilization becomes more complex, it will require much more effort and skill for us to cope with this increasing vulnerability to disorder (entropy) and failure. But to count on perfect design, skill, efficiency, or reliability in any human enterprise is folly. In addition, all human works, no matter how perfect as self-contained engineering creations, are vulnerable not only to such natural disasters as earthquakes, storms, droughts, and other acts of God,* but also to

---

* This fact alone makes it unlikely that the requisite degree of nuclear safety can ever be achieved, especially given the human propensity to build extensively in natural flood plains, known earthquake zones, and other spots liable to natural disasters.

deliberate disruption by crackpots, criminals, terrorists, and military enemies. Nevertheless, despite the patent impossibility of achieving any such thing, modern society seems to be approaching a condition in which nothing less than perfect planning and management will do. Some will object to such a strong statement of the problem, so let us examine several of the arguments that purport to dismiss this concern.

It is sometimes said that the probability of any one of these disastrous events happening is so low as not to be worth worrying about. Of course, humanity must run some risk in order to reap the fruits of technology, but dismissing the problem in this fashion betrays a potentially fatal misunderstanding of the laws of probability, for an apparently low probability of accident may be illusory. First, as we noted in the discussion of reactor safety in Box 10, whether a risk is large or small depends greatly on how many sources of risk there are. That is, if the chances of some kind of reactor accident are 1 in 1000 per reactor year, then 1 accident a year is a certainty (on average) if there are 1000 reactors in operation. Because we already do so many things that have some potential, however small, of altering the climate or unleashing other disasters, we should not be complacent about the apparently highly improbable. Second, some risks are essentially incalculable. There is no way, for example, to estimate the degree of danger to nuclear installations in developing countries from fanatical political terrorists cunning enough to outwit all the safety devices and security procedures. Third, we cannot afford to relax even when the probabilities are truly small, for the million-to-one shot is equally likely to occur at the first event, at the millionth, or well beyond the millionth. If the result of failure is potentially catastrophic, then we are simply engaged in playing a highly recondite version of Russian roulette. As game theorists have shown, a course of action that risks very serious loss is unlikely to be sound, no matter how attractive the potential gain; a prudent strategist limits his risks even if this strategy also limits his gains.

Some believe that we shall soon achieve a level of material and systems reliability far above what we are now capable of; the space program is often cited in support of this belief. However, although the space program is certainly a triumph of technical engineering, most of the problems we are called on to solve are not pure engineering problems. They contain a host of social and other "soft" factors that make them conceptually and practically several orders of magnitude more difficult than the space program. Moreover, this claim conveniently overlooks the fiery death of three astronauts, the near disaster of Apollo Thirteen, and the Challenger disaster, to mention only the American space program. In addition, we have neither the money nor the labor power to turn all our technological acts into a simulacrum of a moon shot. The nuclear industry is a much more realistic model of what we can expect, but as we have seen, despite far greater than average attention to safety and fail-safe design strategies, its safety record is far from perfect. The death toll from Chernobyl alone, as we have noted, is expected to be 70,000 people.

In sum, even massive amounts of money, enormous effort, and supreme technological cleverness can never guarantee accident-free operation of technological devices, and it is indeed strange that technologists—discoverers of the infamous Murphy's Law, which sardonically states that "If something can possibly go wrong, it will"—should so often assume that they can make their creations invulnerable to acts of God or foolproof in normal operations. Indeed, the array of potential ecological and societal disasters confronting a civilization that increasingly depends on the smooth and errorless operation of technological systems should give any prudent individual pause. It is not just that incredible accidents can still happen, as is well illustrated by the fate of the *Titanic,* whose designers believed it to be unsinkable. Rather, we are deliberately adopting new technologies in full awareness that they are by no means "unsinkable."

In fact, the supertanker may be an even better metaphor for modern technological society than the bulldozer (see Box 17). These massive oil barges are maritime disasters looking for a place to happen. The eco-disaster caused by the wreck of the *Torrey Canyon,* not a particularly large supertanker by current or projected standards, was surely a taste of things to come. Any doubt on this score was removed when the *Exxon Valdez* disaster occurred—and that was only the most visible of supertanker spills, which now occur with regularity. Supertankers are fragile. They are cheaply built to minimum standards and in such a way as to flout scandalously nearly all the canons of good seaworthiness established over centuries of experience (Mostert 1974). Their thin and over-stressed hulls are not equal to all the challenges of the sea; they lack the ability to maneuver or stop within any reasonable distance; and they have only a single boiler and a single crew, so that even routine failures leave them helplessly adrift with as much thermal energy in their tanks as is stored in a fair-sized hydrogen bomb.* Like the monstrous supertanker, a highly technological society appears fated to exist on the thinnest of safety

---

* In August 1990, after a year of highly publicized oil spills, Congress passed legislation requiring new supertankers to be double hulled and mandating the eventual retrofitting of existing supertankers (Chasis and Speer 1991, p. 21).

margins,* and there is abundant evidence that such a small margin will eventually prove insufficient. To proceed on the assumption that we can achieve standards of perfection hitherto unattained would be an act of technological hubris exceeding all bounds of prudence.

The End of "Endless" Technological Growth

The important question is not "Can we do it?" in the narrow technological sense. Rather, we must ask, "Can we do all the things we have to do at once, given shortages of money, labor power, and other factors potentially in least supply? At what cost and at what risk? Will we do it? Will we do it in time, given lack of foresight and the very human tendency to wait for a crisis?"

What this array of questions suggests is that, even if the problems of exponential growth seemingly yield to abstract analysis and technological solution, it is possible that they will not be solved simply because we are too human and fallible to deal with them in the real world. In short, exponential technological growth is a false hope, for it can never be the endless process optimists seem to anticipate: Even in the shorter term, technological solutions pose problems of management that can be surmounted only with great difficulty, if at all.

This judgment certainly does not mean that all technological solutions are anathema. Indeed, to counter single-minded technological optimism with an equally single-minded neo-Luddite hostility to technology in all its forms is absurd, for a nontechnological existence is impossible. The questions at issue are what kind of technology is to be adopted, and to what social ends it is to be applied. The whole subject of technology needs to be demythologized so that we have a realistic view of what technology can, what it cannot do, and what its costs are.

The basic features of a valid alternative technology have already been identified (Box 18). Unlike current bulldozer–supertanker technology, it would be based on ecological and thermodynamic premises that are

---

* The U.S. Nuclear Regulatory Commission, attempting to revive the moribund nuclear industry in the United States, exemplifies this willingness to take increased risks. The NRC has been trying, administratively, to eliminate public hearings heretofore required when a reactor is finished and about to go on line. An adverse Court of Appeals decision in November 1990 temporarily frustrated this effort, but both the NRC and the Bush administration are lobbying Congress to eliminate the hearings by legislation (Wasserman 1991, p. 656). The NRC is also expected to promulgate new rules providing for the renewal of reactor licenses past the end of a reactor's natural 40-year-life.

compatible with the coexistence of humans and nature over the long term. As a consequence, it would necessarily eschew merely quantitative progress, striving instead to maximize general human welfare at minimum material cost. Such an alternative (or "soft," "appropriate," "low-impact," "intermediate") technology is certainly possible; that it would also be desirable is a theme we shall return to in Chapter 8.

Even under the most optimistic assumptions, the kinds of alternative technologies under discussion probably cannot support affluence as we in the richest countries have come to define it, so a certain lowering of social sights is called for. In fact, extensive social changes are inevitable. One of the major attractions of the technological fix as a response to the problems of exponential growth is that it appears to avoid the need for awkward social change. In other words, reliance on technological growth makes possible the continuation of business as usual. But as we have seen throughout our discussion of ecological limits, business as usual cannot continue under any circumstances, no matter what one assumes about our civilization's technological response, because a multitude of political, social, and economic issues lie concealed within nearly all aspects of the environmental crisis. The limits to technological growth that we have identified make it even clearer that the essence of the solution to the environmental crisis must be political in the sense specified in the Introduction. We shall explore these thorny issues (particularly the political side effects of continued technological growth) in Part II.

An Overview of Ecological Scarcity

What Is Ecological Scarcity?

Ecological scarcity is an all-embracing concept that encompasses all the various limits to growth and costs attached to continued growth that were mentioned above. As we have seen, it includes not only Malthusian scarcity of food but also impending shortages of mineral resources, biospheric or ecosystemic limitations on human activity, and limits to the human capacity to use technology to expand resource supplies ahead of exponentially increasing demands (or to bear the costs of doing so).* We have seen diminishing returns, which have overtaken not only agricul-

---

* A complete definition of ecological scarcity ought properly to include the social costs attached to continued technological and industrial growth, the economic problems of coping with the physical aspects of scarcity, and certain other sociopolitical factors that will be dealt with in Part II.

18

Alternative Technology

All forms of alternative, or "soft," technology share certain characteristics. First and foremost, they are closely adapted to natural cycles
and processes, so pollution is minimized and as much of the work as
possible is done by nature. Second, they are based primarily on renewable, "income" flows of matter and energy such as trees and solar radiation rather than on nonrenewable, "capital" stocks such as rare ores and
fossil fuels. Third, the first two characteristics encourage the revival of
some labor-intensive modes of production. Fourth, these three together
imply the creation of a "low-throughput" economy, in which the per
capita use of resources is minimized and long-term thermodynamic
and social costs are not ignored for the sake of short-term benefits.
Fifth, all of these seem to point to technologies that are smaller, simpler,
less dependent on a specialized technical elite, and therefore more
decentralized with respect both to location and to control of the means
of production. Finally, among the possible social side effects of such alternative technologies are greater cultural diversity, reduced liability to
misuse of technology by individuals and nations, and less overall
anomie and alienation once individuals have greater control over their
own lives than they do under the current technological dispensation.

Naturally, one way to achieve these goals would be to renounce
modern science and technology entirely and revert to a low-technol-
ogy, pre-modern agrarian society, but the proponents of alternative
technology are not urging a return to some imaginary paradise of pristine closeness to nature. They propose instead a creative blend of the
most advanced modern science and technology with the best of the
old, pre–Industrial Revolution "polytechnics" (Mumford 1970). Yet at
the same time, alternative technology is indeed profoundly anti-
technological, for it is diametrically opposed to autonomous technological growth of the kind that has produced an ecological crisis. Perhaps
technology has not exerted a determining influence on modern society,

tural production but every other economic activity as well; the limits to
the efficiency of pollution control and of energy conversion, the need to
mine ever-thinner ores to get the same useful quantity of metals, the need
to pour ever-more money and energy into the maintenance of the basic
technological infrastructure, and so on. Instead of being able to do ever

as some of its more extreme critics maintain, but it is quite evident that during the last 300 years, society has adapted to technology rather than vice versa. In seeking to reverse this situation and bring the technological process under full social control, alternative technology poses a challenge to the current order that is in the broadest sense primarily political, not scientific or technical. (Indeed, most of the essential components of a viable alternative technology, such as solar power, are already known or invented and merely require development; the process of changeover could therefore be quite fast, unlike the Industrial Revolution, which was retarded by the slow pace of invention.) Thus, although alternative technology is technically feasible and could be installed without unacceptable social costs, its adoption will require a revolutionary break with the values of the industrial era.

A major unanswered question is how high the material standard of living will be. Unfortunately, the answer depends largely on how many people there are. It is abundantly clear that "soft" technology is able to provide an ample sufficiency of material well-being to very large numbers of people. On the other hand, it cannot support the materialistic profligacy now enjoyed by the richest one-fifth of humanity. Humanity's affluent economies have emitted two-thirds of the greenhouse gases, three-fourths of the sulfur and nitrogen oxides, most of the world's hazardous wastes, and 90% of the world's chlorofluorocarbons. It is not possible under any scheme for the world to live like today's Americans, for before it could happen, the planet would be laid to waste. One rough estimate is that alternative technology could support a world population of about 1 billion people at the current standard of living of Norway or the Netherlands (de Bell 1970, p. 154). Human resourcefulness may establish that this is a gross underestimate. But a world maintained in ecological balance with its resources by means of alternative technology will likely contain fewer people than it does now. And those people will have to be more frugal and contribute more physical labor for their affluence than does the richest one-fifth of humanity today.

more with ever less or to substitute one resource for another indefinitely, as economists often claim is possible, we shall have to spend more money, energy, and social effort to obtain the same quantity, or even a diminished quantity, of useful output. Furthermore, most proposed technological solutions to the problems of growth call for more materials (often materials of a very particular and scarce type), create more pollution (or demand more technological solutions to control it), require more energy, and absorb more human resources. Thus the costs of coping with each additional increment of growth rise inexorably and exponentially.

We have also seen that, in general, all sectors are interacting and interdependent, so that on the one hand, the combination of sectoral micro-problems creates an almost overwhelming macro-problem, while on the other hand, the solutions to the macro-problem (as well as those to most of the separate micro-problems) depend on the questionable availability of a host of factors that *may* be in *least* supply. Thus problems exacerbate each other. Also, the solution to one micro-problem is often inconsistent with the solution to other micro-problems or is dependent on the solution of still another problem, which depends in turn on the solution to a third problem, and so on. Nothing less than a coordinated strategy that takes into account the full ensemble of problems and their interactions can hope to succeed.

Thus, stating that ecological scarcity will one day bring growth to a halt is much more than merely asserting that the earth is finite and that growth must therefore cease some day in the future. Ecological scarcity is indeed ultimately grounded on the physical scarcity inherent in the earth's finitude, but it is manifested primarily by the multitude of interacting and interdependent limits to growth that will prevent us from ever testing the finitude of the biosphere and its resources. In fact, as we shall see, ecological scarcity has already begun to restrain growth.

The overall course of industrial civilization as it responds to ecological scarcity is illustrated graphically in Figure 3-3 by the familiar sigmoid or logistic growth curve. In the period between *A* and *B*, the ecological and other resources necessary for growth are present in abundance (at least potentially), and splendid and accelerating growth ensues, as it has during the last 300 years or more. Eventually, however, resources are no longer abundant enough to support further growth, and technological ingenuity can no longer postpone the day of reckoning. At this point of inflection (*C*), deceleration begins; in the narrow transition zone (*B* to *D*), which is approximately one doubling period wide about the point of inflection, considerable further growth due to momentum occurs, but the ecological abundance that fueled accelerating growth begins to disappear, and the first warning signs of ecological scarcity are quickly succeeded by various negative feedback pressures that start to choke off further growth. Beyond the brief transition period these pressures build up quite rapidly, and deceleration continues until equilibrium (*E*) is attained. The zone of transition is therefore the most critical section of the growth curve. The entire changeover from accelerating to decelerating growth occurs in a very brief time, especially

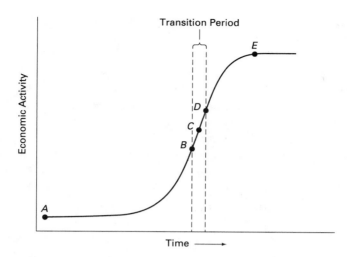

FIGURE 3-3 Growth curve of industrial civilization: *A,* steady state (beginning of accelerating growth); *B,* end of unrestrained growth (beginning of transition period); *C,* point of inflection (beginning of deceleration); *D,* end of transition period; *E,* terminal steady state.

compared to the seemingly infinite period of growth that precedes it, during which the very idea of limits or scarcity, except as temporary challenges to ingenuity, seems ludicrous.

Thus ecological scarcity becomes evident only once the curve is within the transition zone. This being the case, the mere fact that so many aspects of ecological scarcity have been discussed and debated at great length should be ample evidence that industrial civilization is near or past the point of inflection and confronts the prospect of deceleration to a steady state. Yet, in fact, the controversy continues. As we noted in the Introduction, the time factor is the crux of the debate over the limits to growth, so let us examine in greater detail the question of how far away industrial civilization is from the proximate and ultimate limits to growth.

How Far Away Is Ecological Scarcity?

The evidence is overwhelming that we have entered the transition zone. People can impressionistically observe rising pollution problems, not only in industrial nations but in many over-crowded and over-urbanized developing countries. These were the first signs of thermodynamic bills coming due. Since then, specialists have observed degradation of all three of the biological systems on which the world's economy depends: croplands, forests, and grasslands.

Croplands provide feed, food, and many raw materials that industry uses. Forests provide fuel, lumber, paper, and many other products. Grasslands are the source of meat, milk, leather, and wool. As of 1986, 11% of the earth's land was cropland, 31% was forest, and 25% was pasture. The rest of the earth's land surface had little biological activity; it either was desert or was paved over for human use. Since 1981, the amount of land reclaimed for crops has been offset by an equal amount no longer suitable for agriculture or paved over. The amount of grassland worldwide has declined, as overgrazing turns it into pasture. Forests have been shrinking for centuries and, in the 1980s, at a rapidly accelerating rate. The combined area of the three biologically productive areas has been shrinking since the 1980s, whereas the earth's biological wastelands (deserts and paved areas) have been expanding.

Worse, productivity in two of the earth's three biologically productive areas is also down. Throughout the Northern hemisphere, where forest growth rates are measured, trees are growing more slowly. In many areas, whole species of trees and even local forests are dying from acid rain, ozone, and other stresses. Grassland destruction is occurring on every continent, as grazing exceeds the carrying capacity of the land. Even in the United States, a majority of the grassland is in fair to poor condition. As grassland deteriorates, soil erosion accelerates and the capacity to carry livestock is reduced further; eventually, the area turns into a desert. Livestock growers then seek grain from cropland for their animals, putting increased pressure on farmers, whose production of food per capita has not kept up with the increase in human population since 1988.

According to Stanford University biologist Peter Vitousek, humans now appropriate 40% of the land's net primary biological product. Net primary biological product is the amount of energy that primary producers capture via photosynthesis, less the energy they use in their own growth and reproduction. In other words, 40% of the earth's land-based photosynthetic product either is used by humans or has been lost as a result of the alteration of ecosystems by human activity. This means several things. First, as the human impact on the environment increases, other species will find it more difficult to survive. Eventually they will not survive, and human life-support systems will begin to unravel. Second, "eventually" is not so far off. Let us assume a constant level of per capita resource consumption. Then if 5 billion human beings appropriate 40% of the land's NPP, 10 billion human beings will appropriate 80%; before the population got to the projected 14 billion by 2100, humans would have consumed the entire world's net primary biological product, which is impossible. Indeed, even 80% is ecologically impossible; humans cannot survive without the survival of ecosystems made up of other species, most of which would be dead by that point.

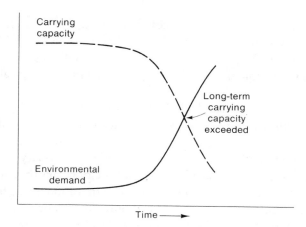

FIGURE 3-4   Growth versus carrying capacity. If growth results in environmental degradation, the carrying capacity is progressively reduced.

As the example of the lily pond makes clear, time is running out. Just as demand for various biological products is increasing to keep up with human population growth and appetites, the carrying capacity of the earth is decreasing with the depletion and degradation of resources (Figure 3-4). Indeed, most ecologists would argue that the carrying capacity has already been exceeded whenever one can observe dangerous levels of pollution, serious ecological degradation, or widespread disturbance of natural balances, all of which are readily observable today. Thus, although precise forecasting is not possible, the available quantitative evidence rather strongly suggests that industrial civilization will be obliged to make an abrupt transition from full-speed-ahead growth to some kind of equilibrium or steady-state society in little more than one generation—and that the process of deceleration has already begun.

The Historical Significance of Ecological Scarcity

The essential meaning of ecological scarcity is that humanity's political, economic, and social life must once again become thoroughly rooted in the physical realities of the biosphere. Scarcity and physical necessity have not been abolished; after a brief historical interlude of apparently endless abundance, they have returned stronger than ever (with political consequences to be taken up in Part II). Because of ecological scarcity, many things that we now take as axiomatic will be inverted in the near future. For example, during the growth era, capital and labor were the critical factors in the economic process; henceforth, land and resources (that is, nature) will be critical. In addition, because the United States, Europe, and Japan—the so-called "haves"—are now living to some extent beyond their ecological means, they may turn into ecological and economic "have nots," while some current "have nots" who are comparatively resource-rich will suddenly become the new "haves." (This transformation is already under way.) All the institutions and values that characterize industrial societies and are predicated on continuous growth will be confronted with ruthless reality tests and revolutionary challenges. Above all, the sudden coming of ecological scarcity means that our generation is faced with an epochal political task. The transition is under way regardless of our wishes in the matter, so our only proper course is to learn how to adapt humanely to the exigencies of ecological scarcity and guide the transition to equilibrium in the direction of a desirable steady-state society.

The great danger from the sudden emergence of ecological scarcity is that we will not respond to its challenges in time. We have already seen that time is probably our scarcest resource; the sheer momentum of growth, the long time constants built into the biosphere, and above all, social response rates that for various reasons lag behind events (and are in any event governed by the factor in least supply) all predispose the world system and most of its subsystems to overshoot (exceed) the level that would be sustainable over the long term. But the inevitable consequence of overshoot is collapse. The trend depicted in Figure 3-4 cannot continue in the real world, for environmental demand can never long exceed the carrying capacity. Figure 3-5 represents the three basic real-world possibilities: (a) smooth convergence on the optimal equilibrium level (which is, as noted above, unlikely); (b) overshoot and collapse with eventual convergence on a relatively high equilibrium level; and (c) overshoot and collapse to a significantly lower than optimal equilibrium level because the carrying capacity has been drastically eroded by the destructiveness associated with the overshoot.

Because the earth's carrying capacity is clearly being depleted and degraded, we are speeding rapidly toward the outcome depicted in Figure 3-5(c), which is highly undesirable for at least three reasons: The suffering and misery created by a large overshoot of the carrying capacity will be enormous. Any large overshoot seems certain to erode the carrying capacity so severely that the surviving civilization will have rather limited material possibilities. And the opportunity to build the basic technological and social infrastructure of a high-level, steady-state society may be irretrievably lost. That is, unless the remaining supplies of non-renewable resources are carefully husbanded and used to make a planned transition to a high-technology steady state, only steady states comparatively poor in material terms will be achievable with the depleted resources left following overshoot and collapse. Thus, although ecological scarcity means that there is no option other than the steady-state society in which people and their demands are in balance with the environment and its resources, the current generation does have a significant say in the type and basic quality of the steady state that will be achieved. The basic policy options are presented graphically in Figure 3-6.

Throughout most of recorded history, the human race has existed in rough equilibrium with its resource base. Growth occurred, if at all, at an infinitesimal pace; even the population of relatively dynamic Europe grew at much less than 1% a year between A. D. 600 and 1600. But then, very suddenly, the Industrial Revolution rocketed the scale of economic activity upward. With the arrival of ecological scarcity, the rocket cannot continue to rise. The first policy option (transition I in Figure 3-6) is an immediate and direct transition to a steady-state civilization relatively affluent in material terms (however frugal it might seem to many now living in the richest countries). If this option is not taken, overshoot must occasion a fall to a significantly lower steady-state level than could have been achieved by carefully planned and timely action (transition II), or even to a level tantamount to a reversion to the traditional premodern agrarian way of life (transition III), so that the entire Industrial Revolution from start to finish will appear as a brief and anomalous spike in humanity's otherwise flat ecological trace, a transitory epoch a few centuries in duration, when it momentarily seemed possible to abolish scarcity.* In short, we stand at a genuine crossroads. Ecological scarcity is not completely new in history, but the crisis we confront is largely unprecedented. That is, it is not a simple repetition of the classic Malthusian apocalypse on a larger scale, in which nothing has changed but the numbers of people, the ruthlessness of the checks, and therefore the greater potential for misery once the day of reckoning comes. The wars, plagues, and famines that have toppled previous civilizations are overshadowed by horrible checks Malthus never dreamt of (such as large-

---

* There is some risk that in trying to make the immediate and direct transition (I in Figure 3-6), we shall achieve a steady-state level somewhat lower than the maximum possible. However, the sacrifice of such a marginal gain seems small compared to the risks attached to overshoot. Moreover, it will always be possible to adjust upward if later experience or further inventions make it feasible; thus the marginal gains will be forgone only temporarily.

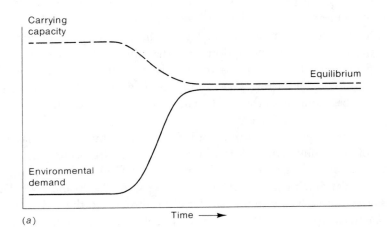

(a)

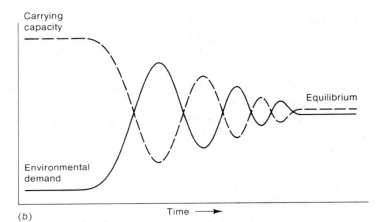

(b)

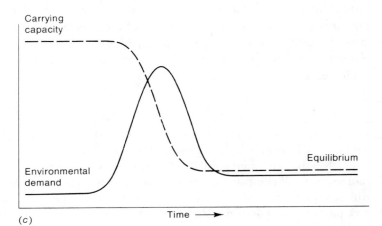

(c)

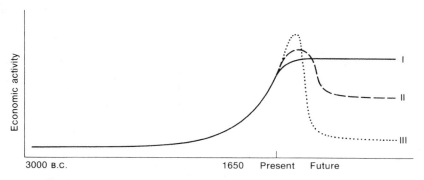

FIGURE 3-6 The ecological history of the world—past, present, and future: I, direct transition to high-level steady state; II, belated transition to somewhat lower-level steady state; III, reversion to pre-modern agrarian way of life.

scale ecological ruin and global radiation poisoning), for these checks are threats to the very existence of the species. On the other hand, we also possess technical resources that previous civilizations lacked when they encountered the challenges of ecological scarcity. Thus in our case a successful response is possible: We can create a reasonably affluent post-industrial, steady-state civilization and avoid a traumatic fall into a version of preindustrial civilization.

This imposing task devolves upon the current generation, and there is no time to lose. Already many trends, such as demographic momentum, cannot be reversed within any reasonable time without Draconian measures. Moreover, as we shall see in Part II, the way ahead is strewn with painful dilemmas. Indeed, nothing can be accomplished without the frustration of many deeply ingrained expectations and the exaction of genuine sacrifices. The epoch we have already entered is a turning point in the ecological history of the human race comparable to the Neolithic Revolution. It will inevitably involve racking political turmoil and an extraordinary reconstitution of the political paradigm that prevails throughout most of the modern world.

FIGURE 3-5 (*Left*) Three scenarios for the transition from growth to maturity: (a) smooth transition to equilibrium with minimal erosion of carrying capacity; (b) overshoot with substantial erosion of carrying capacity; (c) overshoot with drastic erosion of carrying capacity.

The Dilemmas of
Scarcity

4

The Politics of Scarcity

Having explored the general nature and meaning of ecological scarcity, we shall now delve into its political consequences. This chapter examines the basic political dynamics of ecological scarcity; Chapters 5 and 6 assess the specific challenge to the American market system; and Chapter 7 extends the analysis of the preceding three chapters, showing that it applies in all important respects to the rest of the world.

The Political Evils of Scarcity

It was suggested in the Introduction that scarcity is the source of original political sin: Resources that are scantier than human wants have to be allocated by governments, for naked conflict would result otherwise. In the words of the philosopher Thomas Hobbes in *Leviathan* (1651, p. 107), human life in an anarchic "state of nature" is "solitary, poor, nasty, brutish, and short." To prevent the perpetual struggle for power in a war of all against all, there must be a civil authority capable of keeping the peace by regulating property and other scarce goods. Scarcity thus makes politics inescapable.

Presumably, the establishment of a truly just civil authority would eliminate all the political problems that arise from scarcity. With all assured of a fair share of goods, social harmony would replace strife, and people would enjoy long and happy lives of peaceful cooperation. Unfortunately, this has never happened. Although they have certainly mitigated some of the worst aspects of the anarchic state of nature (especially the total insecurity that prevails in a war of all against all), civilized polities have always institutionalized a large measure of in-equality, oppression, and conflict. Thus, in addition to being the source of original political sin, scarcity is also the root of political evil.

The reason is quite simple. For most of recorded history, societies have existed at the ecological margin or very close to it. An equal division of income and wealth, therefore, would condemn all to a life of shared poverty. Not unnaturally, the tendency has been for political institutions to further impoverish the masses by a fractional amount in order to create the surplus that enables a small elite to enjoy more than its share of the fruits of civilized life. Indeed, until recently energy has been so scarce that serfdom and slavery have been the norm—justifiably so, says Aristotle in his *Politics,* for otherwise genuine civilization would be impossible. Except for a few relatively brief periods when for some reason the burden of scarcity was temporarily lifted, inequality, oppression, and conflict have been very prominent features of political life, merely waxing and waning slightly in response to the character of the rulers and other ephemeral factors.

Our own era has been the longest and certainly the most important exception. During roughly the last 450 years, the carrying capacity of the globe (and especially of the highly developed nations) has been markedly expanded, and several centuries of relative abundance have completely transformed the face of the earth and made our societies and our civilization what they are today—relatively open, egalitarian, libertarian, and conflict-free.

The Great Frontier

The causes of the four-century-long economic boom we have enjoyed are readily apparent: the European discovery and exploitation of the New World, Oceania, and other founts of virgin resources (for example, Persian Gulf oil); the take-off and rapid-growth phases of science-based, energy-intensive technology; and the existence of vast reservoirs of "free" ecological goods such as air and water to absorb the consequences of our exploiting the new resources with the new technology. However, the first cause is clearly the most important.

Before the discovery of the New World, the population of Europe pressed hard on its means of subsistence, and as a result, European societies were politically, economically, and socially closed. But with the opening up of a "Great Frontier" in the New World, Europe suddenly faced a seemingly limitless panorama of ecological riches. The land available for cultivation was suddenly multiplied about five times; vast stands of high-grade timber, a scarce commodity in Europe, stretched as far as the eye could see; gold and silver were there for the taking, and rich lodes of other metals lay ready for exploitation; the introduction of the potato and other new food crops from the New World boosted European agricultural production so sharply that the population doubled between

1750 and 1850. This bonanza of found wealth lifted the yoke of ecological scarcity and, coincidentally, created all the peculiar institutions and values characteristic of modern civilization—democracy, freedom, and individualism.*

Indeed, the existence of such ecological abundance is an indispensable premise of the libertarian doctrines of John Locke and Adam Smith, the two thinkers whose works epitomize the modern bourgeois views of political economy on which all the institutions of open societies are based. For example, Locke (1690, paras. 27–29) justifies the institution of property by saying that it derives from the mixture of a man's labor with the original commons of nature. But he continually emphasizes that for one man to make part of what is the common heritage of mankind his own property does not work to the disadvantage of other men. Why? Because "there was still enough and as good left; and more than the yet unprovided could use" (para. 33). His argument on property by appropriation is shot through with references to the wilderness of the New World, which only needed to be occupied and cultivated to be turned into property for any man who desired it. Locke's justification of original property and the natural right of a man to appropriate it from nature thus rests on cornucopian assumptions. There is always more left; society can therefore be libertarian.

The economics of Adam Smith rests on a similar vision of ecological abundance. In fact, Smith is even more optimistic than Locke, for he stresses that the opportunity to become a man of property (and therefore to enjoy the benefits of liberty) now lies more in trade and industry than in agriculture, which is potentially limited by the availability of arable land. Indeed, says Smith, under prevailing conditions, simply striking off all the mercantilist shackles on economic development and permitting a free-for-all, laissez-faire system of wealth-getting to operate instead would generate "opulence," which would in turn liberate men from the social and political restrictions of feudalism. Smith's *The Wealth of Nations* (1776) is therefore a manifesto for the attainment of political liberty through the economic exploitation of the found wealth of the Great Frontier.

The liberal ideas of Locke and Smith have not gone unchallenged, but with very few exceptions, liberals, conservatives, socialists, communists, and other modern ideologists have taken abundance for granted

---

* Of course, the idea of individualism antedated the discovery of the New World, but before that time there had been little opportunity for its concrete expression. However, once the boom permitted it to be expressed, individualism became the basis for almost all of the most characteristic features of modernity: self-rule in democracy, self-enrichment in industrial capitalism, and self-salvation in Protestantism.

and assumed the necessity of further growth. They have disagreed only about how to produce enough wealth to satisfy the demands of hedonistic, materialistic "economic" men and about what constitutes a just division of the spoils. Karl Marx was even more utopian than either Locke or Smith, for he envisioned the eventual abolition of scarcity. He merely insisted that, on grounds of social justice, the march of progress be centrally directed by the state in the interest of those whose labor actually produced the goods.

But the boom is now over. The found wealth of the Great Frontier has been all but exhausted. And technology is no real substitute, for it is merely a means of manipulating *what is already there* rather than a way of creating genuinely new resources on the scale of the Great Frontier. (Moreover, as we saw in Part I, technology is encountering limits of its own.) Thus a scarcity at least as intense as that prevailing in the premodern era, however different it may be in important respects, is about to replace abundance, and this will necessarily undercut the material conditions that have created and sustained current ideas, institutions, and practices. Once relative abundance and wealth of opportunity are no longer available to mitigate the harsh political dynamics of scarcity, the pressures favoring greater inequality, oppression, and conflict will build up, so that the return of scarcity portends the revival of age-old political evils, for our descendants if not for ourselves. In short, the golden age of individualism, liberty, and democracy (as those terms are currently understood) is all but over. In many important respects, we shall be obliged to return to something resembling the premodern, closed polity. This conclusion will be reinforced by a more detailed exploration of the political problem of controlling the competitive overexploitation of resources that has produced the ecological crisis.

The Tragedy of the Commons

It has been recognized since ancient times that resources held or used in common tend to be abused. As Aristotle said, "What is common to the greatest number gets the least amount of care" (Barker 1962, p. 44). However, the dynamic underlying such abuse was first suggested by a little-known Malthusian of the early nineteenth century, William Forster Lloyd (cited in Hardin 1969, p. 29), who wondered why the cattle on a common pasture were "so puny and stunted" and the common itself "bare-worn." He found that such an outcome was almost inevitable.

People seeking gain naturally want to increase the size of their herds. Because the commons is finite, the day must come when the total number of cattle reaches the carrying capacity; the addition of more cattle will cause the pasture to deteriorate and eventually destroy the resource on which the herders depend. Yet even though this is the case, it is still in the rational self-interest of each herder to keep adding animals to his herd. Each reasons that his personal gain from adding animals outweighs his proportionate share of the damage done to the commons, for the damage is done to the commons as a whole and so is partitioned among all the users. Worse, even if he is inclined to self-restraint, an individual herder justifiably fears that others may not be. They will increase their herds and gain thereby while he does not, in spite of his having to suffer equally from the resulting damage. Competitive overexploitation of the commons is the inevitable result.

The same dynamic of competitive overexploitation applies to any "common-pool resource," the economist's term for resources held or used in common.* A classic illustration is the oil pool. Unless one person or organization controls the rights to exploit an oil pool or the owners of the rights can agree on a scheme of rational exploitation, it is in the interest of each user to extract oil from the common pool as fast as he or she possibly can; in fact, failure to do so exposes the individual owner to the risk that others will not leave him or her a fair share. Thus, in the early boom days of the American oil industry, drillers competed with each other to sink as many wells as possible on their properties. The resulting economic and political chaos was remedied by the establishment of state control boards that surveyed the pools and then allotted each owner a quota of production for each acre of oil-bearing land. Oil was thereby transformed from a common pool resource to private property, and exploitation proceeded thereafter in a largely rational and conflict-free manner.

The dynamic of the commons is particularly stark in the case of oil, for one person's gains are another's losses. But even resources that could be exploited cooperatively to give a sustained yield in perpetuity are subject to the same dynamic. Fisheries are a prime example. At first there was abundance enough for all to exploit the resource freely. Conflicts occurred, but their impact was local. Fishing a little farther away or

---

* We are grateful to Margaret McKean for suggesting the term *common-pool resource* for the term *common-property resource* used in the previous edition of this book. She argues that "if resources become property only when human beings attach rights and duties to them, then the problematic resources are non-property, not property." She also points out that the "tragedy of the commons"—that is, tragic overuse—occurs only when human beings do not figure out a way to attach rights and duties to these resources in order to solve the problems of subtractibility or rivalness.

improving techniques were alternatives to fighting over the limited resources in any particular area. In time, however, even the vastness of the ocean began to be more or less fully exploited, and people reacted just as they did in the early days of the oil business, overexploiting and destroying fisheries. In response, coastal nations have privatized parts of the fishing common by declaring complete economic sovereignty over the oceans within 200 miles offshore, so that all the benefits of the nearby fishery would flow to their nationals.* One potential benefit of such "privatization" is that it then becomes possible, within these zones, for collective units such as fishing co-ops, to ban access and to internalize the important externalities. But fishing "wars" and other political conflicts over marine resources remain common in the open ocean. There, fishing operations have increased in scale and technical virtuosity, just as early oil drillers sank dozens of wells on a tiny piece of land. Technological progress in the fishing industry has produced gigantic floating factories, which use driftnets and other ultramodern techniques to catch fish and can or freeze them on the spot, thus eliminating the time that must be spent returning to port in traditional fishing.† The result, not surprisingly, has been relentless competitive overexploitation and an alarming general decline in fish stocks.

Pollution also exemplifies the self-destructive logic of the commons, for it simply reverses the dynamic of competitive overexploitation without altering its nature: The cost to me of controlling my emissions is so much larger than my proportionate share of the environmental damage those emissions cause that it will always be rational for me to pollute if I can get away with it. In short, it profits me to harm the public. (It does not pay me to benefit the public either; see Box 19.)

---

* The unratified Law of the Sea Convention (1982) recognizes offshore 200-mile exclusive economic zones (EEZs) for coastal nations (Corson 1990, p. 145).

† Japanese fishing boats cover more than 500 square miles of Pacific Ocean waters with over 2 million miles of driftnets. The nets sweep up everything that swims into them, depleting the stock of desired species and causing the death or injury of 70% of the catch, including sea mammals, which are unwanted. Korea and Taiwan also driftnet in other areas of the Pacific. The United Nations General Assembly in 1989 called for a moratorium on high-seas driftnetting by 1992, but unfortunately, the resolution allows nations to exempt themselves if they practice unspecified marine conservation measures. Japan, after first announcing that it would exempt itself, recently indicated that it would comply with the moratorium.

Unfortunately, virtually all ecological resources—airsheds, watersheds, the land, the oceans, the atmosphere, biological cycles, the biosphere itself—are common-pool resources. For example, the smoke from factories or the exhaust gases from automobiles cannot be confined so that their noxious effects harm only those who produce them. They harm all in the common airshed. Even most resources that seem to be private property are in fact part of the ecological commons. The logging company that cuts down a whole stand of trees in order to maximize its profits contributes to flooding, siltation, and the decline of water quality in that watershed. And if enough loggers cut down enough trees, even climate may be altered, as has occurred many times in the past. Now that the carrying capacity of the biosphere has been approached, if not exceeded, we are in serious danger of destroying all ecological resources by competitive overexploitation. Thus the metaphor of the commons is not merely an assertion of humanity's ultimate dependence on the ecological life-support systems of the planet; it is also an accurate description of the current human predicament.

In short, resources that once were so abundant that they were freely available to all have now become ecologically scarce. Unless they are somehow regulated and protected in the common interest, the inevitable outcome will be the mutual ecological ruin that the human ecologist Garrett Hardin (1968) has called "the tragedy of the commons." We need to apply much more widely the same kind of social rules and political controls that have traditionally governed the use of grazing lands and other commons in the past (although these controls have not always been sufficiently strong to avert partial or even total destruction of a resource).

A Hobbesian Solution?

Beyond telling us that the answer to the tragedy is "mutual coercion, mutually agreed upon by the majority of the people affected"—by which he means social restraint, not naked force—Hardin avoids political prescription. However, he does suggest that unrestrained exercise of our liberties does not bring us real freedom: "Individuals locked into the logic of the commons are free only to bring on universal ruin; once they see the necessity of mutual coercion, they become free to pursue other goals." By recognizing the necessity to abandon many natural freedoms we now believe we possess, we avoid tragedy and "preserve and nurture other and more precious freedoms." There are obvious dangers in a regime of "mutual coercion," but without restraints on individuals, the collective selfishness and irresponsibility generated by the logic of the commons will destroy the spaceship, so that any sacrifice of freedom by

19

The Public-Goods Problem

The public-goods problem is the obverse of the commons problem. Just as the rational individual gains by harming fellow members of the common, he loses by benefiting them with a public or collective good. At best, he gets only a small return on his investment; at worst, he is economically punished. For example, the good husbandman cannot survive in a market economy; if he maintains his soil while his neighbors mine theirs for maximum yields, sooner or later he must either abandon farming or become a subsistence farmer outside the market. He cannot afford to benefit posterity except at great personal sacrifice. Similarly, although a socially responsible plant owner might wish to control the pollution emanating from her plant, if she does it at her own expense whatever her competitors do, then the plant owner is at a competitive disadvantage. Thus the tragedy of the commons, in which the culprit gets all the benefits from transgressing the limits of the commons but succeeds in relegating most of the costs to others, is turned around. Those who try to benefit the common good soon discover that, while they pay all the costs, the other members of the community reap virtually all of the benefits.

Of course, a producer could try to persuade consumers to pay premium prices for his products as a reward for his virtue. But he

the crew members is clearly the lesser evil. After all, says Hardin, "injustice is preferable to total ruin," so that "an alternative to the commons need not be perfectly just to be preferable" (Hardin 1968, pp. 1247–1248).

Hardin's implicit political theory is in all important respects identical to that of Thomas Hobbes in *Leviathan* (1651). Hardin's "logic of the commons" is simply a special version of the general political dynamic of Hobbes's "state of nature." Hobbes says that where men desire goods scarcer than their wants, they are likely to fall to fighting. Each knows individually that all would be better off if they abstained from fighting and found some way of equitably sharing the desired goods. However, they also realize that they cannot alter the dynamics of the situation by their own behavior. In the absence of a civil authority to keep the peace, personal pacifism merely makes them easy prey to others. Unless all can be persuaded or forced to lay down their arms simultaneously, nothing

would be unlikely to find many buyers for products that, however "virtuous," were no better than the cheaper ones of his competitors. Another conceivable solution would be for the manufacturer who intended to control pollution to take up a collection from all those affected. After all, if his or her pollution is harmful to them, they should be willing to pay something to reduce or eliminate it. However, even if the considerable practical difficulties of organizing such a scheme were overcome, it would almost certainly fail, for people are unlikely to contribute voluntarily to pollution reduction or to the production of any other kind of public good in optimal amounts. The reason is simple: It is entirely rational for individuals to try to make others pay most or all of the costs of a public good that benefits everyone equally; thus the good is never available in optimal quantity under market conditions. For example, no government can subsist on voluntary tax payments. If external defense, internal order, rules for economic competition, public health, education, and other public goods are to be produced in quantities that are rationally desirable for the society, then taxes must be compulsory on all its members. Similarly, if ecological public goods such as clean air and water and pleasant landscapes are to be provided in reasonable mounts, it will be only as a result of collective decisions. Thus, just as for the tragedy of the commons, the answer to the public-goods problem is authoritative political action.

can prevent the war of all against all. The crucial problem in the state of nature is thus to make it safe for men to be reasonable, rather than merely "rational," so that they can share peacefully what the environment has to offer. Hobbes's solution was the erection, by a majority, of a sovereign power that would constrain all men to be reasonable and peaceful—that is, Hardin's "mutual coercion, mutually agreed upon by the majority of the people affected."*

In the tragedy of the commons, the dilemma is not so stark as it is in the state of nature (political order is not at stake), but it is in many ways much more insidious, for even without evil propensities on the part of any person or group, the tragedy will occur. In the case of the village

---

* For a fuller discussion of the virtual identity of analyses and prescriptions in Hardin and Hobbes, see Ophuls 1973.

common, the actors can hardly avoid noticing the causal relationship
between their acts and the deterioration of the commons, but in most
cases of competitive overexploitation, individuals are not even aware of
the damage that their acts are causing, and even if they are aware, their
own responsibility seems infinitesimally dilute. Thus, to bring about the
tragedy of the commons it is not necessary that people be bad, only that
they not be actively good—that is, not altruistic enough to limit their
own behavior when their fellows will not regularly perform acts of public
generosity. That people are in fact not this altruistic is confirmed daily by
behavior we all see around us (and see Schelling 1971).

A perfect illustration of the insidiousness of the tragedy of the
commons in operation is the situation of the inhabitants of Los Angeles
vis-à-vis the automobile:

> Every person who lives in this basin knows that for twenty-five years he
> has been living through a disaster. We have all watched it happen, have
> participated in it with full knowledge just as men and women once went
> knowingly and willingly into the "dark Satanic mills." The smog is the
> result of ten million individual pursuits of private gratification. But there
> is absolutely nothing that any individual can do to stop its spread. Each
> Angeleno is totally powerless to end what he hates. An individual act of
> renunciation is now nearly impossible, and, in any case, would be mean-
> ingless unless everyone else did the same thing. But he has no way of
> getting everyone else to do it. He does not even have any way to talk about
> such a course. He does not know how or where he would do it or what
> language he would use. (Carney 1972, pp. 28-29).*

As this example clearly shows, the essence of the tragedy of the
commons is that one's own contribution to the problem (assuming that
one is even aware of it) seems infinitesimally small, while the disad-
vantages of self-denial loom very large; self-restraint therefore appears to

---

* People do know that by "mutual coercion, mutually agreed upon by the
majority of the people affected," Los Angeles can control its air pollution. In
1989 the Southern California Air Quality Management District (AQMD)
proposed an ambitious antipollution plan that, if fully implemented, could
reduce smog-producing emissions in the Los Angeles area by 70%
(Elmer-Dewitt 1989, p. 65). But because the plan requires changes in the way
people live and conduct business, it has been difficult to obtain mutual
agreement. The AQMD has been forced to abandon several of its proposals, as
critics contend variously that the plan is too expensive, too burdensome on
them (higher downtown parking fees are an example), or that companies will
relocate, resulting in a loss of jobs (Matthews 1991, p. A21).

be both unprofitable and ultimately futile unless one can be certain of universal concurrence. Thus we are being destroyed ecologically not so much by the evil acts of selfish people as by the everyday acts of ordinary people whose behavior is dominated, usually unconsciously, by the remorseless self-destructive logic of the commons.

The tragedy of the commons also exemplifies the political problem that agitated the eighteenth-century French political philosopher Jean-Jacques Rousseau, who made a crucial distinction between the "general will" and the "will of all." The former is what reasonable people, leaving aside their self-interest and having the community's interests at heart, would regard as the right and proper course of action. The latter is the mere addition of the particular wills of the individuals forming the polity, based not on a conception of the common good but only on what serves their own self-interest. The tragedy of the commons is simply a particularly vicious instance of the way in which the "will of all" falls short of the true common interest. In essence, Rousseau's answer to this crucial problem in *The Social Contract* is not much different from Hobbes's: Man must be "forced to be free"—that is, protected from the consequences of his own selfishness and shortsightedness by being made obedient to the common good or "general will," which represents his real self-interest. Rousseau thus wants political institutions that will make people virtuous.

It therefore appears that if under conditions of ecological scarcity, individuals rationally pursue their material self-interest unrestrained by a common authority that upholds the common interest, the eventual result is bound to be common environmental ruin. In that case, we must have political institutions that preserve the ecological common good from destruction by unrestrained human acts. The problem that the environmental crisis forces us to confront is, in fact, at the core of political philosophy: how to protect or advance the interests of the collectivity when the individuals who make it up (or enough of them to create a problem) behave (or are impelled to behave) in a selfish, greedy, and quarrelsome fashion. The only solution is a sufficient measure of coercion (see Box 20). According to Hobbes, a certain minimum level of ecological order or peace must be established; according to Rousseau, a certain minimum level of ecological virtue must be imposed by our political institutions.

It hardly need be said that these conclusions about the tragedy of the commons radically challenge fundamental American and Western values. Under conditions of ecological scarcity, the individuals, possessing an inalienable right to pursue happiness as they define it and exercising their liberty in a basically laissez-faire system, will inevitably produce the ruin of the commons. Thus the individualistic basis of society, the concept of inalienable rights, the purely self-defined pursuit of happiness, liberty as

20

Coercion

The word *coercion* has a nasty fascist ring to it. However, politics is a means of taming and legitimating power, not dispensing with it. Any form of state power is coercive. A classic example is taxation, which is nowhere voluntary, for as the theory of public goods (see Box 19) tells us, the state would starve if it were. Assuming a reasonable degree of consensus and legitimacy, coercion means no more than a state-imposed structure of incentives and disincentives that is designed to advance the common interest. Even Locke's libertarian political theory does not proscribe coercion: If the common interest is threatened, the sovereign must do whatever is necessary to protect it. Nevertheless, unlike Hobbes, Locke does try to set up inviolable spheres of private rights that the sovereign may not invade, and he also demands that power be continually beholden to consent of the governed. The difference between Hobbes and Locke on the matter of coercion is one of degree, with Locke demanding more formal guarantees of limits on the sovereign's power than Hobbes believes are workable. In short, coercion is not some evil specter resurrected from an odious past. It is an inextricable part of politics, and the problem is how best to tame it and bend it to the common interest.

Some aspire to do away with power politics and state coercion entirely by making people so virtuous that they will automatically do

maximum freedom of action, and the laissez-faire principle itself all become problematic. All require major modification or perhaps even abandonment if we wish to avert inexorable environmental degradation and eventual extinction as a civilization. Certainly, democracy *as we know it* cannot conceivably survive.[*]

---

[*] As we know it today, our current political system is essentially statist; the federal government is a bureaucratic and electoral behemoth, beholden to organized and monied interests, dedicated to the satisfaction of human appetite at the expense of nature. Such a "democracy" cannot survive. A genuine democracy that is fundamentally Jeffersonian and Thoreauvian in spirit and practice, however, can survive (see the discussion of these points in Chapter 8 and the Afterword).

what is in the common interest. In fact, this is precisely what Rousseau proposes: small, self-sufficient, frugal, intimate communities inculcating civic virtue so thoroughly that citizens become the "general will" incarnate. However, this merely changes the locus of coercion from outside to inside—the job of law enforcement is handed over to the internal "police force" of the superego—and many liberals (for example, Popper 1966) would argue that this kind of ideological or psychological coercion is far worse than overt controls on behavior. Nevertheless, political education cannot be done away with entirely, for without a reasonable degree of consensus and legitimacy, no regime can long endure. Thus it is again a question of balance. Hobbes and Rousseau, for example, would both agree that law enforcement and political education must be combined, however much they might disagree on what proportion of each is fitting.

Political coercion in some form is inevitable. Failing to confront openly the issues it raises is likely to have the same effect repression has on the individual psyche: The repressed force returns in an unhealthy form. By contrast, if we face up to coercion, full political awareness will dispel its seeming nastiness, and we shall be able to tame it and make it a pillar of the common interest. (The next box suggests a way of taming Leviathan, and Chapter 8 discusses the politics of a steady-state society in more general terms.)

This is an extreme conclusion, but it seems to follow from the extremity of the ecological predicament that industrial humanity has created for itself. Even Hobbes's severest critics concede that he is most cogent when stark political choices are faced, for self-interest moderated by self-restraint may not be workable when extreme conditions prevail. Thus theorists have long analyzed international relations in Hobbesian terms, because the state of nature mirrors the state of armed peace existing between competing nation-states that are obedient to no higher power. Also, when social or natural disaster leads to a breakdown in the patterns of society that ordinarily restrain people, even the most libertarian governments have never hesitated to impose martial law as the only alternative to anarchy. Therefore, if nuclear holocaust rather than mere war, or anarchy rather than a moderate level of disorder, or destruction of the biosphere rather than mere loss of amenity is the issue, the extremity of Hobbes's analysis fits reality, and it becomes difficult to avoid his conclusions. Similarly, although Rousseau's ultimate aim was the creation of a democratic polity, he recognized that strong sovereign power (a "Legislator," in Rousseau's language) may be necessary in certain circumstances, especially if the bad habits of a politically "corrupt" people must be fundamentally reformed.

Altruism Is Not Enough

Some theorists hope or assert that attitudinal change will bring enough major changes in individual behavior to save a democratic, laissez-faire system from ecological ruin. However, except in very small and tightly knit social groups, education or the inculcation of rigid social norms is not sure proof against the logic of the commons. Apparently, it is simply not true that, once they are aware of the general gravity of the situation, a large number of people will naturally moderate their demands on the environment. A number of studies have shown that even the individuals who are presumably the most knowledgeable and concerned about population growth evince little willingness to restrain their own reproductive behavior (Attah 1973; Barnett 1971; Eisner et al. 1970). How much can we expect of most ordinary citizens? The problem is that in order to forestall the logic of the commons, people in overwhelming numbers must be prepared to do positive good *whether or not* cooperation is universal. And in a political culture that conceives of the common interest as being no more than the sum of our individual interests, it seems unlikely that we can prudently count on much help from unsupported altruism (this is not to say that people cannot be educated to be ecologically more responsible than they are at present).

In any event, even the most altruistic individual cannot behave responsibly without full knowledge of the consequences of his or her acts—and such knowledge is not available. If even the experts fiercely debate the pros and cons of nuclear power or the effects of a particular chemical on the ozone layer, using highly abstruse analytical techniques and complex computer programs that only the specialist can fully understand, how is the ordinary citizen to know what the facts are? An additional problem is time. High rates of change and exponential growth are accompanied by a serious lag in public understanding. For example, it seems to take two to four generations for the ideas at the frontier of science to filter down to even the informed public. We have still not completely digested Darwin, much less Einstein and quantum mechanics. How reasonable is it to expect from the public at large a sophisticated ecological understanding any time soon, especially when the academic, business, professional, and political elites who constitute the so-called attentive and informed public show little sign of having understood, much less embraced, the ecological world view? (As noted in the foreword, children do seem to embrace an ecological world view when it is taught to them. This is an encouraging development, but it is not yet known how many curriculums include ecology or how many children will retain their world view as adults.)

Others pin their hopes for a solution not on individual conscience but on the development of a collective conscience in the form of a world view or religion that sees humanity as the partner of nature rather than its antagonist. This attitude will undoubtedly be essential for our survival in the long term, because without basic popular support, even the most repressive regime could hardly hope to succeed in protecting the environment for long. However, mere changes in world view are not likely to be sufficient. Political and social arrangements that implement values are indispensable for turning ideals into actuality. For example, despite a basic world view profoundly respectful of nature, the Chinese have severely abused and degraded their environment throughout their very long history—more, ironically, than the premodern Europeans, who lacked a philosophy expressive of the same kind of natural harmony. Thus Chinese ideals were not proof against the urgency of human desires that drives the tragic logic of the commons.*

It appears, therefore, that individual conscience and the right kind of cultural attitudes are not by themselves sufficient to overcome the short-term considerations that lead people to degrade their environment. Real altruism and genuine concern for posterity may not be entirely absent,

---

* Some (for example, Reich 1971) would protest that our age is different and that a genuinely new consciousness is emerging. This view cannot simply be brushed aside, for substantial changes in values are clearly occurring in some segments of American society, and out of this essentially religious ferment, great things may come. For example, the "back to the land" movement has been much ridiculed, but its symbolic reaffirmation of our ties to the earth has already had a far from negligible impact on the larger society. Nevertheless, that these new values will become universal in the future appears to be essentially a matter of faith at this point. Past hopes for the emergence of a "new man" have been rudely treated by history, so it is difficult to be optimistic.

but they are not present in sufficient strength to avert the tragedy. Only a government with the power to regulate individual behavior in the ecological common interest can deal effectively with the tragedy of the commons.

To recapitulate, the tragic logic of the commons is sustained by three premises: a limited commons, cattle that need ample grazing room to prevent the commons from becoming "bare-worn," and rational, self-seeking herdspeople. If any one of these premises is removed, the tragedy is averted. As we have already seen, the Great Frontier in effect removed the first premise for nearly 400 years. It was precisely this that allowed John Locke, whose political argument is essentially the same as that of Hobbes in every particular except scarcity in the state of nature, to be basically libertarian, whereas Hobbes is basically authoritarian. Thanks to the Great Frontier, Locke and Smith found that there was so much abundance in the state of nature that a Hobbesian war of all against all was unlikely; every person could take away some kind of prize, and competition would be socially constructive rather than destructive, with the "invisible hand" producing the greatest good for the society as a whole. Thus government was required only to keep the game honest—a mere referee, needing only modest powers and minimal institutional machinery—and individuals could be left alone to pursue happiness as they defined it without hindrance by society or the state.

The frontier is gone now, and we have encountered the limits of the commons. However, the physical disappearance of the frontier was for a long time mitigated by technology, which allowed us to graze more cows on the same amount of pasture. Now we have reached the limits of technology: The cows are standing almost shoulder to shoulder, many are starving, and the manure is piling up faster than the commons can absorb it. All that remains is to alter the rational, self-seeking behavior of the individuals and groups that use the commons. This must be done by collective means, for the dynamic of the tragedy of the commons is so powerful that individuals are virtually powerless to extricate themselves unaided from its remorseless working. Our political institutions must indeed force us to be free.

Legislating Temperance

That we must give our political authorities great powers to regulate many of our daily actions is a profoundly distasteful thought. We tend to see political systems that do not bestow our kind of political and economic liberties as "totalitarian," a word that brings to mind all the evil features of past dictatorships. But even Hobbes, no matter how firm his conviction in the necessity of absolutism, certainly did not have Stalinesque tyranny in mind. Hobbes makes clear that order in the commonwealth is not the goal but is rather the means without which the fruits of civilization cannot be enjoyed. The sovereign power is to procure the "safety of the people... But by safety here is not meant a bare preservation but also all other contentments of life which every man by lawful industry, without danger or hurt to the commonwealth, shall acquire to himself" (Hobbes 1651, p. 262). And it is part of the task of the sovereign power to actively promote these "contentments of life" among its subjects. Furthermore, Hobbes will not countenance tyranny. The sovereign power must rule lawfully, give a full explanation of its acts to its subjects, and heed their legitimate desires. Through wise laws and education, the subjects will learn moral restraint. Also, the sovereign power is not to be a dictator regulating every action of the citizen: it does not "bind the people from all voluntary actions" but only guides them with laws that Hobbes likens to "hedges...set not to stop travelers, but to keep them in their ways" (p. 272). Thus many different styles of rule and of life are compatible with his basic analysis.

Similarly, Hardin makes it clear that the problem is to "legislate temperance," not to institute iron discipline. He acknowledges that this may require the use of administrative law, with the consequent risk of abuse of power by the administrators. However, he believes that the application of his formula of "mutual coercion, mutually agreed upon by the majority of the people affected," would be an adequate defense against bureaucratic tyranny, for we would be *democratically coercing ourselves* to behave responsibly (Hardin 1968, p. 1247).

The question of political will is therefore crucial. Given a basic willingness to restrain individual self-seeking and legislate social temperance, social devices acceptable to reasonable persons and suited to a government of laws could readily be found to serve as the "hedges" that will keep us on the path of the steady state.* For example, law professor Christopher Stone (1974) proposes giving natural objects, such as trees,

---

* Merely increasing the power of the state is no solution. As will be shown in Chapter 7 (and contrary to the opinion of many), mere socialism is not a real solution to the tragedy of the commons. That is, giving the state ownership of the means of production is not very useful if the state is committed to economic expansion, for the same ecologically destructive dynamic operates within a socialist economic bureaucracy as in the capitalist marketplace (see Heilbroner 1974 on this point).

mountains, rivers, and lakes, legal rights (comparable to those now enjoyed by corporations) that could be enforced in court.

However, although the socioeconomic machinery needed to enforce
a steady-state political economy need not involve dictatorial control over
our everyday lives, it will indeed encroach upon our freedom of action,
for *any social device that is effective as a hedge will necessarily prevent us from
doing things we are now free to do or make us do things we now prefer not to do.*
It could hardly be otherwise: If we can safely squeeze no more cattle onto
the commons, then we herders must be satisfied either with the herds we
now possess or, more likely, with the lesser number of cattle that the
commons can tolerate ecologically over the long term. The solution to
the tragedy of the commons in the present circumstances requires a
willingness to accept less—perhaps much less—than we now get from
the commons. No technical devices will save us. In order to be able
mutually to agree on the restraints we wish to apply to ourselves, we must
give up the exercise of rights we now enjoy and bind ourselves to
perform public duties in the common interest. The only alternative to
this kind of self-coercion is the coercion of nature—or perhaps that of an
iron regime that will compel our consent to living with less.

Technology's Faustian Bargain

Given this unpalatable conclusion, the seductive appeal of technological
optimism is apparent: If adjusting human demands to the available ecological
resources will entail a greater degree of political authority, then let us by all
means press on with the attempt to surmount the limits to growth
technologically. Thus, to the extent that technologists concede the necessity
of a steady state, they aim at a "maximum-feasible" steady state of technological superabundance in which we will use our alleged mastery of
inexhaustible energy resources to evade ecological constraints, instead of
learning to live frugally on flow resources such as solar energy.* As we have
seen, the barriers to success in such an enterprise are enormous, but for the
sake of argument, let us put aside all questions of practicality and ask instead
what would be the political consequences of implementing these kinds of
technological solutions to ecological scarcity.

---

* In reality, a maximum-feasible steady state is a virtual contradiction in terms,
for squeezing the maximum out of nature runs contrary to basic ecological
principles. Only a life lived *comfortably* within the circle of natural
interdependence merits the designation "steady state." But the technological
optimists customarily talk as though there were no possible model of the steady
state other than the maximum-feasible one.

Alvin Weinberg, who was for many years director of the Atomic Energy Commission's Oak Ridge National Laboratory, has been a leading spokesman for the technological fix, especially nuclear power. Indeed, he has castigated environmentalists for proposing "social fixes" to ecological problems; he argues that technological solutions are "more humane" because they do not "disrupt the economy and...cause the human suffering that such disruption would entail" (Weinberg 1972b). Yet Weinberg himself admits that the specific technological solution he proposes comes with a truly monstrous social fix firmly attached! Because nuclear wastes will have to be kept under virtually perpetual surveillance, and because nuclear technology places the most exacting demands on our engineering and management capabilities,

> We nuclear people have made a Faustian bargain with society. On the one hand, we offer...an inexhaustible source of energy [the breeder reactor].... But the price that we demand of society for this magical energy is both a vigilance and a longevity of our social institutions that we are quite unaccustomed to [Weinberg 1972a, p. 33].

Part of this price is politically ominous:

> In a sense, what started out as a technological fix for the energy-environment impasse—clean, inexhaustible, and fairly cheap nuclear power—involves social fixes as well: the creation of a permanent cadre or priesthood of responsible technologists who will guard the reactors and the wastes so as to assure their continued safety over millennia [Weinberg 1973, p. 431].

Expanding on the "priesthood" theme, Weinberg tells us that because "our commitment to nuclear energy is assumed to last in perpetuity," we will need "a *permanent* cadre of experts that will retain its continuity over immensely long times [but this] hardly seems feasible if the cadre is a national body," for "no government has lasted continuously for 1,000 years." What kind of organization does possess the requisite continuity?

> Only the Catholic Church has survived more or less continuously for 2,000 years or so...The Catholic Church is the best example of [the International Authority] I have in mind: a *central authority that proclaims* and to a degree *enforces* doctrine, maintains its own long-term social stability, and has connections to every country's own Catholic Church [cited in Speth et al. 1974, emphasis added].

In proposing such a technological "priesthood," Weinberg appears to be a true heir of the French utopian social philosopher Claude Henri

Saint-Simon (1760–1825), one of the earliest prophets of technocracy, who believed that it was humanity's mission to transcend nature with technology. Distressed by the disruptive social effects of technology within a bourgeois, laissez-faire political economy, Saint-Simon aspired to create a stable, organic civilization such as that of the Middle Ages, but with science as its religion. To this end he proposed the creation, on the model of the Catholic Church, of a scientific priesthood that would both dispense political justice and promote the economic wealth of society. Saint-Simon stressed social planning, the necessity for authority based on scientific expertise, the subordination of the individual to the needs of society as determined by the experts, and the integration of society and technology—all themes that emerge in the writings of modern technological visionaries.

By whatever name it comes to be called, technocratic government is likely to be the price of Weinberg's Faustian bargain. It will not be formally voted in, of course, but will emerge in a series of small but fateful steps as we follow what seems to be the line of least resistance through our environmental problems. Indeed, critics were alarmed by the civil-rights implications of the safeguards proposed by the Atomic Energy Commission in its draft environmental-impact statement on plutonium recycling. These included the establishment of a federal police force for the protection of plutonium plants and shipments, the extension of current military security-clearance procedures to include all the civilians who might have access to plutonium, and generally increased police powers to cope with the security requirements of a plutonium-based power economy (Speth et al. 1974). The United States abandoned its breeder reactor program in 1984, but the fact remains that there may be no way to ensure the social stability—indeed, the near-perfect social institutions—necessary for an era of nuclear power except with an engineered society under the direction of a technocratic priesthood.

A Pact with the Devil?

It is not nuclear technology alone that offers a pact with a devil who will in the end claim our political souls. Few technological optimists are as candid as Weinberg about the political implications of the solutions they propose, but technocracy has been looming on the horizon for some time. Harrison Brown, a scientist who foresaw most of today's ecological concerns almost four decades ago, predicted that the instability of industrial society would become greater as development proceeded. This and other organizational requirements, he said, will make ever-greater social control necessary, so that "it is difficult to see how the achievement of stability and the maintenance of individual liberty can be made compatible" (Brown 1954, p. 255). Buckminster Fuller, one of the most visionary of the supertechnologists, states plainly that those who run "Spaceship Earth" cannot afford to make "concessions to the non-synergetic thinking (therefore the ignorantly conditioned reflexes) of the least well advised of the potential mass customers [that is, the average citizen]" (Fuller 1968, p. 367). Numerous other writers of varying persuasions see the same trend: more technology means greater complexity and greater need for knowledge and technical expertise; the average citizen will not be able to make a constructive contribution to decision making, so that "experts" and "authorities" will rule perforce; and because accidents cannot be permitted, much less individual behavior that deviates from technological imperatives, the grip of planning and social control will of necessity become a stranglehold (Bell 1973; Chamberlin 1970; Heilbroner 1974).

Thus, the danger in the Faustian bargain lies in the mounting complexity of technology and with the staggering problems of managing the response to ecological scarcity, for these problems will require us to depend on a special class of experts in charge of our survival and well-being: a "priesthood of responsible technologists."

Democracy versus Elite Rule: The Issue of Competence

One of the key philosophical supports of democracy is the assumption that people do not differ greatly in competence, for if they do, effective government may require the sacrifice of political equality and majority rule. Indeed, under certain circumstances democracy *must* give way to elite rule. As the eminent political scientist and democratic theorist Robert Dahl points out, in a political association whose members "differ *crucially* in their competence, such as a hospital or a passenger ship, a reasonable man will want the most competent people to have authority over the matters on which they are most competent" (Dahl 1970, p. 58, emphasis added). In other words, the more closely one's situation resembles a perilous sea voyage, the stronger the rationale for placing power and authority in the hands of the few who know how to run the ship.

Ecological scarcity appears to have created precisely such a situation. Critical decisions must be made. Although it is true that most of them are "trans-scientific" in that they can only be made politically by prudent people, at least the basic scientific elements of the problems must be understood reasonably well before an informed political decision is possible. However, the average person has neither the time to inform himself or herself nor the requisite background for understanding such complex technical problems. Moreover, many people are simply not intelligent enough or well enough educated to grasp the issues, much less the important features of the problems. Even highly attentive and competent specialists do not always understand the problems fully. Even when they do (or claim to), they can almost always be found on both sides of any major question of public policy. (The dispute over nuclear-reactor safety is a prime example, with Nobelists lining up both for and against nuclear power.) Thus, even assuming that the politicians and people understand the issue well enough to ask the right questions, which experts should they listen to? Can they understand what the experts are saying? If we grant that the majority of the people probably will not understand and are therefore not competent to decide such issues, is it very likely that the political leaders they select will themselves be competent enough to deal with these issues? And even if they are, how can these leaders make authoritative decisions that impose heavy present costs or that violate popular expectations for the sake of future advantages revealed to them only as special knowledge derived from complicated analysis, perhaps even as the Delphic pronouncements of a computer?

Such questions about the viability of democratic politics in a super-technological age propel us toward the political thought of Plato. In *The Republic,* the fountainhead of all Western political philosophy, Plato argued that the polity was like a ship sailing dangerous waters. It therefore needed to be commanded by the most competent pilots; to allow the crew, ignorant of the art of navigation, to participate in running the vessel would be to invite shipwreck. Thus the polity would have to be run by an elite class of guardians, who would themselves be guided by the cream of this elite, the philosopher–kings. As the quotation from Dahl suggests, to the extent that Plato's analogy of the ship of state approximates reality, his political prescriptions are difficult to evade. This is precisely why, from Aristotle on, those who have favored democratic rather than oligarchic politics have concerned themselves with keeping the political community small enough and simple enough so that elite rule would not be necessary for social survival. The emerging large, highly developed, complex technological civilization operating at or very near the ecological margin appears to fit Plato's premises more and more closely, foreshadowing the necessity of rule by a class of Platonic guardians, the "priesthood of responsible technologists" who alone know how to run the spaceship.

Such a development has always been implicit in technology, as the ideas of Saint-Simon suggest, but the need for it has become unmistakable in a crowded world living close to the ecological limits, for only through the most exquisite care can we avert the collapse of the technological Leviathan we are well on the way to creating. C. S. Lewis observed that "What we call Man's power over Nature turns out to be a power exercised by some men over other men with Nature as its instrument" (Lewis 1965, p. 69), and it appears that the greater the technological power, the more absolute the political power that must be yielded up to some people by the others. Thus we must ask ourselves whether continued technological growth will not merely serve to replace the so-called tyranny of nature with a potentially even more odious tyranny of people. Why indeed should we deliver ourselves over to a "priesthood of responsible technologists" who are merely technical experts and may well lack the excellence of character and deep philosophical understanding that Plato insists his guardians must possess in order to justify their rule? In fact, why accept the rule of even a genuinely Platonic elite possessed of both wisdom and expertise when all history teaches us that the abilities, foresight, and goodwill of mortal people are limited and imperfect? The technological response to ecological scarcity thus raises profound political issues, in particular one of the most ancient and difficult political dilemmas—*quis custodiet ipsos custodes?* "Who will watch the guardians themselves?"

Technology and the Path to a Brave New World

Modern humanity has used technology along with energy to try to transcend nature. We have seen that it cannot be done; nature is not to be transcended by a biological organism that depends on it. Worse, any attempt to do so will have momentous political and social consequences. Far from protecting us from painful and disruptive social changes, as the technological optimist is wont to claim, continued technological growth is likely to force such changes on us. We are, in fact, in the process of making the Faustian bargain without ever having consciously decided to do so. As a result, we appear to be traveling down the road to total domination by technique and the machine, to the "Brave New World" that Aldous Huxley (1932) warned was the logical end point of a hedonistic, high-technology civilization.*

Technology may not be inherently evil, but it does have side effects, and it does exact a social price. Moreover, in the hands of less-than-per-

---

* All the techniques of social control and biological manipulation forecast in Huxley's dystopian novel are being invented today in our laboratories (Cohen 1973; Delgado 1969; Holden 1973; Kass 1971, 1972; Skinner 1971). And well before these developments occurred, Huxley (1958) was himself appalled to witness in his own lifetime much of what he had imagined as taking place six or seven hundred years in the future.

21

Taming Leviathan: Macro-constraints and Micro-freedoms

The only escape from the political dilemma of ecological scarcity—authoritative rule or ecological ruin—is indicated in the Epigraph: If people exercise sufficient self-control of their passions, the fetters of external authority become unnecessary. Unfortunately, political history suggests that the level of moral restraint and altruism to be expected from the members of large, complex, mass societies is limited at best. These virtues are even less likely to be found in industrial civilization, for its citizens have been brought up to believe that satisfying their hedonistic wants is not only legitimate but positively virtuous. Besides, in complicated and highly interdependent societies, even the most willing citizen would not know how to be ecologically virtuous without a large amount of central direction and coordination. In other words, unless we return to face-to-face, simple, decentralized, small-community living—which may be a desirable long-term goal (so Chapter 8 will argue) but is hardly a short-term possibility—we are stuck with the problem of making authority palatable and protecting ourselves from those who would abuse their ecological guardianship.

Traditional political theory has proposed many answers to this problem. However, one basic principle stands out: If self-restraint is inadequate, macro-constraints are vastly to be preferred to micro-constraints, for the psychological differences between them are crucial. That is, limitations on our freedom that are indirect, remote, and impersonal are

fect human beings, technology can never be neutral, as its proponents too often claim; it can only be used for good or evil. Thus technological fixes are dangerous surrogates for political decisions. There is no escape from politics. As a consequence of ecological scarcity, major ethical, political, economic, and social changes are inevitable whatever we do. The choice is between change that happens to us as a "side effect" of ever-more-stringent technological imperatives and change that is deliberately selected as compatible with our values.

Unfortunately, at this point even total renunciation of technology as dangerous to our democratic health would not enable us to avoid all the

preferable to those that are direct, proximate, and personal. In the former case, the limitations become an almost invisible part of "the way things are," instead of obvious impositions. For example, modern humans feel generally free despite their nearly total submission to such powerful but faceless forces as technological change and the marketplace; the feudal peasant, by contrast, was so bound up in a web of direct personal obligations that he felt much less free, even though this web of obligations may have been in important respects less tyrannical in practice than the impersonal forces to which modern humanity is obliged to submit. Putting the matter more abstractly, the contemporary political philosopher Isaiah Berlin (1969) has defined freedom as the number of doors open to a person, how open they are, and upon what prospects they open. All other things being equal, then, the widest number of meaningful options brings the maximum of freedom; macro-freedom is the sum of the micro-freedoms available to us. Because the destruction of the commons leaves us with few meaningful options, some of the doors now available to us must be partly or even completely closed, but if we wish to preserve a sense of freedom, then this should be done in ways that limit the micro-freedoms, or close the doors of daily life, as little as possible.

Thus an effective way of making authority acceptable is to impose macro-constraints that encourage the behavior necessary to maintain a steady-state society but to leave individuals with a relative abundance of micro-freedoms that, when added up, give them an overall sense of freedom. How such a steady-state society might be "designed" will be discussed in Chapter 8.

political dilemmas described above. During the transition to any form of steady state one can envision, it would be imperative to minimize pollution and use resources as efficiently as possible, and this probably would mean greater centralization and expert control in the short term, even if the long-term goal is a technologically simple, decentralized society favorable to a democratic politics.

Even beyond the transition period, whether a steady-state society can be democratic is at least questionable. A society cannot persist as a genuine democracy unless the people in their majority understand technology and ecology well enough to make responsible decisions. And

22

The Ecological Contract

The Great Frontier and the Industrial Revolution unleashed forces that eventually destroyed the medieval political synthesis, which was based generally on the Heaven-ordained hierarchy of the "great chain of being" and specifically on the "divine right of kings." Changing economic conditions gradually transferred *de facto* political power from monarchs, priests, and nobles to the enterprising middle classes. Although at first the bourgeoisie acquiesced in continued autocratic rule and aristocratic patronage, it eventually tired of supporting what it came to see as unproductive social parasites; it overthrew the *ancien régime* and embarked on democratic self-rule, the only form of government that could be intellectually and practically reconciled with its new sense of individualism. Such major transfers of power must be theoretically and morally legitimated, and the "social contract" theory of government was devised to fulfill this need.

In essence, the theory of the social contract says that individuals are not part of a preexisting hierarchy to which they must unquestioningly adapt but rather are free to decide how they wish to be ruled. It is thus primarily concerned with how free and equal individuals (starting from an anarchic "state of nature") can come together to erect political institutions that will preserve their individual rights to the fullest extent yet also promote the social harmony they need to enjoy these rights in

although the technology of a frugal steady state should be more accessible to the average person's understanding than current technology is, the same may not be true of the ecological knowledge on which the steady-state society will have to be based. Intuition and common sense alone are of little help in understanding the counterintuitive complexity of the human ecosystem—and nowhere else can a little knowledge be so dangerous. Thus, although not intrinsically mysterious, ecology is esoteric in the sense that only those whose talents and training have equipped them to be the "specialists in the general" discussed in the Introduction are likely to possess the kind of competence that would satisfy Dahl's "reasonable man." The ecologically complex steady-state society may therefore require, if not a class of ecological guardians, then at least a class

peace. Ironically, the device of the social contract was used by Hobbes to provide secular support for monarchy, starting from the individualistic, hedonistic, and materialistic premises of the bourgeois world view. However, as it was later developed by Locke and Rousseau, the social contract became the foundation for popular sovereignty and liberal democracy (even Marxism has very deep roots in Rousseau's thought). The untrammeled individual was now king.

As a product of the Great Frontier, the theory of the social contract is fundamentally cornucopian: Nature's abundance being endless and inexhaustible, one has only to solve the problem of achieving social harmony through a just division of the spoils. Nature is thus external to politics. But these cornucopian premises have become as anomalous in an age of ecological scarcity as the divine right of kings was in the era of the Great Frontier and the Industrial Revolution. Ecology and politics are now inseparable; out of prudent self-restraint, if for no other reason, a valid political theory of the steady state will be obliged to give the same weight to ecological harmony as to social harmony. Thus, just as it was the task of the seventeenth- and eighteenth-century political philosophers to create the social-contract theory of government to take account of the new socioeconomic conditions and justify the political ascent of the bourgeois class, so it will be the duty of the next generation of philosophers to create an "ecological-contract" theory promoting harmony not just among humans, but also between humanity and nature.

of ecological mandarins who possess the esoteric knowledge needed to run it well. Whatever its level of material affluence, the steady-state society will not only be ostensibly more authoritarian and less democratic than the industrial societies of today (the necessity of coping with the tragedy of the commons would alone ensure that), but it may also be more oligarchic as well, with full participation in the political process restricted to those who possess the ecological and other competencies necessary to make prudent decisions.*

---

* In Chapter 8, we present the conditions for an ecological democracy which could avoid these consequences.

Hard Political Realities and a New Paradigm

In summary, scarcity in general erodes the material basis for the relatively benign individualistic and democratic politics characteristic of the modern industrial era. Ecological scarcity in particular seems to engender overwhelming pressures toward political systems that are frankly authoritarian by current standards, for there seems to be no other way to check competitive overexploitation of resources and to ensure competent direction of a complex society's affairs in accordance with steady-state imperatives. Leviathan may be mitigated but not evaded (see Box 21).

Ecological scarcity thus forces us to confront once again, perhaps in a particularly acute form, the hard realities and cruel dilemmas of classical politics, from which four centuries of abnormal abundance have shielded us. As a result, we shall have to reexamine fundamental political questions in the light of ecology and construct a new steady-state paradigm of politics based on ecological premises instead of on the individualistic, hedonistic, materialistic, and anthropocentric premises of bourgeois "social contract" theory (see Box 22). The alternative is to let the shape of the steady-state paradigm be decided for us by accepting the outcome of current trends toward technocracy.

Given current political values, this may not seem like much of a choice. However, the one sure thing is that current values and institutions will not be able to endure unchanged. Moreover, as we shall see in Chapter 8, the latitude of choice is wider than might be suspected; indeed, the crisis of ecological scarcity might actually be turned into a grand opportunity to build a more humane and genuinely democratic post-industrial society. In the next two chapters, we shall explore specific features of the American political economy to determine how well it is likely to cope with the challenges of ecological scarcity.

5

The American Political Economy I: Ecology Plus Economics Equals Politics

Having discussed the politics of scarcity in general, we now turn to the particulars of the American situation. As difficult as it sometimes is to keep economics and politics separate, especially in this country, we shall discuss the economic aspects of political economy in this chapter and take up the more political aspects in the next. However, both chapters share an approach different from that taken in by most critiques of the American system. We are not interested here in whether the system falls short of the democratic ideal of freedom, equality, and justice but only in whether it is likely to be able to surmount without fundamental change the challenge of ecological scarcity. To this question both chapters give essentially the same answer: Ecological scarcity undercuts the basic laissez-faire, individualistic premises of the American political economy so that current institutions are incapable of meeting the challenges of scarcity. What is needed is a new paradigm of politics.

Market Failures and Social Costs

As noted in the Introduction, at least some critics of the limits-to-growth argument rely heavily on the market price mechanism to ensure a smooth, gradual transition to the steady state whenever it becomes necessary. They believe that as the costs of fuels and materials rise owing to scarcity, and as the costs of pollution control increase, further growth will become uneconomic; the steady state will therefore be ushered in automatically by market processes. In fact, however, in its current form the competitive market system is an environmental villain—part of the problem that must be solved, rather than the solution. Let us examine how the market fails to deal appropriately with common-pool resources, resource depletion, and other aspects of ecological scarcity.

According to Adam Smith, self-interested participants in a competitive marketplace will be unwittingly led to promote the common good by the "invisible hand" of the market. That is, with consumers and producers acting rationally to maximize their own gain, the market will automatically allocate resources with greatest efficiency and generate a maximum of individual and social prosperity. Thanks to the invisible hand, self-seeking individuals, despite the lack of any intention to do so, will benefit their fellows as they enrich themselves. Smith therefore argued for a laissez-faire, competitive market system of economics.

As we have seen, the premise of abundance necessary to support Smith's contention has vanished. Thanks to ecological scarcity, rational self-seeking individuals, despite the lack of any intention to do so, *harm* their fellows as they attempt to enrich themselves. As steady-state economist Herman Daly (1973, p. 17) aptly puts it, the invisible hand has turned into an "invisible foot" that threatens to destroy the common good with pollution and other "external diseconomies" or "externalities," the economist's terms for the social costs of production that are not accounted for in the price mechanism. In fact, the problem of the invisible foot is simply the economic version of the commons problem discussed in the preceding chapter. Individuals rationally seeking gain (or at least non-loss) are virtually compelled by the logic of the marketplace commons to make economic micro-decisions that are aggregated by the invisible foot into an ecological macro-decision that is increasingly destructive for the society as a whole—and therefore, paradoxically, for the individual as well. Thus an unregulated, competitive, laissez-faire market system, in which all have access to the economic commons and in which common-pool resources are treated as free goods, has produced a tragedy of the commons: the overuse, misuse, and degradation of resources on which we all depend for ecological health and economic wealth.

Other properties of a free-for-all system of wealth getting strongly reinforce its tendency to destroy the commons. For one thing, market decisions are inevitably short-sighted, because the economic value of the future is understated or "discounted." Future values are usually discounted at the interest rate available to a prudent investor; at a 7%

interest/discount rate, the investor would just as soon have $100,000 now as $800,000 in 30 years. Why? Because if he or she invests the $100,000 at 7%, it will be worth $800,000 in 30 years. The investor will have the same amount of money and will have run little or no risk to get it. Similarly, at the same interest/discount rate, a resource that 30 years from now will be worth $800,000 has a present value of no more than $100,000. In fact, for all practical purposes, costs and benefits more than 20 years in the future are discounted to zero; owing in part to such additional factors as the prevailing rate of return on capital, it is a rare economic decision maker whose time horizon extends more than 10 years into the future. Thus critical ecological resources that will be essential for our well-being even 30 years from now not only have no value to rational economic decision makers, but scarcely enter their calculations at all. They are therefore likely to make decisions that irreversibly deplete or destroy vital resources (especially since each decision maker realistically fears that his or her own self-restraint would simply hand over to another the opportunity for profit). Thus, as Karl Marx put it a century ago, the watchword of market capitalism is *"Après nous le déluge,"* as entrepreneurs strive to maximize current benefits at the expense of the future.

An additional problem is that although the market price mechanism handles incremental change with relative ease, it tends to break down when confronted with absolute scarcity or even marked discrepancies between supply and demand. In such situations (for example, in famines), the market collapses or degenerates into uncontrolled inflation, because the increased price is incapable of calling forth an equivalent increase in supply.* In a famine, supply and demand are eventually brought into balance by death, emigration, or the *deus ex machina* of relief efforts—that is, by physical readjustments, not by the price mechanism. Thus the market is unlikely to preside over a smooth and trouble-free transition to a steady state, for the crisis of ecological scarcity involves absolute physical scarcities (lack of food, water, time, or human physiological tolerance for poisons) that mere money can remedy, if at all, only in part (and certainly not indefinitely or all at once). Indeed, shortages leading to rising prices

---

* To use the economist's terms, the market is splendid at coping with relative scarcity, shifting the burden of scarcity so that it is least uncomfortable (for example, by inducing substitution of one resource for another), but it is incapable of dealing with absolute scarcity except by raising prices in general—that is, through inflation. There is reason to suspect that this is the underlying cause of the inflation existing in the industrialized world.

may simply increase the incentives to exploit remaining resources heed-
lessly in a desperate attempt to meet current demand. Rising prices, then,
are not likely to induce timely and appropriate responses to ecological
scarcity, and they will certainly not preserve resources from exhaustion
and degradation. In fact, they may simply intensify the pressures that are
producing the current mode of ecological overshoot.

There are other reasons why the market may fail to respond smoothly and appropriately to the price signals generated by ecological scarcity.
For one thing, scarcity tends to induce competitive bidding and preemp-
tive buying, which lead to price fluctuations, market disruption, and the
inequitable or inappropriate distribution of resources. For example, unwarranted fears that oil supplies might be scarce after Iraq invaded Kuwait
in 1990 caused speculative bidding and major rises in the price of oil,
resulting in oil shortages in Eastern Europe and the developing world,
whose economies could not bear the increased costs. Similarly, despite an
alleged timber shortage in the continental United States (used by logging
interests as an argument against controls on ecologically destructive
practices), the primeval forests of Alaska, one of the few remaining large
sources of high-grade timber in the United States, are being intensively
logged for export to Japan, instead of being preserved for our own future
needs (Harnik 1973).

Economists also assume that consumers will respond in a reasonably
elastic fashion to rising prices due to ecological scarcity. However, this is
by no means obvious. For one thing, many consumer decisions are based
on factors other than price. For instance, very great differentials in cost are
not enough to lure most drivers out of their cars into mass transit, because
such factors as prestige and convenience are more important to the
consumer than mere price. Thus only prohibitive increases in cost would
be likely to reduce significantly the private ownership and use of
automobiles. In addition, prior investment decisions may lock consumers
into using a specific resource, regardless of price. Homeowners and
industries that use natural gas for space or process heat, for example,
cannot easily switch to substitute forms of energy in the short term, no
matter what happens to the price of natural gas. Even very high prices,
then, may not be sufficient to keep the consumption of ecologically
damaging goods and the use of non renewable resources at a level that is
socially optimal.

Additional problems of a more structural nature abound. In the first
place, in a market economy—where the market is *the* economic tool, not
just one among others—all the incentives of producers are toward
growth and the wasteful use of resources. It is in the interest of producers
to have a high-throughput economy characterized by high consumption

through product proliferation and promotion, rapid obsolescence, and the like. It is just not economically advantageous for a producer to make an indestructible, easily repaired, inexpensively operated car. If consumers were perfectly rational—that is, if they acted solely according to their economic advantage—they could no doubt oblige producers to turn out nothing else, but we know very well that consumers are not completely rational (about cars least of all) and that producers do everything in their power to exploit this irrationality to boost sales (for example, by using ads that play upon consumers' social and sexual insecurities). By comparison, the incentives to satisfy needs with minimum inputs of energy and material and the lowest real or long-term cost are quite weak, as is well exemplified by the entrepreneurial flight from passenger rail transportation. In their pursuit of economic advantage, producers can be expected to promote higher consumption and in general to exploit every opportunity to profit by not counting the ecological costs. And of course the growth orientation of the private sector is reinforced by the government, which uses its taxing, spending, and monetary powers to promote prosperity and full employment.

Conversely, producers lack significant market incentives to respond alertly and appropriately to many of the problems created by ecological scarcity. For example, it is simply not in the interest of oil companies or electric power companies to promote alternatives to the current fossil-fuel-based energy economy or to the centralized system of power production and distribution. In fact, for the purely "economic" person, the best of all possible worlds would be one in which people are almost literally dying for lack of what only he or she can supply. It is therefore entirely rational for entrepreneurs to let scarcity reach uncomfortable levels before innovating or bringing new resources to market. Thus the market price system is unlikely to favor far-sighted, much less public-spirited, investment decisions or to promote ecologically sound alternatives to current technologies, especially because some of the logical alternatives, such as alternatively fueled automobiles, could reduce the dependence of consumers on producers. At the very least, producers are likely to wait until demand builds up, and they are ensured of large profits before they invest heavily in such alternatives as fusion, which may require a great deal of time and money to develop to the point of commercial viability. Thus there are major structural obstacles to innovation and investment that will seriously impede response to the pressures of ecological scarcity (particularly in regulated monopoly industries, where real market competition does not exist). At best, market solutions will lag well behind the rapidly developing real-world problems of ecological scarcity.

In short, an unregulated market economy inevitably fosters accelerated ecological degradation and resource depletion through ever-higher levels of production and consumption. Indeed, given the cornucopian assumptions on which a market system of economics is based, it could hardly be otherwise; both philosophically and practically, a market economy is incompatible with ecology.*

The New Economics Is Mostly Politics

If the market in its current form has so many serious environmental liabilities, *which is not disputed by the vast majority of economists,* why do we rely so heavily on it to save us from the consequences of ecological scarcity? The answer seems to be that although they tend to talk as if it were already an accomplished fact, those who put forward this solution are really talking about a market that does not yet exist—a market whose price mechanism has been thoroughly overhauled to eliminate at least some of the liabilities we have noted. In short, those who argue for the market as an economic solution to ecological scarcity are actually urging a political solution, for all of these reforms will require explicit and deliberate social decisions, as a brief review of the proposed changes will indicate.

A number of economic devices have been proposed to mitigate or eliminate the degradation of common-pool resources and to promote the provision of public goods, such as clean air, or the careful husbandry of nonrenewable resources. These devices are principally administrative fiat, the creation of property rights in common-pool resources, effluent or pollution taxes, the auctioning of pollution rights, severance taxes on the use of resources, and the creation of "public markets." Although their technical merits are debated by economists, there is general agreement on the main outlines of a market solution. For example, governments can simply forbid emissions above a certain level, which is the current U.S. policy with respect to automobiles, but economists tend to believe that direct administrative controls are cumbersome and inefficient (in the

---

* As the discussion of a "thermodynamic" economy (in Chapter 2) indicated, ecological economics is grounded on real physical flows rather than on money. An ecological theory of economics would therefore resemble the premodern "physiocratic" or nature-based economic theories that, subsequent to the opening up of the Great Frontier, were eclipsed by the capital- and labor-based economic theories inspired by Adam Smith. Other aspects of an ecological economics will be discussed in Chapter 8 in connection with stewardship and "right livelihood."

23

Determining the Optimal Level of Pollution

The costs of cleaning up the last remaining pollutants is very high.

Optimal level of pollution

Total social costs

This curve is the sum of the two bottom curves.

Units of social cost

Social cost of cleaning up pollution

Social costs of pollution

As pollution increases, the social costs of pollution rise rapidly.

Lowest social cost

Units of pollution

Nature can handle low levels of pollution. If we remove these pollutants, the costs of cleaning up are higher than their social costs.

If we allow more pollution than the optimal level, the social costs of pollution is higher than the cost of cleaning it up.

Economists calculate the estimated optimal level of pollution by plotting a curve reflecting the estimated social costs of pollution against a curve reflecting the social costs of cleaning it up. The sum of the two curves is the total social cost. The lowest point on this third curve is the optimal level of pollution. Source: Miller 1990, p. 581.

economic sense) in that they are not likely to provide the necessary amount of control at the least cost. Alternatively, governments can award environmental property rights to persons and corporations, thus creating a framework for bargaining, negotiation, exchange, and, if all else fails, litigation (none of which can take place effectively over property that nobody owns). According to this view, once the commons is made into private property, market forces and the legal system would work to preserve rather than degrade it (see Box 24). However, economists fear that giving away property rights in the commons to producers would be a tremendous windfall for them and that the dispersed citizens would not be able to organize effectively to buy amenity. And they fear that if the public at large had individual "amenity rights" (so that a producer would have to buy pollution rights from each person affected by its effluent), the result would be economic paralysis. While recognizing that both these devices may be effective and even necessary in some cases, environmental economists generally favor restructuring the system of market prices with various taxes that would oblige producers to "internalize" (that is, incorporate into prices) the environmental costs of production.

In principle, this is a simple and just solution. Government agencies could calculate the public damage caused by the wastes of a producer and

24

Marketing Pollution Rights

One way to control pollution is to sell rights that allow pollution up to the estimated optimum level and then to allow the polluters to trade these rights. The 1990 Clean Air Act experiments with a market approach, setting a cap on sulfur dioxide emissions in the year 2000. Thereafter, if a utility wants to increase its emissions, it must pay another utility to reduce its emissions by an equivalent amount. Congress hopes that this market mechanism will achieve the desired level of sulfur dioxide reduction at least cost. To take a simple example, if it costs utility A $10 million to achieve a given level of pollution control, but costs utility B $5 million to do the same, A may be able to negotiate an agreement with B to pay it $7.5 million dollars if B agrees to spend $5 million of these dollars to offset A's excess pollution. Both utilities "make" $2.5 million on the transaction and, assuming good faith by both parties, achieve the desired pollution control.

then charge that individual or company the appropriate amount as an effluent or pollution tax. Similarly, the government could levy severance taxes designed to promote more rational use of resources, especially virgin nonrenewable resources.* The producer would internalize these new costs of doing business, which would be reflected in the price it charged for its products, so the consumer would pay the real cost of the product—that is, not only the costs of production but also the associated ecological and social costs. Those who ultimately benefited from the use or consumption would justly pay what they should and would thus be more likely to make responsible market decisions, such as reducing consumption if the price were too high. Moreover, if the level of effluent or severance tax were carefully set, the producer would have an incentive to make its operations more efficient in an ecological and social sense— that is, cleaner and thriftier. If it did so, it could pay less tax, lower its prices, and win customers away from less efficient competitors. The theoretical outcome of the well-administered internalization of environmental costs is thus a transformed market that responds readily to the pressures of ecological scarcity.

Unfortunately, this conceptually simple and reasonably equitable approach is far from easy to implement. Some externalities can be assigned a price with relative ease—for example, extra laundry bills or house painting attributable to air pollution from a nearby factory. However, most cannot be readily quantified. Consider the health effects of air pollution. It is almost impossible to know who suffers, to what degree, and from what amount of which agents. Besides, what is the economic cost of life? Of a reduced life span? Of the risk of contracting emphysema? Of being forced by smog to stay inside? What criteria can be used to establish severance taxes on the use of nonrenewable resources—or even on many renewable ones, for that matter, biologists often being uncertain what level of exploitation will provide a sustained yield? How do we put a price on the externalities involved in the unlikely but possible disaster, such as a serious nuclear accident? How do we decide whether irreversible development that forecloses future options, such as building houses on prime farmland, should be undertaken at all, much less what price should be assigned to it?

More generally, what about such social costs of industrial growth as increased commuting time and mental stress, to say nothing of the further

---

* Alternatively, the government could auction off limited pollution or exploitation rights, in effect letting the market establish the tax. This would save much of the cost of information gathering and tax calculation but, among other drawbacks, would give an advantage to those with the greatest market power.

25

Technology Assessment

Just as laissez faire in economics generates environmental costs, so too "laissez innover," or the unrestrained freedom to innovate technologically, generates social costs. In both cases these costs have reached unacceptable levels, and remedies are being sought. If internalization is the economists' answer to environmental costs, "technology assessment" is the technologists' cure for the social costs of laissez innover. The aim of both is the same—to make it no longer possible for individuals and groups to make micro-decisions that produce a macro-decision inimical to the common interest. As might be expected, therefore, technology assessment encounters the same kinds of problems as internalization, but in a more acute form, because the issues involved are much broader and even harder to analyze (especially in terms of quantifiable costs and benefits) than those involved in internalization. Thus decisions about the introduction of new technology are inevitably more political (or "trans-scientific," to use the technologists' term) than those confronted by environmental economists.

The heart of the difficulty is something already familiar to us from previous discussion. As Allen Kneese, a leading environmental economist and expert on cost–benefit analysis, says,

erosion of organic community life? How do we decide whether depletion of resources in the present should proceed if it impoverishes future generations, and to what extent should the cost of future depletion be discounted? We have no way of knowing at present. Indeed, even with perfect information, the economists could not answer most of these questions, for they involve political, social, and ethical issues, not the issue of efficient resource allocation that neoclassical, marginalist economics was designed to handle. They are "trans-economic" questions.*

---

* To take but one example of the difficulty of such assessments, a 1984 EPA cost–benefit analysis of a proposed revision of the Clean Air Act determined that the net benefits over the net costs ranged from −$1.4 billion to +$110 billion, depending on what values were assigned to human life, human health, and a cleaner environment (Miller 1990, p. 583).

It is my belief that benefit–cost analysis cannot answer the most important policy questions associated with the desirability of developing a large-scale, fission-based economy. To expect it to do so is to ask it to bear a burden it cannot sustain. This is so because these questions are of a deep ethical character. Benefit–cost analyses certainly cannot solve such questions and may well obscure them [Kneese 1973, p. 1].

Unfortunately, almost all discussion of technology assessment completely overlooks this point. Just like the majority of economists, who evince an almost religious faith in the tools of neoclassical economics and the technical efficacy of the market mechanism, technologists tend to look upon technology assessment as an exercise in pure cost–benefit analysis that will avoid rather than require political and ethical decisions. But if the criteria and methods of analysis are narrowly technical and economic, then there is little doubt that the Faustian bargain, for example, will be found cost-effective. In its current form, therefore, technology assessment may indeed obscure rather than illuminate the most important questions connected with continued technological innovation.

Thus "the new economics is mostly politics" (Wildavsky 1967). Even though economic analysis and the market itself, which is a highly effective mechanism for efficiently and automatically allocating resources and for sending signals to economic decision makers, both have a definite contribution to make, society will have to use noneconomic criteria to decide which trade-offs are to be made between production and other elements of the quality of life. (The assessment of technology poses a similar problem; see Box 25.) In effect, because economics itself cannot produce what economists call a "social welfare function," a means of assigning monetary values to non-market goods and bads, such a function will have to be invented politically (see Box 26). Moreover, even an overhauled price system will not eliminate all the liabilities we have discussed, so that in any event, an expansion of direct government intervention in the economic process will be

26

Assigning Prices to Environmental Goods

Economists have given thought to how to assign prices to "externalities," or ecological goods and services having no market value. For example, the value of beautiful scenery or a national park or spotted owls can be estimated by asking people the maximum they would be willing to pay to see them, even if they don't actually have to pay it. Likewise, in doing cost-benefit analyses, economists have used the technique in determining how much people might be willing to pay to avoid loss of income and medical costs associated with illness or some medical symptoms.

But the validity of the technique varies with the subject matter. Among other things, when assessing the value of such "goods" as spotted owls or a species of frog, the people give answers that depend on how well informed they are. The technique also fails to incorporate long-term goals because future generations can't bid in the hypothetical market.

Economists have also given thought to how to incorporate uncertainty into the market system. One possibility is to require companies to post a bond equal to the current best estimate of the greatest potential environmental harm their product or service could cause. The bond would pay interest and would be returned if the company could show that its product would not harm the environment. On the other hand, if the product did cause harm, the bond money would be used to repair the damage and to compensate its victims. The technique would require polluters to assume the risks that society now assumes and would give them an incentive to minimize environmental harms.

But this proposal faces substantial political obstacles. Polluters have been powerful enough to obtain legal caps on their environmental liabilities (such as the financial limits to which nuclear power producers are subject in nuclear accidents) and to force society to assume environmental and health risks (for instance, pesticides of dubious safety are permitted in the marketplace until the government proves they are harmful).

needed (for example, to subsidize vital but risky research and development on alternative technologies). Thus, as even leading exponents of the market strategy acknowledge, our political system will be handed the uncomfortable and unwanted burden of making decisions hitherto left to the invisible hand. This means that open collective decision making on a scale never before attempted by our political institutions—as well as much more efficient, innovative, and timely government action in general—will be essential to the success of the market strategy as a means of responding to ecological scarcity.

The Political Costs of Internalization

Because it leans so heavily on politics, the market strategy of coping with ecological scarcity must also be evaluated in political terms. This will raise further questions about how well it can be expected to succeed in the real world of American politics, as opposed to the abstract world of economic analysis.

In the first place, letting the market adjust supply and demand can have such painful social and political consequences that governments have usually gone to considerable lengths to prevent the free play of market forces. Recession, for example, makes exceedingly bad politics. So will internalization, for to the extent that a market strategy is effective environmentally, it is bound to cause social and economic distress. The internalization of environmental costs usually means that people have to pay for what they used to get free. Pollution control, for instance, often does not make a production process more efficient, just more expensive; similarly, severance and pollution taxes increase costs without increasing the quality of goods. The result is a general rise in prices, and the standard of living is eroded by inflation—exactly what economic theory predicts must happen in the face of absolute scarcity.* Furthermore, as is well known, a rise in prices affects income groups selectively. The poor, those with fixed incomes, and in fact all who are not in a position to pass on the increased costs suffer disproportionately. In addition, if a policy of internalization is faithfully implemented, the price of owning certain highly desired but ecologically damaging goods, such as the automobile, seems likely to rise the fastest and highest. This would cause serious disruption of the economy, which depends heavily on mass ownership of the automobile. It would also intensify the maldistributive tendency noted above, with potentially explosive political results if large numbers of people are priced out of car ownership. In any case, ecological scarcity will have painful and disruptive effects on the economy, producing a

---

* In the United States, real wages have declined since 1972, and productivity has been stagnant for about a decade. Both trends are partially attributable to the effects of rising real resource costs, especially energy resources, which are rising as a result of their decreasing energy output/input ratios and the costs of internalizing some of their environmental harms.

lower standard of living as we usually conceive of it. Thus the question is whether it is desirable for the inevitable economic consequences of ecological scarcity to be distributed in a relatively laissez-faire fashion by the invisible hand or by a planned economic contraction designed to mitigate these side effects. This will be a prime political issue confronting the American polity as a consequence of ecological scarcity, so we shall return to it in the next chapter.

Making the Invisible Hand Visible

An additional political problem is the very openness and explicitness of the internalization process, for it will make public what has hitherto been hidden. As its name implies, one of the characteristic features of the invisible hand is that the workings of the economic process are largely concealed, and responsibility for the economic macro-decision that results from the summation and integration of many small economic micro-decisions is diffused. The outcome is due to "the market," not to any particular person or act. This greatly favors the entrepreneurs, who can pursue their own private profit while ignoring the public costs their actions impose on others. The result is "development." Capitalism is thus an economic system founded on hidden social costs, in which development (at least as we have experienced it) would not have occurred if all the costs had been counted in advance. For example, the gentry and the bourgeois clearly did very well by the Industrial Revolution, while the urban and rural masses suffered greatly from the disruptive side effects of development; one sees essentially the same process being repeated today in the countries of the Third World. We also saw in Chapter 3 that the petrochemical industry would not exist in its present form* if the public costs of its pollution had to be paid for by the industry. The extreme resistance of the atomic energy establishment to a full and frank debate on the merits of nuclear power suggests just how threatening full disclosure of costs and benefits is to those who have hitherto been able to hide behind the invisible hand of dispersed, laissez-faire decision making. In brief, honesty and "progress" may not be compatible.

That openness and explicitness are not welcomed by most policy makers is evident from the brief history of the National Environmental Policy Act of 1969 (NEPA). The Act's purpose was, in part, to "promote

---

* Society would undoubtedly want a small petrochemical industry to produce a few valuable products for which there are no substitutes (drugs, artificial hearts, and the like) even if, as is true today, the industry generated nearly as many pounds of hazardous wastes as pounds of product produced.

efforts that will prevent or eliminate damage to the environment and biosphere and stimulate the health and welfare of man." Section 102 of that Act requires all federal agencies to prepare an environmental-impact statement (EIS) on any of their activities that have a significant effect on the quality of the human environment. The statement must include the purpose and need for the proposed action, its probable environmental impact, and the impact of possible alternatives. In other words, the Act requires a full public accounting of the costs and benefits of any Federal action that might have environmental consequences that decision makers and the interested public should be made aware of. (Thirty-six states have also adopted legislation requiring similar statements for governmental— and even some private—projects likely to have a significant environmental impact.) The Act has been interpreted to be only procedural; that is, it established no criteria of acceptability for a proposed action to go forward. The United States Supreme Court ruled in 1980 that the National Environmental Policy Act requires agencies only to "consider" the environmental impacts of its projects. Under the decision, an agency can report that a proposed activity will cause the sky to fall by 2000 and still proceed with the project once it has satisfied the procedural requirements of the Act. Nevertheless, lower courts, in response to environmentalist suits, have forced government agencies to live up rather strictly to the requirement that a full-blown environmental-impact statement be prepared.

As a result, a number of projects have been abandoned when it appeared that an honest public accounting would be too damaging; others have been temporarily shelved for redesign to remove some of the more glaring ecological liabilities; and still others have been held up by the extensive paperwork and inter-agency consultation needed to complete an EIS. Furthermore, the statements have made splendid ammunition for environmental defenders during the regulatory proceedings that are often needed before a project can be carried out. The nuclear power industry, already saddled with relatively heavy procedural and legal requirements involved in site location and reactor licensing, was especially hard hit by the limited public accountability generated by EIS statements, but the Army Corps of Engineers and other development-oriented agencies have also found living up to their obligations under Section 102 to be a troublesome burden.

As the full consequences of Section 102 became apparent, an administrative and legislative backlash set in. In part, this was due to the admittedly expensive and time-consuming process of preparing a draft report, circulating it to interested agencies for criticism, and so forth. However, most of the distress was due precisely to the openness and explicitness of the process, which gives potential opponents (usually a public-interest or environmental action group, but occasionally a rival agency) an opportunity to prevent a project from being carried out. For the sponsoring agency and its legislative and industrial constituency, especially pork-barreling members of Congress, such an outcome is horrible to contemplate, for projects bring prestige, profits, political favor, and many other benefits. Predictably, there was an attempt to weaken NEPA legislatively. When this failed, essentially the same result was achieved by an administrative *modus vivendi,* supported by the United States Supreme Court.*

Instead of treating the process of preparing an EIS as an opportunity to improve the quality of their decision making, sponsoring agencies have turned it into *pro forma,* paperwork compliance with the requirements of Section 102 (Krieth 1973). The Supreme Court has sanctioned this practice, calling the NEPA's requirements "essentially procedural." The court has gone so far as to approve an agency's refusal to consider new and better information once it has completed its own EIS. For example, after the Army Corp. of Engineers issued an EIS concerning a proposed dam on a tributary of the Rouge River in Oregon, the Oregon Department of Fish and Wildlife and the federal government's own Soil Conservation Service found evidence that the environmental impact would be worse than the Army's EIS had predicted. The Army refused to consider the new information or supplement its own EIS, and the Supreme Court sanctioned that refusal. "When specialists express conflicting views," the court said, "an agency must have discretion to rely on the reasonable opinions of its own qualified experts" to support its preferred option. This is so "even if, as an original matter, a court might find contrary views more persuasive" (*Marsh v. Oregon Natural Resources Council* 109 S.Ct. 1851, 1861 (1989).

Fully aware of this minimal reading of NEPA, reviewing agencies often give EISs a perfunctory review, thus saving themselves the money and the staff time that would have had to go into a genuine study, and at the same time chalking up a political favor that can be collected when their own pet projects are passed around for criticism. Because only a few especially controversial projects are important enough to attract the attention of overworked and underfinanced public-interest groups, the vast majority of impact statements glide through with only the most

---

* The Supreme Court has consistently supported the executive branch's views of NEPA. An indication of the court's approach can be seen in the statistics: From 1975 to 1991, the court reviewed 11 NEPA cases, and in all 11, it narrowed the scope of the act. In the same time period, environmental groups asked the Supreme Court to review cases they had lost in lower courts 25 times; the Supreme Court refused to consider their appeals in all 25 cases.

cursory review. In fact, even some highly controversial projects receive pitifully inadequate reviews. Agencies also seek to avoid the larger substantive issues their projects raise by using, as much as possible, narrow and exclusive definitions of *project* and *environmental impact*. For example, when the Department of the Interior engaged in a regional study of leases, mining rights, rights of way, and the distribution of scarce water resources in connection with its plans to approve coal strip mining in the Northern Great Plains, the Court of Appeals ruled that the Interior Department should decide whether to prepare an EIS for the region before any individual strip-mining projects started. But the Supreme Court reversed the decision of the Court of Appeals, accepting the Interior Department's claim that it had no "program" or "proposal" for the region on which to prepare an EIS at that time.

Finally, in the case of the Trans-Alaska Pipeline, a case that arose before the Supreme Court had established that the NEPA is essentially procedural, NEPA was simply set aside. When the Department of the Interior, despite intense White House pressure, could not produce an EIS favorable to the pipeline yet plausible enough to stand up in court, the administration simply persuaded Congress to pass legislation exempting the pipeline project from full compliance with NEPA requirements (Carter 1973; Odell 1973). In Chapter 2, we discussed the environmental havoc spawned by that decision.

Thus, although NEPA has caused some ill-conceived projects to be aborted, it has not resulted in full, open, and explicit environmental accountancy, for agencies have fulfilled only the letter of the law (as minimally interpreted by the Supreme Court) while evading the spirit, and in one important instance even the letter of the law was deliberately set aside. The reason is simple: No project sponsor or developer wants to let it be known that its gain may be the community's loss. Full disclosure of the kind of information needed to internalize the costs of production and make intelligent decisions on future development is deeply threatening to the industrial order and to a political and economic system that has thrived on the invisibility of the invisible hand. One must therefore question whether the openness and explicitness needed to make internalization work can be achieved with our current institutions. At the very least, there will have to be some painful readjustments in our economic mode of life, which is based on hidden costs, and in the administrative and political process allied with it.

Into the Political Cockpit

Internalization would be resisted by important economic interests precisely to the degree that it was effective. Two leading proponents of this approach explain why:

> A system of pollution charges…would establish the principle that the environment is owned by the people as a whole and that the polluters must pay for the privilege of using part of the environment for waste disposal. Such massive transfers of "property rights" and the wealth they represent seldom occur without political upheaval (Freeman and Haveman 1972).

Implicit in any effective program of internalization or environmental management, therefore, is a deliberate reversal of a 200-year-old bias in favor of development and growth. Toward this end, the government will be required to take jealously guarded privileges away from, and impose heavy new obligations on, *some of the most important and powerful actors in the political system.* For example, companies extracting virgin materials now benefit from a substantial tax break in the form of a depletion allowance; in the new scheme of things, not only will this depletion allowance be taken away but additional pollution and severance taxes will be imposed. It is apparent that producers have little incentive to cooperate in a program of internalization, no matter how well and fairly administered, that threatens drastically to curtail both profits and power.

The battle over stricter emission and efficiency (milage) standards for automobiles offers a foretaste of the kind of long political struggle that would become pervasive as Washington tried to force producers to internalize ecological and social costs. Vital issues are at stake. Oil companies and the automobile industry see a threat in higher-priced gasoline, alternative fuels, and smaller automobiles. They believe that if the current trend is allowed to proceed to its logical conclusion, automobiles will eventually be luxury items again, with devastating effects on profits. Labor, too, sees the threat implicit in controls on pollution, and even consumers do not desire goods that are more expensive but no better in terms of utility. In short, internalization will inhibit if not prevent continued growth, and nobody really wants material growth to end. At the very least, everybody wants somebody else to bear the costs of restructuring the economy.

With important interests so clearly opposed, it seems doubtful that a program of internalization can be fully, fairly, and efficiently administered. At best, there will be considerable lag as time-consuming political battles are fought out in the legislature and the courts, and economists agree that substantial delay in internalizing costs will be fatal to the market strategy. At worst, internalization will be only partially (as well as belatedly) applied, and primary reliance on politically more feasible but economically less efficient measures, such as direct regulation, will shackle the market with a jumble of controls that will increase the irrationality of the price mechanism without removing any of its environmental liabilities.

The market strategy thus seems likely to founder on politics, for the attempt to reform the price mechanism will transfer from the invisible hand to the highly visible political realm decisions on matters of critical importance to major interests in the society. If the invisible hand must be made visible and obedient to some explicit conception of the common interest, then this will inevitably bring about a basic change in the character of American government, which has relied heavily on the market mechanism. Laissez-faire economics has been, in effect, a surrogate for politics. With its demise, government seems likely to turn into a political cockpit for competing economic interests fighting desperately for survival in an age of ecological scarcity.

Farewell to "Economic Man"

The most fundamental concepts the system has habitually used to frame its decisions are being called into question. A laissez-faire market system of economics reflects the values of "economic man," and these values themselves, not just laissez-faire institutions, have become pernicious. For example, to a purely economic person there is no higher value than the individual wants of those living today; in pursuit of these wants it is economically optimal to keep growing until one further increment of growth will precipitate ecological catastrophe (Pearce 1973). The ultimate consequence of such a policy of ecological brinkmanship would, of course, be ruin. But who cares about ultimate consequences? In fact, any purely economic person *must* ignore the interests of posterity, for it has no agent he or she can bargain with in a market place and nothing of economic value to offer the person of today. It is an economic fact that posterity never has been, and never will be, able to do anything for us. Posterity is, therefore, damned if decisions are made "economically."

Thus the current paradigm of political economy, which is based on the primacy of economics, must give way to one based primarily on politics. The market will remain an essential tool for performing vital economic tasks, but it will cease to be the dominant mode of allocating social values. Similarly, economics, instead of being the master science it has been since the beginning of the Industrial Revolution, will be reduced to the more modest but still important role of handmaiden to ecological politics, supporting the material goals of the polity and managing the human ecological household in a way that respects the laws of nature and the long-term interests of humanity.

6

The American Political Economy II: The Non-politics of Laissez Faire

The preceding chapter showed that the invisible hand is no longer to be relied on for social decisions; we shall be obliged to make explicit political choices in order to meet the challenges of ecological scarcity. This is an embarrassing conclusion, for we Americans have never had a genuine politics—that is, something apart from economics that gives direction to our community life. Instead, American politics has been but a reflection of its laissez-faire economic system.

The Political Functions of Economic Growth

From our earliest colonial beginnings, rising expectations have been a fundamental part of the American credo, each generation expecting to become richer than the previous one. Thanks to this expectation of growth, the class conflict and social discontent typical of early nineteenth-century Europe were all but absent in America; politics was accordingly undemanding, pragmatic, and laissez-faire. Thus, said Alexis de Tocqueville in his classic study of American civilization *Democracy in America,* we were indeed a "happy republic."

Growth is still central to American politics. In fact, it matters more than ever, for the older social restraints (the Protestant ethic, deference, isolation) have all been swept away. Growth is the secular religion of

American society, providing a social goal, a basis for political solidarity, and a source of individual motivation. The pursuit of happiness has come to be defined almost exclusively in material terms, and the entire society—individuals, enterprises, the government itself—has an enormous vested interest in the continuation of growth.

The Economic Basis of Pragmatic Politics

Growth continues to be essential to the characteristic pragmatic, laissez-faire style of American politics, which has always revolved around the question of fair access to the opportunity to get on financially. Indeed, American political history is but the record of a more or less amicable squabble over the division of the spoils of a growing economy. Even social problems have been handled by substituting economic growth for political principle, transforming non-economic issues into ones that could be solved by economic bargaining. For example, when labor pressed its class demands, the response was to legitimize its status as a bargaining unit in the division of the spoils. Once labor had to be bargained with in good businesslike fashion, compromise, in terms of wages and other costable benefits, became possible. In return for labor's abandonment of uncompromising demands for socialism, others at the economic trough "squeezed over" enough for labor to get its share. Similarly, new political demands by immigrants, farmers, and so on were bought off by the opportunity to share in the fruits of economic growth. The only conflict that we failed to solve in this manner was slavery and its aftermath, and it is typical that once the legitimacy of black demands was recognized in the 1960s, the reflex response was to promote economic opportunity via job training, education, "black capitalism," and fair hiring practices—that is, the wherewithal to share the affluence of the envied whites. If blacks prosper economically, says our intuitive understanding of politics, racial problems will vanish.

As a political mode, economic reductionism has many virtues. Above all, it is a superb means of channeling and controlling social conflict. Economic bargaining is a matter of a little more or a little less. Nobody loses on issues of principle, and even failure to get what you want today is tolerable, for the bargaining session is continuous, and the outcome of the next round may be more favorable. Besides, everybody's share is growing, so that even an unfair share is a more-than-acceptable bird in the hand. Most people understand that in a growth economy, individuals or groups have more to gain from increases in the size of the enterprise as a whole than from any feasible change in distribution. Furthermore, people have gotten what was of primary interest to them—access to income and wealth—and with their chief aim satisfied, they were able to repress desires for community, social respect, political power, and other values that are not so easily divisible as money.

This characteristic style of conflict resolution presupposes agreement on the primacy of economics and a general willingness to be pragmatic and to accept the bargaining approach to political and social as well as economic issues. Unfortunately, the arrival of ecological scarcity places issues on the political agenda that are not easily compromisable or commensurable, least of all in terms of money. Trade-offs are possible, of course, but environmental imperatives are basically matters of principle that cannot be bargained away in an economic fashion. Environmental management is therefore a role for which our political institutions are miscast, because it involves deciding issues of principle in favor of one side or another rather than merely allocating shares in the spoils. Worse, a cessation or even a slowing of growth will bring opposing interests into increasingly stark conflict. Economic growth has made it possible to satisfy the demands of new claimants to the spoils without taking anything away from others. Without significant growth, however, we are left with a zero-sum game, in which there will be winners and losers instead of big winners and little winners. Especially in recent years, growth has become an all-purpose "political solvent" (Bell 1974, p. 43), satisfying rapidly rising expectations while allowing very large expenditures for social welfare and defense. Without the political solvent of growth to provide quasi-automatic solutions to many of our domestic social problems, our political institutions will be called on to make hard choices about how best to use relatively scarce resources to meet a plethora of demands. More important, long-suppressed social issues can now be expected to surface— especially the issue of equality.

Ecological Scarcity versus Economic Justice

To state the problem succinctly, growth and economic opportunity have been substitutes for equality of income and wealth. We have justified large differences in income and wealth on the grounds that they promote growth and that all members of society would receive future advantage from current inequality as the benefits of development "trickled down" to the poor. (On a more personal level, economic growth also ratifies the ethics of individual self-seeking: You can get on without concern for the fate of others, for they are presumably getting on too, even if not so well as you.) But if growth in production is no longer of overriding impor- tance, the rationale for differential rewards gets thinner, and with a cessation of growth it virtually disappears. In general, anything that diminishes growth and opportunity abridges the customary substitutes for equality. Because people's demands for economic betterment are not likely to disappear, once the pie stops growing fast enough to accommodate their needs, they will begin making demands for redistribution.

Even more serious than the frustration of rising expectations is the prospect of actual deprivation as substantial numbers of people get worse off in terms of real income as a result of scarcity-induced inflation and the internalization of environmental costs. Indeed, the eventual consequence of ecological scarcity is a lower standard of living, as we currently define it, for almost all members of society. One does not need a gloomy view of human nature to realize that this will create enormous political and social tension. It is, in fact, the classic prescription for revolution. At the very least, we can expect that our politics will come to be dominated by resentment and envy—or "emulation," to use the old word—just as it has many times in the past in democratic polities.

To make the revolutionary potential of the politics of emulation more concrete, let us imagine that the current trend toward making automobile ownership and operation more expensive continues to the point where the car becomes once again a luxury item, available only to "the carriage trade." How will the average person, once an economic aristocrat with his or her own private carriage but now demoted to a scooter or a bicycle, react to this deprivation, especially in view of the fact that the remaining aristocrats will presumably continue to enjoy their private carriages?

Of course, such an extreme situation is probably a long way off (although many would be priced out of the market today if all the social costs attributable to the automobile were internalized). Yet it is toward such a situation that the rising costs due to ecological scarcity are pushing us. Already, in striking contrast to the not-too-distant past, the price of a detached house in the most populous areas of the country is more than the average family can afford to pay. Also, as the cost of food and other basic necessities continues to increase, less disposable income will be left for the purchase of automobiles and other highly desired goods. In sum, deprivation is inevitable, even in the short term.

This point has not been lost on advocates for the disadvantaged, who have already protested vehemently against the regressive impact of even modest increases in the cost of energy (through increased gasoline taxes, for example) and goods.* More generally, they fear that lessened growth

---

* For every $1.00 increase in the price of oil, about 78,000 jobs are lost in the United States. Yet if the gasoline tax were adjusted to pay the cost of all public subsidies to the automobile, Americans would have to pay $4.50 per gallon of gas (Schaeffer 1990, p. 15).

will tend to restrict social mobility and freeze the status quo, or even turn the clock back in some areas, such as minority rights.

The political stage is set, therefore, for a showdown between the claims of ecological scarcity on the one hand and socioeconomic justice on the other. If the impact of scarcity is distributed in a laissez-faire fashion, the result will be to intensify existing inequalities. Large-scale redistribution, however, is almost totally foreign to our political machinery, which was designed for a growth economy and which has used economic surplus as the coin of social and political payoff. Thus the political measures necessary to redistributing income and wealth such that scarce commodities are to a large degree equally shared will require much greater social cooperation and solidarity than the system has exhibited in the past.

They will also require greater social control. Under conditions of scarcity, there is a trade-off between freedom and equality, with perfect equality necessitating almost total social control (as was attempted in Maoist China). However, even partial redistribution will involve wholesale government intervention in the economy and major transfers of property rights, as well as other infringements of liberty in general, that will be resisted bitterly by important and powerful interests.

Thus either horn of the dilemma—laissez faire or redistribution—would toss us into serious difficulties that would strain our meager political and moral resources to or beyond capacity. American society is founded on competition rather than cooperation, and scarcity is likely to aggravate rather than ameliorate the competitive struggle to gain economic benefits for oneself or one's group. Similarly, our political ethic is based on a just division of the spoils, defined almost purely in terms of fair access to the increments of growth; once the spoils of abundance are gone, little is left to promote social cooperation and sharing. As Adam Smith pointed out, the "progressive state" is "cheerful" and "hearty"; by contrast, the stationary state is "hard," the declining state "miserable" (Smith 1776, p. 81). How well will a set of political institutions completely predicated on abundance and molded by over 200 years of continuous growth cope with the "hardness" of ecological scarcity?

The Non-politics of Due Process

This dilemma is only a specific instance of a more general problem. In many areas, the American government will be obliged to have genuine policies—that is, specific measures or programs designed to further some particular conception of the public interest. This will require radical changes, because in our laissez-faire political system, ends are subordinated to political means. In other words, we practice "process" politics as opposed to "systems" politics (Schick 1971). As the name implies, process politics emphasizes the adequacy and fairness of the rules governing the process of politics. If the process is fair, then, as in a trial conducted according to due process, the outcome is assumed to be just—or at least the best that the system can achieve. By contrast, systems politics is concerned primarily with desired outcomes; means are subordinated to predetermined ends.

The process model has many virtues. Keeping the question of ends out of politics greatly diminishes the intensity of social conflict. People debate the fairness of the rules, a matter about which they find it relatively easy to agree, and they do not confront each other with value demands, which may not be susceptible to compromise. However, by some standards, the process model hardly deserves the name of politics, for it evades the whole issue of the common interest simply by declaring that the "will of all" and the "general will" are identical. The common interest is thus, by definition, whatever the political system's invisible hand cranks out, for good or ill.

Of course, we have found that pure laissez-faire politics, like pure laissez-faire economics, produces outcomes that we find intolerable, but our instinct has always been to curb the social costs of laissez faire by reforms designed to preserve its basic features: We check practices that prevent the efficient or fair operation of the market rather than converting to a planned economy; we promote equal opportunity rather than redistributing wealth or income. Planning with certain ends in mind does take place in such a political system. Each separate atom or molecule in the body politic (individuals, corporations, government agencies, advisory commissions, and supreme courts) plans in order to maximize its own ends, and the invisible hand produces the aggregated result of action on these private plans. But the central government does not plan in any systematic way, even though its ad hoc actions—VA and FHA home loans, tax breaks for homeowners, and the like—do in a sense constitute a "plan" for certain outcomes—in this case, suburban sprawl.

In reality, "the American political system" is almost a misnomer. What we really have is congeries of unintegrated and competitive subsystems pursuing conflicting ends—a non-system. And our overall policy of accepting the outcome of due process means that in most particulars we have non-policies. Now, however, just as in economics, the externalities produced by this laissez-faire system of non-politics have become unacceptable. Coping with the consequences of ecological scarcity will require explicit, outcome-oriented political decisions taken in the name of some conception of an ecological, if not a political and social, common interest. What likelihood is there of this happening?

Who Dominates the Political Marketplace?

Critics of the American political system almost never question the necessity (or superiority) of process politics. If bad outcomes are generated, it must be because powerful interests dominate the political marketplace and prevent the will of the majority from being fully and fairly translated into outcomes. There has been, say the critics, a wholesale expropriation of the public domain by private interests (Lowi 1969; McConnell 1966). Nevertheless, although much of this criticism is incontrovertible, the general preferences of the American people are in fact quite well reflected in political output. People want jobs, economic opportunities, and a growing economy. Indeed, to the extent that the system has had a guiding policy goal at all, it has been precisely to satisfy the rising expectations of its citizens. Even if special interests have benefited disproportionately from the measures taken to promote this end, most of the benefit has been transmitted to the vast majority of the population. The problem, then, is not that our political institutions are unresponsive to our wills but that what we desire generates the tragedy of the commons.

Naturally, to the extent that our government is largely a brokerage house for special interests, the situation is much worse, because such interests have an even bigger stake in continued economic growth. But within a process system of politics, government decisions that consistently favor producer over consumer interests are all but inevitable, for the political marketplace is subject to the public-goods problem (Box 19). For example, those who have a direct and substantial financial interest in legislation and regulation are strongly motivated to organize, lobby, make campaign contributions, advertise, litigate, and so forth in pursuit of their interests. By comparison, the great mass of the people, who will be indirectly affected and whose personal stake in the outcome is likely to be negligible, have very little incentive to organize in defense of their interests. After all, the "right" decision may be worth $10 million to General Motors but will cost each individual only a few pennies. Thus those who try to stand up to special interests on environmental issues find themselves up against superior political resources all across the board.

The gross political inequality of profit and nonprofit interests is epitomized by the favorable tax treatment accorded the former. By law, tax-deductible donations cannot be used for lobbying or other attempts to influence legislation (for example, by advertising). Thus the nonprofit organizations that depend very heavily on donations are severely handicapped; if they lobby, they undercut their financial support. Businesses, by contrast, can deduct any money spent for the same purpose from their taxable income and pass on the remaining expense in the form of higher

prices. The public, both as consumers and as taxpayers, therefore sub-
sidizes one side in environmental disputes. Moreover, the law is self-
protecting, for public-interest groups cannot even lobby to have it
changed without losing their tax-exempt status.

Thus the outcome of the process of American politics faithfully reflects
the will of the people and their desire for economic growth. However, just
as in the economic marketplace, the public suffers from certain negative
externalities as a result of the inordinate political power of producer interests;
political power tends to be used to ratify and reinforce, rather than counter-
mand, the decisions of the economic market. In sum, the American political
system has all the drawbacks of laissez faire, wherein individual decisions add
up to an ecologically destructive macro-decision, as well as a structural bias
in favor of producers that tends to make this macro-decision even more
destructive of the commons than it would otherwise be.

The Ecological Vices of Muddling Through

The logic of the commons is enshrined in a system of process politics
obedient to the demands of both consumer and producer for economic
growth. The ecological vices of this system are further intensified by the
decision-making style characteristic of all our institutions—disjointed
incrementalism or, to use the more honest and descriptive colloquial
term, "muddling through."

Incremental decision making largely ignores long-term goals; it
focuses on the problem immediately at hand and tries to find the solution
that is most congruent with the status quo. It is thus characterized by
comparison and evaluation of marginal changes (increments) in current
policies, not radical departures from them; by consideration of only a
restricted number of policy alternatives (and of only a few of the
important consequences for any given alternative); by the adjustment of
ends to means and to what is "feasible" and "realistic"; by serial or
piecemeal treatment of problems; and by a remedial orientation in which
policies are designed to cure obvious immediate ills rather than to bring
about some desired future state. Moreover, analysis of policy alternatives
is not disinterested, for it is carried out largely by partisan actors who are
trying to improve their bargaining position with other partisan actors.

Muddling through is therefore a highly economic style of decision
making that is well adapted to a pragmatic, laissez-faire system of politics.
Moreover, it has considerable virtues. Like the market itself, disjointed
incrementalism promotes short-term stability by minimizing serious conflict
over ultimate ends, by giving everybody something of what they want, and
by bringing bargained compromises among political actors, satisfying their

needs reasonably well at minimumal intellectual and financial cost. At the same time, it promotes the consensus and legitimacy needed to support public policy. It is also basically democratic; like the economic market, it reflects the preferences of those who participate in the political market (assuming that all legitimate interests can participate equally, which is not always the case). Disjointed incrementalism is also conservative in a good sense: It does not slight traditional values, it encourages appreciation of the costs of change, and it prevents overly hasty action on complex issues. It may also avoid serious or irreversible mistakes, for an incremental measure that turns out to be mistaken can usually be corrected before major harm has been done. Under ideal circumstances, disjointed incrementalism therefore produces a succession of policy measures that take the system step by step toward the policy outcome that best reflects the interests of the participants in the political market.

Unfortunately, muddling through has some equally large vices. For example, it does not guarantee that all relevant values will be taken into account, and it is likely to overlook excellent policies not suggested by past experience. In addition, disjointed incrementalism is not well adapted to handling profound value conflicts, revolutions, crises, grand opportunities, and the like—in other words, any situation in which simple continuation of past policies is not an appropriate response. Most important, because decisions are made on the basis of immediate self-interest, muddling through is almost guaranteed to produce policies that will generate the tragedy of the commons. It is perfectly possible to come up with a series of decisions that all seem eminently reasonable on the basis of short-term calculation of costs and benefits and that satisfy current preferences but that yield unsatisfactory results in the long run, especially because the future is likely to be discounted in the calculation of costs and benefits. In fact, that is just how we have gotten ourselves into an ecological predicament. Thus the short-term adjustment and stability achieved by muddling through is likely to be achieved at the expense of long-term stability and welfare.

A perfect illustration of the potential dangers of muddling through is our approach to global warming. As a result of millions of separate decisions made by industry and individuals, 6 billion tons of carbon dioxide are emitted into the atmosphere each year, and emissions are increasing by 3% annually. Yet no real congressional debate has occurred on whether to control these private decisions in order to reduce carbon emissions. Even worse, the executive branch blithely ignores the problem and advocates a more aggressive pursuit of the traditional energy and growth policies that have brought about the rise in carbon dioxide emissions. As a result, we go on unwittingly pursuing business as usual, making short-term calculations of costs and benefits, and bring upon ourselves the greenhouse effect almost by default.

Indeed, in its purest form, muddling through *is* policy making by default instead of by conscious choice—simply an administrative device for aggregating individual preferences into a "will of all" that may bear almost no resemblance to the "general will." Unfortunately, the contrasting synoptic, or outcome-oriented, style of decision making cannot be fully achieved in the real world because of limits to our intellectual capacities (even with computers), lack of information (plus the cost of remedying it), uncertainty about our values and conflicts between them, and time constraints, as well as many lesser factors. Moreover, in its pure form, synoptic decision making could lead to irreversible and disastrous blunders, obliviousness to people's values, and the destruction of political consensus. Thus some measure of muddling through is a simple administrative necessity in any political system.

However, we Americans have taken muddling through, along with laissez faire and other prominent features of our political system, to an extreme. We have made compromise and short-term adjustment into ends instead of means, have failed to give even cursory consideration to the future consequences of present acts, and have neglected even to try to relate current policy choices to some kind of long-term goal. Worse, we have taken the radical position that there can be no common interest beyond what muddling through produces. In brief, we have elevated what is an undeniable administrative necessity into a philosophy of government, becoming in the process an "adhocracy" virtually oblivious to the implications of our governmental acts and politically adrift in the dangerous waters of ecological scarcity.

Disjointed incrementalism, then, provides an almost sufficient explanation of how we have proceeded step by step into the midst of ecological crisis and of why we are not meeting its challenges at present. As a normative philosophy of government, it is a program for ecological catastrophe; as an entrenched reality with which the environmental reformer must cope, it is a cause for deep pessimism. At the very least, the level or quality of muddling through must be greatly upgraded, so that ecology and the future are given due weight in policy making. But goal-oriented muddling through comes close to being a contradiction in terms (especially within a basically democratic system). Moreover, incrementalism is adapted to status-quo, consensus politics, not to situations in which policy outcomes are of critical importance or in which the paradigm of politics itself may be undergoing radical change (Dror 1968, especially pp. 300-304; Lindblom 1965; Schick 1971, especially p. 158). Thus steering a middle course will be difficult at best, and it may not be possible at all during the transition to a steady-state society.

Policy Overload, Fragmentation, and Other Administrative Problems

Disjointed incrementalism is not the only built-in impediment to an effective response to ecological scarcity. In the first place, the growing scale, complexity, and interdependence of society make the decision-making environment increasingly problematic, for the greater the number of decisions (and, above all, the greater the degree of risk they entailed), the greater the social effort necessary to make them. Given the size and complexity of the task of environmental management alone, especially with the declining margin for ecological or technological error, there would be a danger of administrative overload. But the crisis of ecological scarcity is only one crisis among many—part of a crisis of crises that will afflict decision makers in the decades ahead (Platt 1969). An allied crisis of priorities also impends, as burgeoning demands for environmental cleanup, more and better social services, and so on compete for the tiny portion of government resources remaining after the "fixed" demands of defense, agricultural supports, and other budgetary sacred cows are satisfied, so that decision makers will simply lack sufficient funds to act effectively across the board. (In the United States, this has been true throughout the 1980s and early 1990s.) In addition, there may be critical shortfalls in labor power, especially technical and scientific labor power. In short, the problems are growing faster than the wherewithal to handle them, and political and administrative overload is therefore a potentially serious problem for the future, if not right now.

A second serious problem is fragmented and dispersed administrative responsibility. The agency in charge of decisions on air pollution, for example, usually has no control over land-use policy, freeway building, waste disposal, mass transit, and agriculture! Also, some elements of policy are handled at the federal level, whereas others belong to the state and local governments; the boundaries of local governments, especially, have no relationship to ecological realities. As a result, it frequently happens that one agency or unit of government works at cross-purposes with another, or even with itself, as in the old Atomic Energy Commission, which was charged with both nuclear development and radiation safety.* Furthermore, each agency has been created to perform a highly

---

* The Nuclear Regulatory Commission, whose mission is protection of the public from nuclear and radiation hazards, in practice also promotes nuclear energy. It has become, as do most government agencies, the captive of the industry it is charged to regulate.

specialized function for a particular constituency, which leads to a single-mindedness or tunnel vision that deliberately ignores the common interest. In brief, we have as many different policies as we have bureaus and it is difficult to get them to pull together.

A third major defect of our policy-making machinery is that decisions inevitably lag behind events—usually far behind. In part, the problem is that the decision makers' information and knowledge are deficient and out of date. Owing to the complexity and scope of the problems of environmental management, these deficiencies are either impossible to remedy or too costly. Thus, even if they are inclined to be forward-looking, decision makers are virtually obliged to muddle through critical problems with stopgap measures that provoke disruptive side effects. Much the larger part of the time-lag problem, however, is that the procedural checks and balances built into our basically adversary system of policy making can subject controversial decisions to lengthy delays. For example, Congress in 1977 amended the Clean Air Act to protect visibility in large national parks and wilderness areas; it took until 1990, however, for the EPA to issue draft regulations to implement the law. Thereafter, before the EPA issued its final regulations, the White House weakened them, sacrificing two-thirds of the visibility reductions that the EPA had proposed (Rauber 1991, p. 28). The matter may still end up in court. By presidential decree, the White House Office of Management and Budget subjects all EPA regulations to cost-benefit analysis. But the 1977 law requires power plants to install "the best available retrofit technology" to eliminate the air pollution impairing visibility in the parks. Opponents argue that in so far as cost-benefit analysis causes regulations to be issued that do not require the use of the best available technology, the use of such anlysis is illegal. They also argue a proper cost-benefit analysis, in any event, supports the original EPA draft regulations—that OMB simply manipulated the data to weaken them. This example suggests that the best we can expect in most cases is long wars of legal attrition against environmental despoilers. However, the adversary legal system is already having difficulty coping with environmental issues,* and there is some risk that environmental policy making may simply bog down in a morass of

---

* Increased volume is only part of the problem. The traditional legal machinery for redressing civil wrongs, designed for two-party litigation, is having trouble with standing to sue and other issues that crop up in the typical environmental suit, where society as a whole is one of the parties. Also, technology creates new situations faster than the courts can work out precedents, and much of the scientific evidence used in environmental litigation is of a probabilistic and statistical nature that ill accords with the standards of proof traditionally demanded by courts.

hearings, suits, countersuits, and appeals, as government agencies, business interests, and environmentalist groups use all the procedural devices available to harass each other. And even if total stalemate is avoided, there are bound to be significant delays—an ominous prospect now that an anticipatory response to problems has become essential for their solution.

Additional hindrances to effective environmental decision making abound. The narrowly rationalistic norms and *modus operandi* of bureaucracies, for example, are at odds with the ecological holism needed for the task of environmental management. History also shows that regulatory agencies tend to be captured by the interests they are supposed to be regulating, so that they rapidly turn into guardians of special interest instead of public interest. In addition, the institutions charged with environmental management are frequently so beholden to their own institutional vested interests or so dominated by sheer inertia that they actively resist change, employing secrecy, special legal advantages available to government agencies, and other devices to squelch the efforts of critics and would-be reformers (for example, Lewis 1972). In recent years, environmental decision making has been hindered by nonstatutory mechanisms established in the White House. The President's Council on Competitiveness, for example, is a non-statutory body that, after closed meetings with industry, has repeatedly forced the Environmental Protection Agency to rewrite regulations to make them hospitable to industry interests. As of this writing, the Council has successfully forced the EPA to gut four major pro-visions—some say the "pillars"—of the 1990 Clean Air Act (Weisskopf Sept. 1991, p. A1).* These difficulties suggest that the problem is not simply to overcome inertia and vested interest but rather to arrest the institutional momentum in favor of growth created by two centuries of pro-development laws, policies, and practices. This will require across-the-board institutional reform, not merely new policies.

In sum, administrative overload, fragmented and dispersed authority, protracted delays in making and enforcing social decisions, and the institutional legacy of the era of growth and exploitation are likely to obstruct timely and effective environmental policy making.

---

* Industry and administration lobbyists had tried to persuade Congress to adopt their substitutes for all four provisions when the legislation was being considered, but Congress had refused to adopt them. Environmentalists may thus be in a strong position to challenge these regulations as illegal, but even if they prevail, implementation of the law's requirements will be substantially delayed.

How Well Are We Doing?

None of the tendencies and trends we have just considered inspires much optimism that our political institutions at any level are adequate to the challenges of dealing with ecological scarcity. Although the final verdict is not yet in, this conclusion is certainly reinforced by the quality of their performance so far.

Energy policy is a good illustration. Despite a consensus that a coherent national energy policy is absolutely essential to avoid economic and social turmoil, a menacing international trade deficit, and even the compromise of its political independence, the United States has no genuine policy, much less a coherent one. Instead, the past decade has seen almost continual dithering and muddle and devotion to business as usual. In 1989 President Bush called for a long-term comprehensive policy, but what he proposed in 1991 was a mere grab-bag of favored projects of the oil, coal, gas, and nuclear power industries. Among these was more off-shore oil drilling, drilling in environmentally pristine areas, and the doubling of nuclear power capacity by 2030. The President proposed few conservation measures, only a minuscule increase in research on renewable energy, and support only for selected alternative fuels (ethanol and methanol but not for hydrogen, fuel cells, or electric vehicles). He proposed nothing to combat greenhouse gas emissions, except as an incidental consequence of his support for nuclear power.

When the Reagan administration made similar proposals during its years in office, it and Congress fought each other to a stalemate on energy conservation, environmental protection, and the relative support for renewable energy and fuels versus the support for fossil fuel and nuclear energy development. For example, a majority in Congress has thus far rejected oil drilling in the Arctic and other wilderness areas; instead it favors raising automobile fuel-efficiency standards to 40 miles per gallon by 2001, which would save 5 to 10 times the oil expected to be produced by oil drilling in the Arctic. But the Congressional majority is not veto-proof and the result is a stalemate. Continued stalemate, regrettably, sets the stage for an eventual general collapse of our energy economy because of either rising costs of petroleum or intolerable levels of pollution.

Similarly, our political institutions have so far conspicuously failed to meet the challenge represented by the automobile. The decline in air quality was sufficiently alarming to cause Congress to pass the Clean Air Act in 1970. For all its faults, this was a landmark piece of environmental legislation, and acting under the law's authority, government agencies forced emission control on a reluctant automobile industry. However,

Detroit several times succeeded in winning delayed compliance. Moreover, the air-quality standards mandated by Congress in the 1970 Clean Air Act simply could not be achieved through technology alone. Yet when the Environmental Protection Agency tried to impose on key municipalities pollution-control plans that would have penalized or restricted car use (through gas rationing and parking surcharges, for instance), the resulting political ruckus soon forced the EPA into retreat, and all pretense of meeting the original standards was abandoned. At the same time, Congress tried to control emissions through Corporate Fuel Economy Standards (CAFE). A 1975 law required manufacturers to raise the average efficiency of the cars they sold to 27.5 miles per gallon by 1985. Again, the executive branch granted the automobile industry many delays in complying with the law, and by 1990, efficiency standards had reached only 26.5 miles per gallon (which meant that the hoped-for reduction in pollution was nullified by a doubling, since the law was passed, of the number of vehicle miles driven). The 1990 Clean Air Act does tighten emission standards further and hopes to achieve its objectives via technological changes such as the use of reformulated gasoline in the nine most polluted metropolitan regions by 1995. But although the 1990 law will help, cleaning up the air and reducing greenhouse gas emissions cannot be achieved by more stringent emissions standards alone, because improvements in clean-air technology are more than eaten up by growth in the automobile fleet and increases in vehicle miles driven. In short, as was shown in Chapter 3, Americans must simply drive much less than they do now (with much more efficient vehicles) and use mass transit much more. Unfortunately, having allowed the automobile so completely to dominate our lives that to restrict its use would produce instant economic and social crisis, we are repeatedly reduced to the desperate hope that some kind of technological fix will turn up in time to prevent natural feedback mechanisms—extreme price rises, national bankruptcy, intolerable levels of air pollution—from taking matters out of our hands.

Thus in these and other critical areas we are failing to meet the challenges. Everybody wants clean air and water, but nobody wants to pay the price. Nor do we wish to give up the appurtenances of a high-energy style of life or to accept the major restructuring of the economy and society that would be needed to reduce greenhouse-gas emissions significantly. Even modest invasion of sacrosanct private property rights—for example, in the form of vitally needed land-use law—has also proved to be well beyond our current political capacity. In fact, since the beginning of the 1980s there has been considerable backlash and backsliding on environmental issues, leading to relaxed standards and blatant denial of problems. In short, although there has been genuine progress since environmental issues first became a matter for political concern, our political institutions have so far largely avoided the tasks of environmental management and have for the most part done too little too late in those efforts they have undertaken.

As we have seen, the basic institutional structure and *modus operandi* of the American political system are primarily responsible for this. Nevertheless, the lack of courage and vision displayed by the current set of political actors should not escape notice. Neither Congress nor the executive branch has provided real leadership or faced up to crucial issues. To the extent that they have acted, as in the area of pollution control, they have acted faintheartedly or, what is almost worse, expediently rather than effectively. Say what one will about the institutional impediments and the difficulty of the problems, it is hard to conclude that our political leaders are doing the job they were elected to do. But of course, the inability or reluctance of our political officials to act simply reflects the desires of the majority of the American people, who have so far evinced only modest willingness to make minor sacrifices (for example, to support and engage in recycling) for the sake of environmental goals, but no willingness to accept fundamental changes in their way of life (for example, to restrict development to areas where public transit is available, or to support and use public transit and drive less). Our public officials can hardly be expected to commit political suicide by forcing unpopular environmental measures on us. Until the will of the people ordains otherwise or fundamental changes are quite literally forced on us, the best we can expect is piecemeal, patchwork, ineffective reform that lags ever farther behind onrushing events.

The Necessity for Paradigm Change

Our political institutions, predicated almost totally on growth and abundance, appear to be no match for the mounting challenges of ecological scarcity. This is a shocking conclusion about a political system that was once regarded, even by many foreigners, as marvelously progressive. For all its faults, the virtues of the American political system are undeniable: It worked well for nearly two hundred years, and it was eminently just and humane by any reasonable historical standard. Unfortunately, the problems of scarcity that confront the system today are problems that *it was never designed to handle.* Many of its past virtues are therefore irrelevant; what we must now address are its equally undeniable failings in the face of ecological scarcity.

Efforts to patch up the current paradigm of politics with new modes of decision making and planning—or even with new policies—will not succeed. These can only delay, and perhaps intensify, the ultimate break-down. Only a new politics based on a set of values that are morally and practically appropriate to an age of scarcity will do (see Chapter 8). To achieve this new politics will require a revolution even more fundamental than that which created our nation in the first place, for the characteristic features of American civilization, not merely the nature of the regime, must be transformed. A great question stands before the American polity: Will we make the effort to translate our ideals of equality and freedom into forms appropriate to the new age of scarcity, or will we not even try, continuing prodigally to sow as long as we can and leaving the future to reap the consequences? Only time will tell whether the return of scarcity must inevitably presage retrogression to the classical scenario of inequality, oppression, and conflict, but one way or another, we Americans are about to find out what kind of people we really are.

7

Ecological Scarcity and
International Politics

The Comparative Perspective

Our principal focus so far in Part II has been on the American political system, specifically the strong market orientation of its political economy. However, as noted in the Introduction, the United States is only the most extreme version of modern industrial civilization, and the peculiarities of the American version of this civilization ought not to be allowed to obscure the wider implications of the analysis. Some problems may be uniquely American, but most are universal in one form or another. Let us therefore extend the analysis first to other nations and then to the international political arena. We shall find that the basic political dynamics and dilemmas of ecological scarcity discussed in Chapter 4 remain unchanged. Furthermore, much of the specific analysis of American institutions in Chapters 5 and 6 can in fact be applied, with appropriate modifications, to all developed and even many developing countries, capitalist and communist alike, as well as to the world in general. The crisis of ecological scarcity is thus a planetary crisis.

Western Europe

In Chapters 2 and 3, we compared various aspects of Western Europe's environmental situation to the United States. Western Europe's environmental problems are essentially the same in character and magnitude as those of the United States, and its governments on the whole seem to exhibit the same degree of capacity to deal with them. In some ways, however, Western European governments have dealt with their problems more effectively than this country has. For example, Western European governments, with the exception of the United Kingdom, have managed to reduce atmospheric emissions of sulfur oxides, nitrogen oxides, and particulates to lower levels per unit of gross domestic product than has the United States. On the other hand, emissions of nitrogen oxides, volatile organic compounds, and toxic wastes are rising in almost all countries. Surface waters in Western Europe suffer from low oxygen levels, eutrophication from nitrates and phosphates, and high levels of toxic metals. In addition, lakes are dying from acid deposition. Although isolated success stories have occurred (for example, Sweden reduced discharges of toxic metals into its waters from 1300 tons per year in 1972 to 55 tons per year in 1985), Western European waters are generally in the same condition as those in the United States—polluted, with little overall improvement. In certain respects, as a result of its greater density of population and industrial development, Europe's pollution problems are worse than our own. The contamination of the Baltic and Mediterranean Seas and of the Rhine River, heavy oil spillage from tankers and refineries, and the "acid rains" that fall particularly on Scandinavia are only some of the most notorious examples. Europeans also have more threatened species and declining forests than does the United States.

With respect to resources, Western Europe's predicament is clearly much worse. Even taking due account of the temporary respite that development of North Sea gas and oil has brought, Europe's long-term dependence on external sources of energy is far greater than our own; for example, Western Europe has nothing resembling America's vast coal reserves. Similarly, Western European mineral resources are almost negligible compared to actual and potential demand. Perhaps more critically, Western Europe as a whole is a major net importer of food and fiber, and the dependence of many European countries, such as Denmark and the Netherlands, on food imports is overwhelming (both to feed the populace and, ironically, to sustain energy-intensive agricultural systems that are mainstays of their economies). Thus Western Europe is even more overextended ecologically in relation to its own resources than the United States. For us Americans, a major disruption of world trade would cause painful retrenchment, to be sure, but there would be little danger of starvation, and domestic sources of energy would be available in sufficient quantity to keep the economy limping along. Europe does not enjoy such luxury. World trade must continue along established lines or economic collapse threatens.

Western European political systems have the same tendency as the American system to permit activities that degrade the environment. All share the same growth-oriented world view. All have followed the path blazed by the United States toward high mass consumption and, to a somewhat lesser extent, high energy use. All are mass democracies in which political parties compete for favor largely on the basis of how well they can satisfy the material aspirations of the citizenry. In short, having traveled the same basic path in roughly the same manner for the last 250 years, we Westerners have wound up in approximately the same place.

Nevertheless, just as there are some differences in the nature and degree of ecological scarcity, so too there are some significant differences in the potential for political adaptation. For one thing, Europe has had to contend with ecological scarcity in numerous ways even during an era of unparalleled abundance. Not possessing the same cornucopia of found wealth, for example, Europe has never been so profligate with its resources as the United States. For instance, Europeans manage to achieve roughly comparable living standards while using only about half as much energy per capita as Americans. Also, Europeans practice sustained-yield forestry, control land use quite stringently by U.S. standards, support and use public transit more than Americans do, and so on. Thus, both because of necessity and because of a generally less doctrinaire attachment to the principles of laissez faire, there exists in Europe a much greater willingness to accept planning and social controls. Moreover, at least in some quarters, disenchantment with bourgeois acquisition as a way of life has grown markedly. The rise of Green political parties in several countries has introduced an ecological agenda into the political arena. In general, therefore, European nations may cope somewhat better with ecological scarcity than the United States, despite the greater physical challenges they will face.

Japan

Although in terms of ecological scarcity Japan's situation is much more desperate than that of Europe, Japan possesses countervailing political and social advantages over Europe. With about half the population of the United States, a land mass about the size of Montana that is mostly mountainous and poorly endowed with mineral and energy resources, and the second largest economy in the world, Japan is a very tight little island indeed. Prevented from gaining by military means a position of power, respect, and economic security in the international community, the Japanese entered the great postwar international GNP stakes determined to win economically what could not be won by force of arms. They "aped" (their own word) the acquisitive ethic and mass-democratic institutions of the West so effectively that they achieved economic growth of unprecedented intensity and rapidity. This extraordinary "success" earned them notorious pollution problems, such as the mercury poisoning that killed 1900 people and paralyzed or affected thousands more, and a level of dependence on foreign trade and foreign sources of raw materials and fuels that makes them extremely vulnerable to international turmoil and resource scarcities, whether due to natural exhaustion or to artificial restriction by cartels. Japan thus faces ecological scarcity in an extreme form. A serious interruption of oil supplies from the Persian Gulf, a substantial decline in the fish catch, the inability or unwillingness of the United States and other countries to continue to supply vast quantities of food, timber, minerals and other vital products—these and numerous other potential threats could have severe consequences for Japan, which has totally committed itself to the modern way of industrial life and to living far beyond its ecological means.

Beginning in the early 1970s, and especially after the energy crisis of 1973–1974, the Japanese awoke to the fact that they were headed for an ecological precipice. The Japanese government cracked down on pollution with progressively greater severity and has recently moved to conserve energy and control growth in general. An awkward problem for Japanese political leaders, who are largely drawn from the business-oriented, conservative Liberal–Democratic Party (LDP) that has ruled throughout the postwar era, is that the powerful economic interests that are the LDP's main source of support, financial and otherwise, are also the chief polluters and main beneficiaries of growth. On the other hand, there are a number of positive factors. Because of extreme congestion, the cost of living, the cost and importance of educating children, and the cost of caring for the aged (before the government started to do so in 1973), Japanese families have successfully controlled their birth rate in the postwar era. That success was achieved despite the ruling party's efforts to overcome labor shortages by tightening the country's abortion law, prohibiting the sale of birth control pills, and proposing to give "baby bonuses" to women who had a third and fourth child. In addition, the adoption and implementation of environmental regulations have been facilitated by a tradition of government intervention in all areas of economic and social life. The result has been that on many environmental indicators, Japan by the late 1980s was doing better than either the United States or Western Europe. For example, between 1975 and the late 1980s, Japan had reduced its annual sulfur dioxide emissions by more than 60%; the comparable reduction for the United States was only 25%. Japan was the only industrialized country that successfully reduced emissions of nitrogen oxides during that period. Likewise, Japan has been reducing toxic pollution of its waterways more than most European countries and more than the United States.

Eastern Europe and The Soviet Union

The former Soviet bloc is the most interesting and revealing comparative case. Because they were the leading non-market industrialized nations, these countries should seemingly have been exempt, if not from the basic political dynamics of scarcity, at least from most of the failings of American market economics and politics discussed in the preceding chapters. In fact, however, the U.S.S.R. and its former satellites have severe environmental problems and have demonstrated far less capacity to deal with them than the United States and other market-oriented democracies.*

That the Soviet Union and Eastern Europe have serious environment problems is not denied by governmental spokespeople. Cities are blackening and their structures are deteriorating, mountains are teeming with dead trees, crop yields are diminishing, rivers are little more than open sewers, ground-water is polluted, and clean drinking water is scarce. Life expectancy is actually going down in some areas. Statistically, Russian men die 7 to 10 years earlier than men in other developed countries; in Northern Siberia, men die 22 years and women 14 years earlier than men and women in the northern countries of Western Europe (Yablokov et al. 1991, p. C3). People in industrial areas have high rates of cancer and of respiratory, skin, liver, and other diseases. Indeed, some pollution levels are astonishing by Western standards. Coal is the primary fuel in Eastern Europe; it is "dirty" coal, containing high sulfur levels, and is burned with few or no emission controls. Czechoslovakia's soils receive 25 metric tons of emitted pollutants annually per square kilometer, compared to 0.6 tons for Sweden. Twenty million Russians breathe highly polluted air. Many people in the former Soviet bloc have become environmental refugees; they have moved from Prokopyevsk, Nizhny Tagil, Kirishi, Angarsk, and other places in search of clean air and water (Yablokov et al. 1991, p. C3). Industries and municipalities throughout the former bloc commonly discharge their wastes, untreated, into rivers. The Soviet Union treats only 30% of its sewage annually. Many present and former Soviet cities—among them Riga, the capital of Latvia—have no sewage treatment plants at all. 75% of Russia's surface water is too polluted to drink. 50% of Poland's cities don't have sewage treatment plants; 35% of its industries don't treat their wastewater. 65% of Poland's river water is too corrosive (to say nothing of its contamination with sewage and toxins) to be

---

* In the summer of 1991, what had been the Soviet Union appeared to be breaking up into independent or semi-independent countries, with a confederate government at the "center" whose final form has not yet emerged. However, we shall continue to use the term *Soviet Union* to refer to the area that was, until mid-1991, one country.

used by industry. Even automobile pollution, which should be low because the number of cars per capita in these countries is far lower than in the West, is a problem. Soviet bloc cars had no pollution controls. Some were built with two-stroke engines that burn oil and gasoline together to create a sooty, smoking exhaust. Needless to say, with such high levels of air and water pollution, lakes everywhere in the region are expiring; in the Soviet Union, whole seas are polluted and their fish populations dying. A 1989 study showed that 69% of the fresh-water fish in Russia are "extremely contaminated" by mercury-based pesticides (Yablokov et al. 1991, p. C3). Sadly, people are dying too; the Soviet Ministry of Health has itself found that disease and mortality rates are higher in badly polluted areas than in other areas of the Soviet Union.

Eastern European countries also use energy inefficiently—about half as efficiently as does the United States, which itself is only half as efficient as Western Europe and Japan. Moreover, despite a relatively favorable position compared to Europe and America, the U.S.S.R. is not exempt from ecological scarcity with respect to its resources. For example, sizable grain purchases in recent years have made it evident that the Soviet Union's agricultural situation is problematic, even if the prospect of Malthusian starvation is remote. Some of the problem may be attributed to deterioration of the country's soils. Two-thirds of the country's arable land suffers from soil erosion; 1.5 billion tons of topsoil are lost each year. In addition, more than 10% of the U.S.S.R.'s irrigated land has been lost to salinization. Soil deterioration and pollution result in 16 billion rubles worth of crop losses each year in the Soviet Union.

Even the Soviet Union's apparent abundance of domestic energy resources may be illusory, at least in part. Some of its resources are not readily exploitable, for they lie in remote and environmentally forbidding regions; they also may be less substantial than rough estimates had indicated. Finally, they probably cannot be fully exploited without advanced Western technology and Western capital, which Soviet bloc countries desperately seek (and which, at least from public sources in the West, has not been readily available).

Why have communist countries failed so dismally at coping with their environmental problems? Communist propaganda denied the possibility of such failure. After all, it said, the Soviet system is not held in thrall by selfish market interests. Therefore it would easily be able to deal with any environmental problems that cropped up, whereas pollution and other environmental ills in the West were seen as serious emerging "contradictions" (inherent self-destructive forces) that capitalist nations would not be able to overcome. It didn't work out that way. Why?

First, the ideology of growth and belief in the power of technology were even more strongly entrenched in the U.S.S.R. than in the West, so abandoning or even compromising growth in production for the sake of environmental protection or resource conservation was a much more heretical concept. For one thing, as pointed out in the Introduction, the Marxist utopia depends for its achievement on the abolition of material scarcity. Thus to abandon growth would be tantamount to abandoning a utopian promise that had inspired the whole society. Worse, this cherished utopian goal was used to justify many features of Soviet life that seemed to conflict with basic Marxist principles. Soviet leaders, for example, explained the use of differential rewards (as opposed to the true communist principle "to each according to his needs") as a necessary expedient to help build the requisite material and productive base for a utopia of abundance. More important, the "proletarian dictatorship" and "democratic centralism" exercised by the Communist Party were also rationalized with this brand of logic. The loss of such convenient justifications could thus cause awkward political repercussions.

Second, largely as a result of this fundamental ideological bias toward material expansion (but also because of preoccupation with national security) the primacy of narrow economic concerns in policy matters was almost total, and fixation on production to the virtual exclusion of all else made the Soviet elite very resistant to more than token concern for the environment.

Third, although they were employed by the state rather than by private corporations, Soviet economic managers competed with other managers within the basic framework of the national plan, and their reluctance to spend money on nonproductive pollution control, their willingness to foist the external costs of production onto others, and their desire to win promotions by overproducing the quota made them behave, with respect to the environment, just like capitalist managers. Moreover, economic managers in the Soviet Union and Eastern Europe had far greater political power than their Western counterparts. In one respect, the tragic logic of the commons operated even more viciously in those countries: Because not only air and water, but virtually all natural resources, were (thanks to state ownership) treated as free or semi-free goods, there was an even greater tendency on the part of economic managers to use land, energy, and mineral resources wastefully.

Fourth, because government decisions were made in private council by leaders who put production and the vested interests of the state economic bureaucracy first, those concerned about the problems of growth had little opportunity to influence policy as it was being formed; they could only point out the adverse consequences of past policies.

Environmentalists, to the extent that they dared to appear, were vigorous-
ly suppressed in Soviet bloc countries. Only after Chernobyl and Mikhail
Gorbachev's *glasnost* could people participate in environmental protests.

In short, Soviet economic and political institutions seem designed
to produce environmental deterioration and resource depletion more
relentlessly than their American counterparts. The essential reason was
sardonically stated by a leading expert on Soviet environmental policy:
"The replacement of private greed by public greed is not much of an
improvement" (Goldman 1970). But even that was an understatement,
for private greed has been subject to modest, though inadequate,
controls. The irony is that Soviet bloc environmental standards on the
books were stringent; air and water quality standards were even stricter
than those in Western countries. This was because the standards were
promulgated on the basis of public health and did not need to be
compromised in order to accommodate powerful economic interests.
There was only one problem: The powerful economic interests—that
is, powerful government ministries, operating in secret and with no
outside input—saw to it that the regulations were not enforced. The
standards had no chance of prevailing among ministers whose overriding goal was growth.

The communist experience demonstrates that although both Western
and communist economic institutions produce environmental deterioration, the nature of the particular political institutions through which the
"greed" for material growth is translated into economic output makes a
difference. In the final analysis, the greed of powerful economic interests
can be controlled more easily when openness and democracy prevail. The
communist experience also demonstrates that if they are very strongly
committed to growth, highly centralized and effective governments may
wreak more and faster havoc on the environment than even the most
laissez-faire government.

The Third World

The developing or less-developed countries (LDCs)* constituting the
so-called Third World of course differ greatly from each other in many
important respects, but for the purposes of our discussion, little is lost by

---

* Lacking any reasonable alternative, we employ these well-established but, it
seems to us, culturally biased terms in their narrow economic sense. Bhutan (see
Box 27), a country that preserves the ancient and admirable Tibetan culture in
virtually all of its traditional richness, is scarcely undeveloped, fanatical
modernizers to the contrary notwithstanding.

27

Bhutan: Developing Sustainably

Bhutan, alone among modern countries, practices sustainable development. According to Christopher Flavin, (*World Watch*, 1990) although the country's last two kings have supported some economic growth, they have given first priority to the environment. Government officials nationalized the country's forests in 1979 to stop them from being over-exploited. Now, forestry officials permit only selective logging at a rate they believe is sustainable. They have established large wildlife reserves and greenbelts in which no logging is permitted. They have outlawed the export of raw logs. The country generates most of its energy renewably—with wood and hydropower. Government policy favors energy growth by more hydropower and solar energy rather than by fossil or nuclear fuels. Bhutan's population growth has been about 2% per year. Women enjoy a high status. All couples receive population counseling at the time of marriage and after the birth of each child. The government has made contraceptives available at 70 Basic Health Units spread around the country. There are no great disparities of wealth in the country. Most farmers grow their crops on small plots and graze cattle and yaks in the mountains.

Bhutan has escaped the growth ideology common to the rest of the world partly because of geography and partly because of culture. Bhutan is a very isolated Himalayan country; it has been almost entirely inaccessible to the outside world. Its religion, Buddhism, retains a love for nature and all living things. It has few economic tensions and no powerful business interests—no industries left behind by the British. Its people know little about consumer items that people in other cultures crave. Its kings have not been corrupted by business or other powerful interests, and they have left much government decision making to local government units. As the country proceeds to develop, economic and consumer interests that clash with the requirements of sustainability may yet come into being. But so far, Bhutan has retained an exceptional status.

considering them together. In brief, most LDCS, and especially the group of exceptionally poor countries sometimes called the Fourth World, are not sufficiently developed to experience neo-Malthusian ecological scarcity. Instead, they confront ecological scarcity in its crudest Malthusian form: too many people, too little food. Because this core problem, along with its major ramifications, was covered in Part I, no more need be said about it here, except that almost everywhere the difficulties seem greatly to exceed the capacity of current governments in the LDCs to cope with them. Even now, for example, many governments cannot assure all their citizens enough food to maintain life, and the future prospects are grim. However, there are some interesting exceptions to this general picture.

The LDC's run the gamut from virtual non-development to what is usually called semi-development, in which considerable industrialization and modernization coexist with continued non-development, especially in rural areas. In general, countries moving toward semi-development seem to follow established models. Mexico and Brazil, for example, have followed a basically American path (Mexico City has a smog problem worse than that of Los Angeles), and Brazil's treatment of its undeveloped wealth, especially such fragile and irreplaceable resources as the Amazon rain forest, epitomizes frontier economics at its most heedless. On the other hand, Taiwan and South Korea have proceeded more or less along the lines laid down by Japan and have encountered many of the same problems. The environmental problems of developing countries are exacerbated by the burden of their debt to the developed world. In 1989 developing countries paid $77 billion in interest on their debts and $85 billion in principal, paying out $50 billion a year more in debt repayments to wealthy nations than they received in new developmental assistance. This massive transfer of resources from poor to rich not only impoverishes the people of these regions but is also the propelling force behind many environmental disasters—the unsustainable logging of forests, the overgrazing of pasture, the depletion of mineral resources, the overexploitation of fisheries, and the growing of crops (such as coffee) that can earn cash as exports rather than crops that can feed their own famished people.

The International State of Nature

The International Macrocosm

If in the various national microcosms constituting the world political community the basic dynamics of ecological scarcity apply virtually across the board, in the macrocosm of international politics they operate even more strongly. Just as it does within each individual nation, the tragic logic of the commons brings about the over-exploitation of such common-pool resources as the oceans and the atmosphere. Also, the pressures toward inequality, oppression, and conflict are even more intense within the world political community, for it is a community in name only, and the already marked cleavage between rich and poor threatens to become even greater. Without even the semblance of a world government, the solutions of such problems depends on the good will and purely voluntary cooperation of over 170 sovereign states—a prospect that does not inspire optimism. Let us examine these issues in more detail, to see how ecological scarcity aggravates the already very difficult problems of international politics.

The Global Tragedy of the Commons

The tragic logic of the commons operates universally, and its effects are readily visible internationally—in the growing pollution of international rivers, of seas, and now of even the oceans; in the overfishing that has caused a marked decline in the fish catch in some areas, as well as the near extinction of the great whales; and in the impending scramble for seabed resources by maritime miners or other exploiters. There is no way to confine environmental insults or the effects of ecological degradation within national borders, because river basins, airsheds, and oceans are intrinsically international. Even seemingly local environmental disruption inevitably has some impact on the quality of regional and, eventually, global ecosystems. Just as it does within each nation, the aggregation of individual desires and actions overloads the international commons. But, like individuals, states tend to turn a blind eye to this, for they profit by the increased production while others bear most or all of the cost, or they lose by self-restraint while others receive most or all of the benefit. Thus Britain gets the factory output while Scandinavia suffers the ecological effects of "acid rain"; the French and Germans use the Rhine for waste disposal even though this leaves the river little more than a reeking sewer by the time when, downstream, it reaches fellow European Economic Community member Holland.

Even though the problems are basically the same everywhere, the political implications of the tragedy of the commons are much more serious in the international arena. It has long been recognized that international politics is the epitome of the Hobbesian state of nature: Despite all the progress over the centuries toward the rule of international law, sovereign states, unlike the citizens within each state, acknowledge no law or authority higher than their own self-interest; they are therefore free to do as they please, subject only to gross prudential restraints, no matter what the cost to the world community. For example, despite strong pressures from the international community, including a 5-year moratorium on commercial whaling by the International Whaling Commission, Japan and Iceland continue to hunt whales. More than 13,000 whales have been killed since the international community banned whaling. The United States relentlessly spews huge amounts of carbon dioxide into the air commons, despite efforts among other industrialized nations to get an agreement to reduce greenhouse emissions.

In international relations, therefore, the dynamic of the tragedy of the commons is even stronger than within any given nation state, which, being a real political community, has at least the theoretical capacity to make binding, authoritative decisions on resource conservation and ecological protection. By contrast, international agreements are reached and enforced by the purely voluntary cooperation of sovereign nation states existing in a state of nature. For all the reasons discussed in Chapter 4, the likelihood of forestalling by such means the operation of the tragedy of the commons is extremely remote. Worse, just as any individual is nearly helpless to alter the outcome by his or her own actions (and even risks serious loss if he or she refuses to participate in the exploitation of the commons), so too, in the absence of international authority or enforceable agreement, nations have little choice but to contribute to the tragedy by their own actions. This would be true even if each individual state was striving to achieve a domestic steady-state economy, for unless one assumes agreement on a largely autarkic world, states would still compete with each other internationally to maximize the resources available to them. Ecological scarcity thus intensifies the fundamental problem of international politics—the achievement of world order—by adding further to the preexisting difficulties of a state of nature. Without some kind of international governmental machinery with enough authority and coercive power over sovereign states to keep them within the bounds of the ecological common interest of all on the planet, the world must suffer the ever-greater environmental ills ordained by the global tragedy of the commons.

The Struggle Between Rich and Poor

Ecological scarcity also aggravates very seriously the already intense struggle between rich and poor. As is well known, the world today (some forthcoming changes will be discussed in the following section) is sharply polarized between the developed, industrialized "haves," all affluent in a greater or lesser degree and all getting more affluent all the time, and the underdeveloped or developing "have nots," all relatively and absolutely impoverished and (with few exceptions) tending to fall further and further behind despite their often feverish efforts to grow. The degree of the inequality is also well known: The United States, with only 6% of the world's population, consumes about 30% of the total energy production of the world and comparable amounts of other resources; it throws away enough paper and plastic plates and cups to set the table for a worldwide picnic six times a year (Durning 1991, p. 161).

The rest of the "haves," though only about half as prodigal as the United States, still consume resources far out of proportion to their population. Conversely, per capita consumption of resources in the developing world ranges from one-tenth to one-hundredth that in the "have" countries. To make matters worse, the resources that the "haves" enjoy in inordinate amounts are largely and increasingly imported from the developing world. For example, developed nations consume two-thirds of the world's steel, aluminum, copper, lead, nickel, tin, zinc, and three-fourths of the world's energy. Thus economic inequality and what might be called ecological colonialism have become intertwined. In view of this extreme and long-standing inequality (which, moreover, has its roots in an imperialist past), it is hardly surprising that the developing world thirsts avidly for development or that it has become increasingly intolerant of those features of the current world order it perceives as obstacles to its becoming as rich and powerful as the developed world.

Alas, the emergence of ecological scarcity appears to have sounded the death knell for the aspirations of the LDCS. Even assuming (contrary to fact) that there were sufficient mineral and energy resources to make it possible, universal industrialization would impose intolerable stress on world ecosystems. And humans, in particular, could not endure the pollution levels that would result. Already, the one-fifth of the world's population that lives in industrial countries generates most of the world's toxic wastes, two-thirds of the world's greenhouse emissions, three-fourths of the world's nitrogen oxides and sulfur emissions, and 90% of the gases that are already destroying the world's protective ozone layer.

In short, the current model of development, which assumes that all countries will eventually become heavily industrialized mass-consumption societies, is doomed to failure.* Naturally, this conclusion is totally unacceptable to the modernizing elites of the developing world; their political power is generally founded on the promise of development. Even more important, simply halting growth would freeze the current pattern of inequality, leaving the "have nots" as the peasants of the world community in perpetuity. Thus an end to growth and development would be acceptable to the developing world only in combination with a radical redistribution of the world's wealth and a total restructuring of the world's economy to guarantee the maintenance of economic justice. Yet it seems absolutely clear that the rich have not the slightest intention of relieving the plight of the poor if it entails the sacrifice of their own living standards. Ecological scarcity thus greatly increases the probability of naked confrontation between rich and poor.

Who Are Now the "Haves" and the "Have Nots"?

An important new element has been injected into this struggle. The great "resource hunger" of the developed world, and even of some parts of the developing world, has begun to transfer power and wealth to those who have resources to sell, especially critical resources such as petroleum. As a result, the geopolitics of the world is changing.

This process can be expected to continue. The power and wealth of the major oil producers are bound to increase over the next five decades, despite North Sea and Alaskan oil and regardless of whether or not the

---

* The ecologically viable alternative, depicted in Part III, is a locally self-sufficient, semi-developed, steady-state society based on renewable or "income" resources such as photosynthesis and solar energy. Only Bhutan seems self-consciously to be developing sustainably as a matter of principle (see Box 27). Others find themselves unable to see such apparent frugality as a realistic option. All the pressures impel them toward "efficiency," standardization, centralization, and large scale. In addition, because sustainable development does not work when population pressure is extreme and most developing countries are heavily overpopulated, choosing restraint or frugality sometimes implies a willingness to use harsh measures—for example, compulsory abortions to stabilize populations or forced resettlement to save the rainforests. It is not surprising that most leaders prefer to continue in the illusory hope of achieving heavy industrialization. In addition, the lust for status and prestige, the desire for military power, and many other less than noble motives are also prevalent, and the frugal modesty of semi-developed self-sufficiency can do little to satisfy them.

Organization of Petroleum Exporting Countries (OPEC) manages to act in a unified manner.

Some believe that oil is a special case and that the prospect of OPEC-type cartels for other resources is dim (Banks 1974; Mikesell 1974). These assessments may be correct, but it seems inevitable that in the long run an era of "commodity power" must emerge. The hunger of the industrialized nations for resources is likely to increase, even if there is no substantial growth in output to generate increased demand for raw materials, because the domestic mineral and energy resources of the developed countries have begun to be exhausted. The United States, for example, already imports 100% of its platinum, mica, chromium, and strontium; over 90% of its manganese, aluminum, tantalum, and cobalt; and 50% or more of 12 additional key minerals (Wade 1974). However, the developed countries seem determined to keep growing, and assuming even modest further growth in industrial output, their dependence on developing world supplies is bound to increase markedly in the next few decades.* Thus, whatever the short-term prospects for the success of budding cartels in copper, phosphates, and other minerals, the clear overall long-term trend is toward a seller's market in basic resources and therefore toward "commodity power," even if this power grows more slowly and is manifested in a less extreme form than that of OPEC.†

Thus the basic, long-standing division of the world into rich and poor in terms of GNP per capita will eventually be overlaid with another rich–poor polarization, in terms of resources, that will both moderate and intensify the basic split. Although there are many complex interdependencies in world trade—for example, U.S. food exports are just as critical to many countries as their mineral exports are to us—it is already clear

---

* Naturally, there will be short-term exceptions. Developed countries, for example, will remain somewhat independent of Middle Eastern oil supplies while North Sea oil production and Alaskan oil production remain at high levels and as long as other non–OPEC sources of oil exist. On the other hand, one of the reasons for the Persian Gulf war was undoubtedly the desire to keep Kuwaiti oil in friendly hands. The respites from the overall trend toward increasing dependence will be transitory and limited to particular commodities.

† Actually, OPEC-like cartels in other resources might be preferable to a disorganized seller's market. Cartels can be bargained with and integrated into the normal diplomatic machinery, so that the drastic price fluctuations and outright interruptions of supply that cause extreme economic distress are avoided. But the price of stability is higher prices for commodities and increased political power for cartel members.

that the resource-rich nations of the developing world stand to gain greater wealth and power at the expense of the "haves." Already, through nationalization and forced purchase, the OPEC nations have largely wrested control of drilling and pumping operations from the Western oil companies; it is only a matter of time before they expand into other areas of the oil business. In addition, as is already evident, the newly resource-rich are not likely to settle for mere commercial gains. They have long-standing political grievances against other nations—especially the developed nations—that they will try to remedy with their new power. Unfortunately, the defense the industrial powers are most likely to use against unfriendly economic or political moves on the part of any of these countries will be military, as it was in the Persian Gulf war, where the industrial nations have an overwhelming advantage.

This discussion leaves out the majority of poor countries—those without major resources of their own. As they are forced to pay higher prices to resource-rich countries, they will suffer—indeed, they already have suffered—major setbacks to their prospects for development. This is true not only of the hopelessly poor countries of Africa and Asia but also of countries whose development programs have already acquired some momentum.

In sum, world geopolitics and economics are in for a reordering. Western economic development has involved a net transfer of resources, wealth, and power from the current "have nots" to the "haves," creating the cleavage between the two that now divides the world. In recent years, developed nations have added the additional burden of debt repayment, increasing the wealth transferred to them from the "have nots." In the long term, this situation will change; wealth will also be transferred to those nations that have scarce resources. But only the relatively few "have nots" that possess significant amounts of resources will gain; the rest of the poor will become more abject than before. Thus the old polarization between rich and poor seems likely to be replaced by a threefold division into the rich, the hopelessly poor, and the newly enriched—and such a major change in the international order is bound to create tension.

Conflict or Cooperation?

How this tension will play out in the years ahead is hard to say. The danger is that to many of the declining "haves," ill-equipped to adapt to an era of "commodity power" and economic warfare, the grip of the newly enriched on essential resources will seem an intolerable stranglehold to be broken at all costs. At the same time the poor, having had their revolutionary hopes and rising aspirations crushed, will have little to lose but their chains. Thus the world may face turmoil and war on top of ecological scarcity—a horrible prospect, given the ecologically destructive character of modern warfare (see Box 28).

Some, on the other hand, hope or believe that ecological scarcity will have just the opposite effect: Because the problems will become so overwhelming and so evidently insoluble without total international cooperation, nation states will discard their outmoded national sovereignty and place themselves under some form of planetary govern-ment that will regulate the global commons for the benefit of all humanity and begin the essential process of gradual economic re-distribution. In effect, states will be driven by their own vital national interests (which they recognize as including ecological as well as tradi-tional economic, political, and military factors) to embrace the ultimate interdependence needed to solve ecological problems (Shields and Ott 1974). According to this hypothesis, the very direness of the outcome if cooperation does not prevail may ensure that it will.

The pattern, so far, has fallen somewhere between these extremes. War for resources has broken out on occasion; as we have noted, one of the reasons for frequent United States military activity in the Middle East is to keep the control of oil fields and commerce in friendly hands. On the other hand, there has also been considerable talk about cooperative international action to deal with the problems of environmental degrada-tion, and some momentum toward greater cooperation has developed. However, with the possible exception of agreements among the nations of the European Economic Community, which are in a common market, international environmental agreements have been piecemeal and unen-forceable, or what they have required has been that which the most reluctant nation has been willing to concede—measures that are usually inadequate. The international environmental regulatory process thus re-sembles process politics in the American context, with this difference: The players—in this case the polluters—cannot be forced to come to the bargaining table; if they do come, they can't be forced to agree to anything; and if they do agree to something, the agreement cannot be enforced.

An Upsurge in Conference Diplomacy

A look at the record of international environmental agreements thus far concluded reveals only modest accomplishments. By the late 1960s some of the alarming global implications of pollution and general ecological

28

War and Ecocide

War may occasionally be the lesser of evils, but by its very nature it has always been anathema to any reasonable person. To the human ecologist it is doubly horrible. One of the most appalling features of the modern world is the enormous amount of ecological damage and resource wastage that can be attributed to warfare and military preparedness. Resources that should have been used for human welfare (or that should never have been used at all) have been sacrificed to the gods of national security in the jungles and rice paddies of Vietnam and the deserts of the Middle East. But this is only the most obvious wastage. The military consumes prodigious amounts of energy; the Pentagon uses more energy in a year than a mass transit system consumes in 14 years. In just 1 hour, an F–16 fighter jet on a training flight consumes more fuel than an average motorist does in a year (Renner 1991, p. 137). The U.S. military emits more carbon into the atmosphere than the total emissions of Great Britain. It emits 76% of the halons and 50% of the CFCs the United States emits into the stratosphere (Renner 1991, p. 140). It uses more nickel, copper, aluminum, and platinum than the whole developing world combined (Renner 1991, p. 140). The Pentagon also generates more toxic wastes than the five biggest chemical companies combined (Renner 1991, p. 143). We have already discussed

degradation had become widely apparent, and preparations began for the first major international conference on the environment at Stockholm in 1972. Depending on one's point of view, the Stockholm Conference—to give it its proper title, the United Nations Conference on the Human Environment—was either a major diplomatic success or an abysmal failure. On the positive side, the elaborate preparations for the conference (each country had to make a detailed inventory of its environmental problems), the intense publicity accorded the over two years of preliminary negotiations, and the conference itself fostered a very high level of environmental awareness around the globe. Virtually ignored by diplomats in 1969, the environmental crisis had by 1972 rocketed right

the frightening amounts of nuclear toxic waste generated by the U.S. and Soviet military machines.

Even more criminal from an ecological point of view is the increasingly ecocidal nature of modern warfare. Nuclear warfare, of course, is the prime villain, for any substantial number of nuclear explosions would poison world ecosystems and gene pools for untold generations and probably disrupt the structure of the atmosphere enough to cause mass extinctions. (In addition, the widespread dispersal of nuclear materials and technology for so-called peaceful purposes increases the probability of nuclear proliferation and therefore of nuclear war and terrorism.) However, any form of chemical and bacteriological warfare is potentially ecocidal—for example, the use of broadcast herbicides in Vietnam. But even more conventional forms of modern warfare are exceedingly destructive of local ecologies. In Vietnam, for instance, the U.S. military devastated millions of acres of farm and forest with saturation bombing and giant earth-moving machinery. Iraq was bombed back into the nineteenth century in the Persian Gulf war. Moreover, the fires that the Iraqi government set in the oil fields of Kuwait wasted a tremendous amount of oil before they were extinguished.

Of course, armies have employed ecocidal weapons—for example, scorched earth and salted lands—since ancient times. Yet war today, or even armed peace, is far more wasteful of scarce resources and far more destructive to the earth. War has been rightly called the ultimate pollutant of the planet Earth.

up alongside nuclear weapons and economic development as one of the big issues of international politics. The second major achievement of the Stockholm Conference was the establishment of the United Nations Environment Program (UNEP) to monitor the state of the world environment and to provide liaison and coordination between nation states and among the multitude of governmental and non-governmental organizations concerned with environmental matters. Finally, a few preliminary agreements covering certain less controversial and less critical ecological problems, (such as setting aside land for national parks and suppressing trade in endangered species) were reached either at the conference or immediately thereafter.

Despite these acknowledged achievements, environmentalists were by and large rather unhappy with the conduct and outcome of the conference. They were especially disillusioned, for example, by the way the original ecological purity of the conference's agenda was rapidly watered down by pressures from countries of the developing world, who made it plain that they would have nothing to do with the conference unless, in effect, underdevelopment was interpreted as a form of pollution. Moreover, a great part of the proceedings was devoted not to the problems on the agenda, but to the kind of "have" versus "have not" debate discussed above, and routine ideological posturing on such political issues as "colonialism" consumed additional time. Also, cold-war politics refused to take a vacation. Thus the perhaps naively idealistic hope of many that the ecological issue would at last force quarrelsome and self-seeking sovereign nation states to put aside stale old grudges, recognize their common predicament, and act in concert to improve the human condition was completely dashed.

Worse, some of the features of the current world order most objectionable from an ecological point of view were actually reaffirmed at Stockholm, including the absolute right of sovereign countries to develop their own domestic resources without regard to the potential external ecological costs to the world community, and the unrestricted freedom to breed guaranteed by the Universal Declaration of Human Rights. Subsequent environmentally oriented conferences have made no inroads on these fundamental principles, and they have mounted only cumbersome attacks on some forms of environmental degradation.[*]

For example, the U.N. World Population Conference in 1974 somehow managed to end "without producing explicit agreement that there was a world population problem" (Walsh 1974). Another World Population Conference, in Mexico City in 1984, also failed to achieve concrete results. The best the international community has been able to achieve on population policy is that by 1988, 48 heads of government had signed a 1985 statement supporting a world goal of population stabilization. However, no binding targets or concrete program have been agreed to. The United States, which at one time provided leadership and major financing for international population control efforts, has reduced its support for them since 1981.

---

[*] Among the most important U.N. environmental conferences not mentioned in the text were conferences on women (1974), human settlements (1976), water (1977), desertification (1977), renewable energy (1981), human environment (1982), and the ozone layer (1987, 1989).

The first international Law of the Sea Conference took place in 1973. Its supporters wanted to establish an all-encompassing treaty dealing with overfishing, seabed mining, and pollution controls, premised on the view that the oceans are the "common heritage of mankind." They failed. By the time a draft treaty was finally agreed to in 1982, it had carved the oceans into national zones of exploitation for 200 miles out from each nation's coasts; the nation controlling the Exclusive Economic Zone may control who, if anyone, may enter the zone for economic purposes. Only seabeds were declared a "common heritage of mankind," to be mined according to regulations established by an International Seabed Authority. However, even this was too much for the United States and many other industrial countries. The U.S. position is that seabeds should be mined on a first-come, first-served basis, without international regulation. Therefore, the United States has refused to sign or ratify the treaty. Behind these differences in legal position are differing national interests, not just differing views on the best way to protect the ocean environment. The industrial countries have or will soon have the capacity to begin deep seabed mining; they can enrich themselves further, or at least put off the day of their own mineral depletion, with a first-come, first-served approach. The developing world, on the other hand, benefits from a more controlled "common heritage" approach. So nationalism may block adoption of the proposed treaty. Only 40 countries had ratified it by 1989, and 60 must do so for it to go into effect.

Other international agreements affecting the sea are confined to narrower issues, but even so, some have been difficult to implement. MARPOL, the 1973 International Convention for the Prevention of Pollution from Ships, established minimal distances from the land for ocean dumping, limited the dumping of garbage, required ports to provide facilities for receiving trash from incoming ships, and prohibited the dumping of plastics. Only 39 nations had ratified this treaty 18 years later, and it took 14 years for the United States to ratify it. Under the treaty, the U.S. ban on the dumping of plastics took effect in 1989, although the largest source of plastics dumping, the armed forces, will not be brought under the treaty until 1994. The London Dumping Convention of 1972 has won wider support; 63 nations have signed it. It prohibits ocean dumping of heavy metals, specified carcinogens, and radioactive and other hazardous substances. Yet enforcement of the treaty has been spotty; violators are caught and punished with only as much vigor as each nation chooses to muster. In this connection, as we have noted earlier, even ocean treaties and conventions that have nearly universal support, (such as the moratorium on whaling and the prohibition of driftnet fishing) either have loopholes through which nations can jump or, as with whaling, are often openly flouted. Occasionally, another signatory to such a treaty who is angry about such defiance may engage in a trade sanction against a delinquent country. Generally, however, a country's flouting of environmental agreements does not result in the international community's imposing a meaningful penalty.

Still, ocean conventions and treaties have had more results beneficial to the environment than have international agreements concerning hazardous wastes and air pollution. Regarding hazardous wastes, the most the international community has managed to agree to is a prohibition of transboundary shipment of such goods by stealth. A 1989 United Nations draft treaty forbids transboundary movement of waste without notification by the exporter, without the consent of the importer, or with documents that do not conform to the shipment. A 1990 United Nations system of Prior Informed Consent regarding restricted chemicals (primarily pesticides) requires that prospective importing nations be provided with information about the benefits and risks of a chemical before deciding whether to allow it to be imported. Environmentalists had wanted a stronger draft making it illegal for a country to export to another country a chemical that is banned within the exporting country itself. But the United States, which exports large amounts of banned pesticides, led the successful opposition to that proposal.

International air pollution treaties and conventions have also been weak. Developing countries have largely refused to agree to international controls of air pollution, fearing that such controls will impede their pace of development. Most air pollution agreements have been concluded only among Western European nations, sometimes with the United States and/or Britain not going along, although they were invited to sign. Examples include the 1988 conventions to reduce sulfur and nitrogen emissions. The United States and Western European governments have also had several conferences on reducing greenhouse emissions, but the United States has refused to agree to targets to reduce carbon emissions and has frustrated the attempts of European nations to come to a binding agreement among themselves. The best of the international air pollution agreements was the Montreal commitment by industrial countries to phase out their use of CFCs by 2000 (see Chapter 3). But we have seen that this agreement will be too little and too late to avert hundreds of thousands of cancer deaths and millions of cancer cases that will result from ozone depletion.

The forces that prevent strong international environmental agreements are many. First, the spirit of militant nationalism that has animated so much of the history of the postwar world has not abated, except among Western European governments. Indeed, the tendency of the world is to move in the opposite direction, with the nations of Eastern Europe and the Soviet Union breaking up into militant ethnic states. Nation states insist on the absolute and sovereign right of self-determination in use of resources, population policy, and development in general, regardless of the wider consequences. Second, the demand among Third World countries for economic development has, if anything, increased in intensity, and whatever seems to stand in the way, as ecological considerations often do, gets rather short shift. Third, largely because their prospects for development are so dim, the countries of the developing world have begun to press even harder for fundamental reform of the world system (a "new international economic order"). Thus every discussion of such environmental issues as food and population is inevitably converted by those who represent the developing world into a discussion of international economic justice as well, which enormously complicates the process of negotiation. In short, environmental issues have become pawns in the larger diplomatic and political struggle between the nations.

In addition, diplomats, like national leaders, have attempted to handle the issues of ecological scarcity not as part of a larger problematique but piecemeal, so that their interaction with other problems is all but ignored. For example, the World Food Conference was solely concerned with the problem of feeding the hungry and paid virtually no attention to the eventual ecological consequences of growing more food or subsidizing further overpopulation with radically increased food aid. To some extent, therefore, the successes of international conferences that simply try to solve one small piece of the larger problem are as much to be feared as their failures.

If one wished to be optimistic, one could conclude that the world community has taken the first attitudinal and institutional steps toward meeting the challenges of ecological scarcity. A more realistic assessment, however, would be that although modest environmental improvements have been achieved, major impediments to further progress remain. One might even be forced to conclude, more pessimistically, that the world political community as presently constituted is simply incapable of coping with the challenges of ecological scarcity, at least in a timely way.

Planetary Government or the War of All Against All

In short, the planet confronts the same problems as the United States, but in a greatly intensified form. Even before the emergence of ecological scarcity, the world's difficulties and their starkly Hobbesian implications were grave enough. Some saw the "revolution of rising expectations" pushing the world toward a situation in which wants greatly exceeded capacity to meet them, provoking Hobbesian turmoil and violence (Spengler 1969). The world lives under the blade of a deadly Sword of Damocles. The hair holding this environmental Sword has come loose; pollution and other environmental problems will not obligingly postpone their impact while diplomats haggle, so the Sword is already descending toward our unprotected heads. There is thus no way for the world community to put the environmental issue in the back of its mind and go about its business. The crisis of ecological scarcity is a Sword that must be parried, squarely and soon.

The need for a world government with enough coercive power over fractious nation states to achieve what reasonable people would regard as the planetary common interest has become overwhelming. Yet we must recognize that the very environmental degradation that makes a world government necessary has also made it much more difficult to achieve. The clear danger is that, instead of promoting world cooperation, ecological scarcity will simply intensify the Hobbesian war of all against all—with the destruction of the common planet (for purposes of human habitation) the tragic outcome.

Learning to Live
with Scarcity

8

Toward a Politics of the
Steady State

How we are to learn to live with ecological scarcity is the problem that
will dominate the coming decades. However daunting this task must
seem, it is indeed possible to make a transition to a relatively desirable
steady state instead of simply letting nature take its course, which is
certain to lead in the opposite direction. However, we must recognize
that a large measure of devolution or retrogression in terms of our
current values will inevitably follow 400 years of continual evolution
and "progress." But not all the political, social, economic, cultural, and
technological advances of the past four centuries must be abandoned.
Too, the sooner we confront the challenge squarely, the greater the
likelihood of saving the best of this legacy and, what may be more
important, of making a virtue out of necessity. Our actions over the
critical next few decades will therefore either create or preclude a
relatively desirable future for ourselves and our descendants. However,
we offer no concrete or formal solutions to the political dilemmas of
ecological scarcity. There are several reasons for this.

Learning to See Anew

First, the most important prerequisite for constructive change is a new
world view based on, or at least compatible with, the realities of the
human ecological predicament. The ecological crisis is in large part a
perceptual crisis: Most people simply do not realize that they are part of
a delicate web of life that their own actions are destroying, yet any viable
solution will require them to see this. Once such a change in

"paradigm" has occurred—once people have chosen to adopt ecological limitations deliberately as a consequence of their new understanding—then practical and humane solutions will be found in abundance. Indeed, as we have already seen, the essential elements of the steady state are not so hard to discern, and some good work has been done on suitable institutions. But the psychological readiness and political will to adopt them are wanting. Thus "metanoia," or a fundamental transformation of world view, must pave the way for concrete action.

Second, at this juncture any specific set of solutions would immediately be criticized as politically unrealistic. Indeed, how could it be otherwise? Current political values and institutions are the products of the age of abnormal abundance now drawing to a close, so any solutions predicated on scarcity would necessarily conflict with them. Of course, to work "within the system" to prevent further ecological degradation and promote in-cremental change toward the steady state is an essential task deserving great support. But to accept current political reality as not itself subject to radical change is to give away the game at the outset and render the situation hopeless by definition. Indeed, it must be understood that *ultimately politics is about the definition of reality itself.* As John Maynard Keynes pointed out, we are all the prisoners of dead theorists; the ideas of John Locke, Adam Smith, Karl Marx, and all the other philosophers of the Great Frontier in effect define reality for us. Before we can even see what the problem is, we must throw off their bonds they have clamped on our imagination. To put it another way, normal politics is indeed "the art of the possible"; it consists of working as best one can for valued objectives "within the system"—that is, inside the current political paradigm. However, politicking (to give it its true name) is only one part of politics, and the lesser part at that. In its truest sense, *politics is the art of creating new possibilities* for human progress. Because the current system is ecologically defective, we must direct our concrete political ac-tivities primarily toward producing a change of consciousness that can lead to a new political paradigm. Until people at large begin to see a new kind of reality based on ecological understanding, environmental politicking within the system can be only a rear-guard holding action designed to slow the pace of ecological retreat. Rejecting current political realities and relying primari-ly on a change of consciousness may seem utterly impossible to achieve, given what people want and believe today, but it should be remembered that only a little over a century ago it was legal to treat human beings as property. Already, many people are finding our slavish treatment of nature stupid at best and morally repugnant at worst. The events of the decades to come are bound to increase their number. Looking back on us as we ourselves look back on our slave-holding ancestors, our descendants will wonder why it took us so long to come to our senses.

Third, the transition will take several decades. Thus it is not necessary for us to have all the answers to ecological scarcity today. What is essential is for us to begin the disciplined and serious search for such answers now, instead of waiting until the point of panic-stricken extremity. We sometimes forget, for example, that our Constitution was the culmination of several decades of intense and sustained political discussion and action by our founding fathers. We confront a challenge perhaps greater than theirs, and we should not deceive ourselves about the magnitude and duration of the task. Moreover, as this example suggests, no one person, no one work, no one invention can hope to supply more than a small piece of the solution that will eventually emerge. The final result will be a mosaic of many elements, some designed by dint of human effort, others fashioned by the accidents of history. Thus to advance and promulgate specific solutions at this stage might be positively harmful, for such premature closure is likely to deflect us from the much more crucial task of going back to first principles—that is, to politics. Once we have agreed on political "first principles" and the lessons of history, as our founding fathers generally did at Philadelphia, then building the institutional machinery to incorporate them will not be such a difficult task. In sum, before trying to give rebirth to our political institutions, we must first allow time for a proper gestation.

Fourth, the hour is very late. Now that everyone can recognize the evils of ecological scarcity, it is probably much too late for a carefully planned transition to the steady state. Had we prudently listened to earlier warnings and acted appropriately when the environmental crisis first came to light many years ago, we might have devised a comprehensive master plan for the transition. This is no longer possible, for as we have seen, some measure of ecological overshoot (with attendant disruptive side effects that are unpredictable) is virtually foreordained. Besides, we are so committed to most of the things that cause or support ecological evils that we are almost paralyzed; nearly all the constructive actions that could be taken at present (for example, drastically restricting population growth) are so painful to so many people in so many ways that they are indeed totally unrealistic, and neither politicians nor citizens would tolerate them.* Only after nature has mandated certain changes and overwhelmingly demonstrated the advisability of others will it be possible to think in terms of a concrete program of transition. Until then, our

---

* All societies display social fanaticism to some extent. Their first response to threatening doubts is to redouble their efforts to shore up belief in the current paradigm, which is after all a kind of civil religion. Thus our tendency is to react to the challenges of ecological scarcity by ignoring or denying it and mounting more and more desperate efforts to stave off the inevitable changes.

time and effort is better spent laying the scientific and philosophical groundwork so that the moment of ripeness will find us prepared to move rapidly from thought to action.

Finally, in many respects seen and unseen, the process of transition is well under way. It is no accident that many radical critiques of industrial civilization, either grounded in ecology or self-consciously related to it, have been produced in recent years. Nor is it surprising that so many groups and individuals are experimenting with radically different life styles and technologies, many of them avowedly based on ecological principles. Nor that a quasi-religious ferment of self-examination and self-criticism seems to have sprung up throughout the industrial world, leading to new images of people and of human needs and potentials. The raw materials for social transformation are being produced right now, and the process of tearing down the old reality and constructing the new has already begun.* Thus, although this natural transition process is halting, belated, and imperfect, the ill of ecological scarcity is tending to manufacture its own remedy. Certainly, inspiring leadership and a comprehensive theory will be necessary at some point; without them, ordinary people would lack the social vision and sense of direction they need to make constructive personal responses to any social challenge, and the transition would degenerate into a mere muddling through. However, to a very large extent the transition will evolve, instead of being created by theorizing and social planning. The individuals and groups composing the collectivity will more or less willingly seek a viable and attractive set of social answers by responding to the pressures of ecological scarcity in their daily lives. The answers that emerge will then be ratified by theory. (Similarly, colonial Americans had already evolved many features of the distinctively American way of life well before they were formalized in political institutions.)

In short, excessive or premature specificity about the institutions of the steady-state society is either not very useful or a positive hindrance. Again metanoia is the key, for it will almost automatically engender concrete, practical arrangements that are congruent with it.

Nevertheless, a general outline of a solution to the problems of ecological scarcity is implicit in the concept of the steady state. Let us therefore review the essential characteristics of a steady-state or sustainable society.

---

* In his *The Promise of the Coming Dark Age,* the historian L. S. Stavrianos identifies important elements of this process and shows how the "grass" of a new civilization based on decentralized self-management is even now pushing up through the "concrete" of the moribund industrial order.

The Characteristics of the Steady State

It is not possible to specify the structural features of the steady-state society. The great diversity of human societies, which have existed in a virtual steady state throughout most of recorded history, shows that there are many different ways to similar ends. However, any such set of structural features would clearly have to reflect certain basic characteristics of the steady state. In preceding chapters we have discussed in some detail its purely *physical* characteristics: primary dependence on income or flow resources, the maintenance of population levels within the ecological carrying capacity, resource conservation and recycling, generally good ecological husbandry, and so on. Let us now focus on the necessary *sociopolitical* characteristics of any steady-state society, regardless of how it chooses to give them social form. For the reasons given in the last section, the following treatment is tentative. It is designed merely to indicate the general direction we will travel as we move from our current industrial civilization toward the steady state.

Communalism   We have been living in an age of rampant individualism that arose historically from circumstances of abnormal abundance. It seems predictable, therefore, that on our way toward the steady state we shall move from individualism toward communalism. The self-interest that individualistic political, economic, and social philosophies have justified as being in the overall best interests of the community (as long as the growth "frontier" provided a safe outlet for competitive striving) will begin to seem more and more reprehensible and illegitimate as pollution and other aspects of scarcity grow. And the traditional primacy of the community over the individual that has characterized virtually every other period of history will be restored. How far the subordination of individual to community values and interests will have to go and how it will be achieved are for the future to determine. Rigid caste systems and inflexible feudal hierarchies are unlikely to be necessary, but the degree of individual subordination (for example, of property "rights") that will eventually be required would probably seem quite insupportable to many living today.

Authority   As the community and its rights are given increasing social priority, we shall necessarily move from liberty toward authority, for the community will have to be able to enforce its demands on individuals. This prospect may seem alarming, but the historical record does not justify the fear that any concession of political rights to the community must lead to the total subjugation of the individual by an all-powerful state. There is no reason why authority cannot be made strong enough to maintain a steady-state society and yet be limited. *The personal and civil rights guaranteed by our Constitution could be largely retained in an appropriately designed steady-state society.* Nor need the right to own and enjoy adequate personal property be taken away. Only the right to use private property in ecologically destructive ways would have to be checked. Thus authority in the steady state need not be remote, arbitrary, and capricious. In a well-ordered and well-designed state, authority could be made constitutional and limited.

Government   Allied with the foregoing transition will be a movement away from egalitarian democracy toward political competence and status. Because the mere summation of equally regarded individual wants into Rousseau's "will of all" has become ecologically ruinous, we must find ways of achieving the "general will" that stands higher than the individual and his or her wants. To this end, certain restrictions on human activities must be competently determined, normatively justified, and then imposed on a populace that would do something quite different if it was left to its own immediate desires and devices. This can be accomplished in more than one way. Power can be given or allowed to accrue to those who *are fittest to rule* (as in Thomas Jefferson's "natural aristocracy"). However, the creation of a ruling class, no matter how open and well qualified, immediately delivers us into the classic "Who will watch the guardians?" dilemma—and as noted earlier, the greater the scarcity, the greater the likelihood of oppression of the ruled by the ruling class. This dilemma can be avoided, at least in part, by founding the political system by common consent on *a set of values fit to be ruled by*—that is, principles designed to foster the common interest of the steady state instead of the particular interests that would destroy it. If this could be done successfully, government would be respected and there would be a fundamental agreement among citizens on how their communal life was to be regulated. But administrative authority would be decentralized and within the reach of the citizens. Citizens pursuing an ecological ethos would themselves promote and support laws obliging themselves to live within ecological limits, much as citizens enforced the criminal laws in nineteenth-century America (Tocqueville 1835, p. 71). Then the need for the ministrations of a ruling class would be much lessened, and to the extent that a class of Jeffersonian natural aristocrats was still needed to make the system work, it could be subjected to constitutional restraints, also as in the earliest days of the American republic. (See Box 29 for further discussion of this "design criteria" approach.) Nevertheless, once the basis of political values becomes something other than personal self-interest, age-old dilemmas related to the legitimacy of rule immediately arise. Future political theorists will therefore have to overcome the exceedingly difficult problem of legislating the temperance and virtue needed for the ecological survival of a steady-state society without at the same time exalting the few over the many and subjecting individuals to the unwarranted exercise of administrative power or to excessive conformity to some dogma.

Politics   Because the free play of market forces and individual initiative produces the tragedy of the commons, the market orientation typical of most modern societies will have to be strictly governed. If we want a viable and attractive steady-state society, we must determine its basic principles and then put them into effect in either a planned or a designed fashion (see Box 29 for a discussion of the important distinction between the two). Another way of stating this is to say that we must move from non-politics toward politics. Laissez faire is a device for making political decisions about the distribution of wealth and other desired goods automatically and rather non-politically, instead of in face-to-face political confrontation (as happened, for example, in the Greek democracies). As noted in Chapter 6, shifting from a process or non-political mode to an outcome or political mode holds serious dangers, for the political struggle can escalate into revolution and counter-revolution (again as happened in the Greek democracies).* It will thus be necessary for the political and social philosophers of the steady state to discover principles of legitimacy, authority, and justice that will keep the political struggle within reasonable bounds. Yet even if they are successful in this task, they are unlikely to be able to discover a device as effectively non-political as the market for making political decisions. At least at the outset, those who live in the steady state will therefore have to be genuinely political animals in Aristotle's sense, self-consciously involved in designing and planning their community life.

Stewardship   The character of economic life will change totally. Ecology will engulf economics, and we shall move away from the values of growth, profligacy, and exploitation typical of "economic man" toward sufficiency, frugality, and stewardship. The last especially, at least in its minimal form of trusteeship, will become the cardinal virtue of ecological economics. To use the analogy of ecological succession, we shall move

---

* The discussion in Box 29 suggests that there is thus considerable merit in agreeing *politically* on design criteria for the state that will minimize the scope of politics and political decision making thereafter. As a mode of politics, non-politics has considerable merit.

29

Planning Versus Design

There is a subtle and often overlooked but important distinction between planning and design. Both are attempts to achieve a desired
real-world outcome by influencing nature. Although the difference
is sometimes obscure in practice, planning is the attempt to produce
the outcome *by actively managing the process,* whereas design is the attempt to produce the outcome *by establishing criteria to govern the
operations of the process so that the desired result will occur more or less automatically without further human intervention.* Because of the scale and
complexity of human activities, planning inevitably requires large
bureaucracies and active intervention in people's lives. The Soviet
Union's economic planning machinery was perhaps the most
elaborate, but virtually all modern societies (to a considerable extent,
even developing societies) are increasingly pervaded by the apparatus
of planning. As a result, we have all become personally familiar with
the inefficiencies, limitations, and costs of such cumbersome and
bureaucratic social control. Thus the apparent need for even more
planning to cope with the exigencies of ecological scarcity raises the
frightening and repugnant prospect of minute and total daily supervision of all our activities, in the name of ecology, by a ponderous
and powerful bureaucratic machine, a veritable Orwellian Big
Brother.

However, this is not inevitable, for we can adopt a design approach instead of a planning approach to the problematique of
ecological scarcity. By self-consciously selecting and implementing a
set of design criteria aimed at channeling the social process quasi-
automatically within steady-state limits, we can avoid having constantly to plan, manage, and supervise. An example of such a design
criterion comes from social critic Ivan Illich (1974a), who has
proposed an absolute, across-the-board speed limit of 15 to 25 miles
per hour—that is, the speed of a bicycle. Illich believes that adoption
of this single proscription would eliminate most of the worst
ecological and social consequences of high energy use without subjecting individuals to daily bureaucratic regulation. One can debate
the merits of this particular proposal, but it nevertheless illustrates

how powerfully the adoption of a few simple (albeit drastic in terms of current values) design criterion could indeed have major social impacts sufficient to produce a steady-state society without also creating a Big Brother to supervise it. Another well-known example of a design approach to solving environmental problems is economist Kenneth Boulding's proposal for achieving population control with marketable baby licenses; once the basic idea was accepted, the system would operate with minimal bureaucratic supervision, and people would be able to determine for themselves how to respond to the market pressures created by the licensing system (that is, they could have as many children as they wanted by buying additional licenses from those who wanted few or no children) (1964, pp. 135–136). The Clean Air Act of 1990 adapts a design approach with a market option to sulfur dioxide emissions. The Act sets a cap on sulfur dioxide emissions for each utility company. If a utility seeks to increase its emissions over the cap, it must pay another utility to make an equivalent reduction.

It should be evident that the design approach has substantial advantages over planning, a point not lost on our founding fathers, who unconsciously favored a design strategy in establishing our system as a political and economic marketplace governed predominantly by laissez faire. Now, of course, these particular design criteria are inappropriate for our changed circumstances, so they must be exchanged for new ones, but it would seem wise to emulate our founding fathers in their preference for design over planning.

It should also not be forgotten that design is nature's way. As a consequence of certain basic physical laws (the design criteria), natural systems and cycles operate automatically to produce an integrated, harmonious, self-sustaining whole that evolves in the direction of greater biological richness and order, eventually reaching a climax that is the ultimate expression of the design criteria. The essential task of the political and social philosopher of the steady state is therefore to devise design criteria that will be just as effective and compelling as those of nature in creating an organic and harmonious climax civilization but that are neither so ruthless nor so cruel. In other words, what are the humane alternatives to nature's wars, plagues, and famines as design criteria for a steady state?

from pioneer to climax economics; the rapid growth and exploitation of new possibilities typical of the pioneer stage will give way to a state of stable maturity in which maximum amenity is obtained from minimum resources, and energy is devoted primarily to maintenance of the current capital stock rather than to new growth. In short, quality will replace quantity, and husbandry will replace gain, as the prime motives of economic life. Learning to live with scarcity does not mean learning to live without. If the ideal of stewardship were more positively embraced, then, as numerous human ecologists have suggested, economic activities could be designed to "woo the earth" so that it would become a garden yielding beauty as well as ample subsistence (Dubos 1968). Approached in this spirit, the economics of the steady state, no matter how frugal and careful, need not involve joyless self-abnegation on the part of individuals, for they would be participating in what could be a deeply satisfying civilizational task. Yet it must be acknowledged that many people living today might not share this sanguine assessment of the potential for delight and self-fulfillment in a steady-state economy characterized by frugality.

Modesty   In the area of cultural norms the changes are less predictable. However, once the limitations nature imposes on people have become clearer, Faustian striving after power and "progress" should give way to modesty of both ends and means. One human ecologist describes the impending social change as one from tragedy to comedy (Meeker 1974): We shall abandon the tragic hero's deadly serious and angst-ridden quest for greatness and new fields to conquer (which usually ends badly for himself and others) and learn instead cheerfully to enjoy the simple pleasures of ordinary life. The people inhabiting the future steady state could therefore be more relaxed, playful, and content than those living today, who must spend large amounts of energy constantly striving just to keep afloat in the waves of change that inevitably accompany rapid material growth. Moreover, although some profess to see the steady state as tantamount to rigor mortis, once the getting and spending of material wealth have ceased to be the prime determinant of status and self-esteem, the search for social satisfaction and personal fulfillment can turn toward the artistic, cultural, spiritual, intellectual, and scientific spheres—none of which are seriously confined by physical limitations. (Even space programs and other types of "big science" are possible in the steady state, provided the political will exists to expend scarce resources in this fashion.) In sum, there is no intrinsic reason why a steady-state society, despite its material frugality, should suffer from cultural stagnation, nor is there any reason why personal and cultural life should not be at least as rewarding as it is in today's industrial civilization. But the rewards must necessarily be rather different, for the culture of the steady state will certainly be far more frugal and modest than our own.

Diversity   The steady-state society should be less homogeneous and more culturally diverse than our own. As noted previously, the pressures of ecological scarcity urge upon us technological pluralism, some more labor-intensive modes of production, and smaller-scale enterprise adapted to local ecological realities. As a result, populations are likely to be spread more evenly over the land and to be more self-sufficient in the basic necessities. Thus extreme centralization and interdependence, which depend on high levels of energy use, should give way to greater decentralization, local autonomy, and local culture. The extent of the reversion to diversity and decentralization is unpredictable because, barring a total collapse of technological civilization, the continued existence of modern communications is likely to forestall a return to the era when each locality was in effect a little country all its own. In addition, the kind of political and economic arrangements ultimately adopted will largely determine how far this process goes. That is, diversity, decentralization, and local autonomy seem to fit more naturally with some of the political and economic choices mentioned above than with others. For example, a decision in favor of a planned rather than a designed steady-state society would actually be a decision in favor of maximum standardization and central control. Nevertheless, the limitations on energy and material use in any conceivable steady state seem likely to lessen substantially the current high degree of homogenization, centralization, and interdependence.

Holism   Because the kind of one-dimensional thinking that created the crisis of ecological scarcity in the first place will no longer be tolerable, there will be a decisive movement away from scientific reductionism (the assumption inherited from Francis Bacon that nature is to be understood by dissecting it into its smallest constituent parts) toward holism, the contrary assumption that nature is best understood by focusing on the interrelationships that link all parts of the whole. In other words, what has been called the "systems paradigm" will become the dominant intellectual and epistemological mode; biology, or more specifically ecology, will replace physics as the master science. The effects of this intellectual inversion are likely to be profound. For example, embracing holism will tend to make thinkers generalists first and specialists second. More important, however, greater holism would alleviate many current social ills. Reductionist science has left most individuals psychologically adrift—by ruthlessly destroying older world views without putting anything in their place, by fragmenting the corpus of knowledge, and by alienating people from nature. A new synthesis based on a fuller understanding of the total ecology of the planet would go a long way toward making the average person feel at home in the universe once again.

Morality   Finally, the steady-state society will undoubtedly be characterized by genuine morality, as opposed to a purely instrumental set of ethics. It seems extremely unlikely, for example, that a real commitment to stewardship could arise out of enlightened self-interest; it will require a change of heart. But the same could be said about many of the other developments we have outlined. Indeed, the crisis of ecological scarcity can be viewed as primarily a moral crisis in which the ugliness and destruction outside us in our environment simply mirror the spiritual wasteland within; the sickness of the earth reflects the sickness in the soul of modern industrial individuals, whose whole lives are given over to gain, to the disease of endless getting and spending that can never satisfy their deeper aspirations and must eventually end in cultural, spiritual, and physical death. If this assessment is correct, then the new morality of the steady state must involve a movement from matter toward spirit, not simply in the sense that material pursuits and values will inevitably be deemphasized and restrained by self-interested necessity, but also in the sense that there will be a recovery or rediscovery of virtue and sanctity. We shall learn again that canons higher than self-interest and individual wants are necessary for people to live in productive harmony with themselves and with others. Thus the steady-state society, like virtually all other human civilizations except modern industrialism, will almost certainly have a religious basis—whether it be Aristotelian political and civic excellence, Christian virtue, Confucian rectitude, Buddhist compassion, Amerindian love for the land, or an amalgam of these and other spiritual values.

Post-modernity   To sum up, ecological scarcity obliges us to abandon most basic modern values in favor of ones that resemble pre-modern values in many important respects. This does not mean that we shall simply revert to an earlier mode of existence, although this is what could happen if we fail to exercise forethought and self-restraint. Again using the analogy of ecological succession, it can be said that the very success of the industrial stage of civilizational succession has created conditions under which, to avoid a simple relapse into pre-modern civilization, we must move to a new, higher, more mature stage of post-industrial or post-modern civilization that shares many features of earlier civilizations while being something new in world history. Thus the emergence of the steady-state society will in one way or another bring the modern era to a close. It is now for us to decide whether we will accept the challenge implicit in the crisis of ecological scarcity by creating a genuinely post-modern civilization that combines the best of ancient and modern.

The Roots of Wisdom: Political Philosophy

Having seen what some of our choices on the path to the steady-state society might be, we come to the second and harder question: Where shall we find the wisdom to make such fateful choices and to guide us in the momentous enterprise of building a post-modern civilization? There is obviously no straightforward answer to this question, but there are some discernible avenues of approach. One is to make a profound study of the ills of industrial civilization.

A logical starting point in this endeavor is ecological philosophy, the attempt to discover the larger meaning and practical lessons of human ecology. Although it engages throughout in ecological philosophy, this book emphasizes politics, and it must be complemented by the works of other writers who have asked nature how people can live in harmony with it.*

The next step toward mastery of the problem is to study the work of contemporary radical social critics who judge the industrial paradigm from what could be called a post-industrial perspective.† That is, whatever the differences among them, they all examine the proudest successes of industrial civilization, such as science and development, and find them more or less pernicious. Accordingly, they propose not reforms but the creation of an entirely new post-industrial order. (Thus they do

---

* See, for example, Dubos, R. *So Human an Animal* and *A God Within*; Leopold, A. *A Sand Country Almanac*; Shepard, P. and McKinley, D. *The Subversive Science: Essays Toward an Ecology of Man*; Nash, R. *The Rights of Nature: A History of Environmental Ethics*; Rolston, M. *Environmental Ethics: Duties to and Values in the Natural World*; Berry, T. *The Dream of the Earth*; Regan, T. ed., *Earthbound: New Introductory Essays in Environmental Ethics*; Scherer, D. ed., *Upstream/Downstream*; Wenz, P. *Environmental Justice*; and Milbrath, L. *Envisioning a Sustainable Society*.

† See, for example, Bookchin, M. *The Ecology of Freedom: Remaking Society*; Roszak, T. *The Making of a Counter Culture* and *Where the Wasteland Ends*; Maslow, A. *The Psychology of Science* and *The Farther Reaches of Human Nature*; Illich, I. *Deschooling Society*, *Tools for Conviviality*, and *Energy and Equity*; Schumacher, E. *Small Is Beautiful: Economics as if People Mattered*; and Mumford, L. *The City in History* and *The Myth and the Machine*.

not simply repeat old criticisms. Their work looks forward, however much it may sometimes seem to hark back to the concerns of the earliest critics of the Industrial Revolution.) We must ponder unflinchingly the secular heresies of these post-industrial critics in order to liberate ourselves from inherited prejudices.

As important as it is to analyze modern industrial civilization in the light of the crisis of ecological scarcity, it ought to be evident that the questions raised throughout this book are scarcely new but in fact are modern variations on ancient themes. This being the case, once we have understood those things that make us unique, we must expect to receive the greater part of our guidance from the past—particularly from political philosophy, the long and rich tradition of discourse that is concerned precisely with how people can best live in community.

We have already seen that the values of a steady-state society would have to resemble pre-modern values in many important respects, but steady-state values bear a particularly uncanny resemblance to the ideas of the British conservative thinker Edmund Burke, the last great spokesman for the pre-modern point of view. For instance, the major tenet both of ecological philosophy and of Burke is trusteeship or, better yet, stewardship. Burke wrote mainly about humanity's social patrimony rather than its natural heritage, but from the nature of his reasoning it is clear that he meant both: The current generation holds the present as a patrimony in moral entail from its ancestors and must pass it on to posterity—improved if possible, but at all costs undiminished. Beyond this general overriding imperative, almost all of Burke's ideas resonate strongly with those of the ecological philosophers:

General skepticism about the possibility of "progress"

Awareness that the solution to one problem generates a new set of problems

Acceptance of human limits and imperfections

The need for organic change in order to preserve the balance and harmony of the whole social order

The interdependence and thus mutual moral bondage of society

The need to check aggressive self-interest, the contingent and situational nature of morality

The inevitability and desirability of diversity among human beings both within societies and among societies

Progress as a gradual evolution toward what is immanent in a historical society

The social order as part of, or as an outgrowth of, the natural order

Politics as the balancing of many conflicting and equally legitimate claims to achieve for humanity the best possible state given the objective situation

Burke also grasped the profound social implications of the Enlightenment and the Industrial Revolution. He foresaw, for example, that turning the direction of society over to "sophisters, economists, and calculators" (his epithets for the amoral capitalists who typified the new way of accumulating wealth) would destroy community, lead to the atomization of society, and set one person against another in an endless and self-destructive struggle for gain. He also saw that zeal for liberty and equality in the abstract would soon lead to the destruction of all the "little platoons" (that is, the guilds, communes, and other intermediate corporate bodies) intervening between the individual and the state, so that individuals would eventually be left standing alone and defenseless before an all-powerful state that in theory represented their interests but in practice was largely beyond their control. As we have seen, both these issues are closely intertwined with our general analysis of ecological scarcity.

Ecology broadly defined is thus a fundamentally conservative orientation to the world. Indeed, one biologist has called the climax state (the natural analog of the steady state) "a perennial feudal society" (McKinley 1970). However, it by no means follows that we must adopt Burke's political doctrines. Rule by a landed aristocracy would be anachronistic at best and reactionary at worst. Yet in our search for a set of social and political ideas that correspond to an ecological world view, Burke will surely have much to teach us.

Human ecology is also consonant with even older bodies of political thought, such as the classical tradition. In Book Two of his *Republic,* Plato says that although people need tools, some division of labor, and the like—in other words, a modest level of development—in order to live a civilized and humane life, they do not seem to know when to stop. Thus they overdevelop, and the consequences include luxury, vice, class struggle, war, and many other ills. To prevent this, says Plato, we must restrain people with wise rule by philosophers who know that what people desire is not always desirable for them and that true justice requires the establishment of controls to maintain the balance and harmony of the whole. The classical tradition also distrusts technology: Just as excessive or uncontrolled economic development threatens to turn the direction of society over to monied interests and the vagaries of the market, so, too, uncontrolled technological change undermines politics, the rational (in the broadest sense) direction of human affairs, by turning social decisions over to the apparent imperatives of mere things.

The more modern anarchist tradition may also contain valuable lessons, for decentralization, local autonomy, modesty, community, and other characteristics of the steady state seem favorable to developments in this direction. Indeed, to the extent that the environmental movement shares a common political ideology, it is predominantly anarchist. Moreover, the whole issue of a planned versus a designed steady state cuts so close to the central problem of anarchism that it is perhaps the most directly relevant body of theory for many of the critical issues raised in the preceding section.

Western political philosophy taken as a whole contains many valuable lessons. Let us examine briefly two of the most important and obvious ones.

First, it is only a slight exaggeration to say that all political theory teaches the necessity of prudence, which *Webster's Third New International* tells us is a comprehensive term implying "a habitual deliberateness, caution, and circumspection in action," further qualified as (1) "wisdom shown in the exercise of reason, forethought, and self-control," (2) "sagacity and shrewdness in the management of affairs...shown in the skillful selection, adaptation, and use of means to a desired end," (3) "providence in the use of resources," and (4) "attention to possible hazard or disadvantage." As the preceding discussion has amply demonstrated, the behavior of industrial civilization has been imprudent in the extreme. Unlike abundance, however, scarcity is extraordinarily intolerant of lapses in prudence, so this virtue must certainly be a part of the steady-state solution, regardless of the particular doctrinal and institutional form it eventually takes. The lessons of prudence can of course be acquired in the school of hard knocks, but they are perhaps best learned from the great political theorists of the past (as well as the political historians, such as Thucydides and Tacitus, who have traditionally been read along with them) for whom prudence is the cardinal virtue of politics.

A second indispensable political virtue is individual self-restraint. The Epigraph to this book, taken from Burke, explains why in the lucid prose for which he is famous. Reduced to its essentials, his argument states that

Man is a passionate being.

There must therefore be checks on will and appetite.

If these checks are not self-imposed, then they must be applied externally by a sovereign power.

We have seen how this problem has surfaced again and again in our analysis—in the Hobbesian dynamics of the tragedy of the commons, in the consequences of accepting the Faustian bargain of nuclear technology, and so on. The essential political message of this book is that we must learn ecological self-restraint before it is forced on us by a potentially monolithic and totalitarian regime or by the brute forces of nature. We are currently sliding by default in the direction of one (or both) of these two outcomes. Only the restoration of some measure of civic virtue (to use the traditional term) can forestall this fate, and the necessary lessons in virtue are, again, better learned from political philosophy than from personal suffering.

If we are to take political philosophy seriously again, we should broaden our perspective beyond the specifically Western tradition of political thought, for the political history and theory of other civilizations have much to teach us. For example, given the probable nature of the steady-state society, there is much in our own political tradition that seems to favor the revival of something like the classical city state. However, the Western political tradition never satisfactorily resolved the problem of keeping peace between city states. Thus it might be valuable to study the *millet* system of the Ottoman Empire, which granted the widest measure of local autonomy to individual cities and provinces while still providing them with peace and most of the other benefits of a larger political community. On the other hand, it might be argued that reversion to the city state is unrealistic given the numbers of people to be accommodated and the size of the territory to be governed. If so, then the history and political thought of agrarian societies—especially China from the Shang Dynasty to Mao—are worthy of the closest study. Similarly, feudal societies, whose resonance with ecology we have noted, should contain many important lessons; Westerners would do well to go beyond their own medieval history to study Tokugawa Japan, which existed in almost total autarky for several centuries yet supported (albeit frugally) a rather large population at a high cultural level.

However, we must not expect political theory and history to provide us with specific solutions or even neatly packaged object lessons on what not to do. What is essential is that we once again approach politics from a philosophical perspective instead of grasping after easy answers that fit current prejudices. As Ivan Illich (1974b) says on the subject of our modern dependence on "energy slaves," "The energy crisis focuses concern on the scarcity of fodder for these slaves. I prefer to ask whether free men need them." Once—*and if*—we approach the totality of our problem with ecological scarcity from this perspective, asking the questions that really need to be asked, then solutions informed by political wisdom will certainly emerge.

The Roots of Wisdom: Ultimate Values

Political philosophy alone is not enough. The wisdom to ask the right questions comes ultimately from so-called higher values, and all the great theorists of politics invoke them as an essential element in their political arguments. However, to assert the necessity of ultimate values in this day and age is heretical. Because scientific orthodoxy maintains that values have no epistemological standing, any statement that one value is to be preferred to another is therefore scientifically meaningless. But because science is our standard of social reality, value questions must not be socially meaningful either. Similarly, the modern liberal-democratic orthodoxy holds that people have an inalienable right to create their own values; accordingly, any attempt to judge these values or replace them with others in the name of some nebulous ideological concept such as "the common interest" is taken as anti-liberal and ultimately fascist. Thus, to the ideologically committed scientist and democrat, all values are equal, and politics can be no more than the clash of personal and factional interest, moderated only slightly by some minimal ethical conceptions about what constitutes a just division of the spoils. Politics, therefore, comes to be devoted almost exclusively to the utilitarian satisfaction of desire or appetite, which, in the absence of any higher values, necessarily becomes the sole measure of individual and social good. The idea that public authority might exist in part to direct people toward virtuous ends becomes anathema.

Yet wisdom, if only the rough and ready kind acquired by everyday living, tells us that not all values are equal and that virtue matters. In practice, science and democracy alike would be a shambles without the implicit values that govern them; indeed, "science" and "democracy" are themselves high-level values that generate the criteria by which utilitarian political decisions can be made in industrial civilization. We know too that the Protestant faith, even though it was not everywhere established, was the unofficial religion of the Industrial Revolution, providing a transcendental explanation of the human condition as well as justification for acquisitiveness and other bourgeois traits. Our founding fathers were motivated by deep religious faith to set up our political institutions "under God." We used to have, in effect, a positive standard of right and virtue, one that still lingers on in an unconscious and degenerate form. Thus we have had a value-based civic religion all along. We have simply never acknowledged it as such.

Of course, civic religions are never easily changed, and resistance to turning politics once again into more than a mere clash of interests will be very high. We must also recognize that, given the litany of horrors that is human history, the widespread suspicion of values is not without foundation. Too many crimes have been committed by leaders and peoples who were convinced that they had God on their side. Moreover, it is not always an easy task to distinguish genuine needs, the satisfaction of which is essential to human well-being, from mere wants, the fulfillment of which is dispensable without real sacrifice. Nevertheless, however difficult and controversial the task, we have no choice but to search for some ultimate values to inform the construct of a post-modern civilization. What follows is an effort to indicate what these values ought to be, but the discussion is even more condensed, tentative, general, and personal than the previous discussion of political values. It merely suggests that there is already remarkably widespread agreement on what an appropriate set of ultimate values ought to be under any circumstances and that these values favor a certain type of steady state.

We noted earlier that the crisis of ecological scarcity is fundamentally a moral and spiritual crisis. In looking out at the ecological ruin we have made of the earth, we see what manner of people we have become. Worse, the degraded environment so impoverishes us spiritually that we are likely to cause further ecological ruin. But the point has been reached where such a vicious circle can no longer continue without serious consequences for humanity. The earth is teaching us a moral lesson: The *individual virtues that have always been necessary for ethical and spiritual reasons have now become imperative for practical ones.* These virtues were pithily summarized in the fifth century B.C. by the Taoist sage Lao Tzu:

> Nature sustains itself through three precious principles, which one does well to embrace and follow.
> These are gentleness, frugality and humility [Chap. 67].

Implicit in gentleness, frugality, and humility are simplicity and closeness to nature. *Walden,* Henry David Thoreau's famous symbolic critique of an American society rapidly headed in the opposite direction, is an extended sermon on the necessity of natural simplicity as the only way to avoid living the quietly desperate lives of those weighed down by striving for power, possessions, and position. Such simplicity does not mean rejection of all progress, as Thoreau makes clear in his chapter on "Economy."

Though we are not so degenerate but that we might possibly live in a cave or a wigwam or wear skins today, it certainly is better to accept the advantages, though so dearly bought, which the invention and industry of mankind offer. In such a neighborhood as this, boards and shingles, lime and bricks, are cheaper and more easily obtained than suitable caves, or whole logs, or bark in sufficient quantities, or even well-tempered clay or flat stones. I speak understandingly on this subject, for I have made myself acquainted with it both theoretically and practically. *With a little more wit we might use these materials so as to become richer than the richest now are, and make our civilization a blessing. The civilized man is a more experienced and wiser savage* [1854, p. 295, emphasis added].

It is of course quite obvious that development as we know it, in all its complexity, violence, prodigality and pride, is utterly noxious to these fundamental ethical–spiritual principles. The greatest sociologists and political economists would hardly disagree. In his classic work *The Protestant Ethic and the Spirit of Capitalism,* the renowned nineteenth-century German sociologist Max Weber foresaw the spiritual death that awaited an increasingly rationalized, bureaucratized society: "Specialists without spirit, sensualists without heart; this nullity imagines that it has attained a level of civilization never before attained" (cited in Burch 1971, p. 159). John Stuart Mill, one of the ablest and most ardent philosophical defenders of liberty and other bourgeois values, was nevertheless distressed by "the trampling, crushing, elbowing, and treading on each other's heels" that the relentless struggle to "get on" seemed inevitably to produce (1871, p. 748). Mill also foresaw that the long-term consequences of development would be pernicious.

> If the earth must lose that great portion of its pleasantness which it owes to things that the unlimited increase of wealth and population would extirpate from it, for the mere purpose of enabling it to support a larger, but not a better or a happier population, I sincerely hope, for the sake of posterity, that they will be content to be stationary, long before necessity compels them to it [p. 751].

Even Adam Smith, perhaps the person most directly responsible for the materialistic and economic nature of modern civilization, clearly believed that one who pursued wealth was prey to vanity, greed, and other foolish and ignoble motives (1792, III-2). The eminent twentieth-century economist John Maynard Keynes, whose fame and influence ironically rest primarily on his prescriptions for keeping the engine of economic growth in high gear, was still more adamantly opposed to the values of "economic man."[*] Noting that the whole long era of development has "exalted some of the most distasteful of human qualities into the position of the highest virtues," he hoped for its speedy end, so that men and women would once more be

> free...to return to some of the most sure and certain principles of religion and traditional virtue—that avarice is a vice, that the exaction of usury is a misdemeanor, and the love of money is detestable, that those walk most truly in the paths of virtue and sane wisdom who take least thought for the morrow. We shall once more value ends above means and prefer the good to the useful. We shall honour those who can teach us how to pluck the hour and the day virtuously and well, the delightful people who are capable of taking direct enjoyment in things, the lilies of the field who toil not, neither do they spin [1971, p. 192].

It follows from what these writers say (and from similar sentiments expressed by people of every age and tradition) that nothing of real value would be lost if development were to cease. Rather, the likelihood of men and women leading reasonably happy, sane, fulfilled, and harmonious personal lives would be enhanced.[†] Moreover, once the ultimately fruitless and self-destructive quest for ever more private affluence was abandoned, public amenity would be free to grow and to produce all the kinds

---

[*] Paradoxically, Keynes believed that because "foul is useful and fair is not," we could not afford to abandon these values until we were out of "the tunnel of economic necessity" a hundred years hence-that is, until we had abolished scarcity. Even if this were possible, the problem with this qualification is that the tunnel is likely to be endless unless one learns to say, "Enough!" For growth simply produces more mouths and greater wants and is thus self-defeating. Furthermore, even Keynes suggested that economics be radically devalued during our passage through the tunnel.

[†] The available empirical evidence supports the position that economic development is largely irrelevant to personal happiness. Easterlin (1973) shows that people's sense of economic well-being depends primarily on their relative standing. (Thus the American poor, most of whom are quite rich by any historical or comparative standard, nevertheless feel acutely deprived.) The popular demand for more growth is therefore largely motivated by a desire to keep up with (or catch up with) the Joneses. Unfortunately, this is a never-ending pursuit; a few Joneses will always pull ahead of the crowd and inspire emulation, so the package of goods that one needs to feel non-poor grows constantly. Relative equality and distributive justice thus seem more important for individual happiness and well-being than does the absolute level of production.

of cultural riches people have been able to enjoy in the past, even if the gross quantity of production were less than it is today. Indeed, social critic Lewis Mumford argues persuasively that the inhabitants of ancient Pompeii, an ordinary Roman provincial town, enjoyed a quality of life superior in many important respects to that attainable in present-day California (1973, pp. 462–473). Nor should we forget the cultural glory of Athens, Florence, Kyoto, and other ancient centers of civilization whose achievements antedate the Industrial Revolution. Thus development appears to be virtually irrelevant to cultural richness and progress; social arrangements, not wealth in itself, seem to determine the level of social amenity.* In sum, "with a little more wit we might...become richer than the richest now are, and make our civilization a blessing."

The Minimal, Frugal Steady State

The nature of the most desirable type of steady state should now be clear. We saw earlier that the attempt to achieve a high-throughput or maximum-feasible steady-state society involved a Faustian bargain fraught with dire political consequences. Now we see too that the maximum-feasible steady state, which aims at gratifying as far as possible the materialistic and hedonistic appetites of the populace, flies in the face of the lessons to be discovered in political philosophy and in the ethical–spiritual teachings of wise people of every age and tradition. In other words, political and spiritual wisdom alike urge the adoption of the minimal, frugal steady state as the form of a post-industrial society.

Politically, a minimal steady state would, as its name implies, follow the favorite prudential maxim of our founding fathers: "That government is best that governs least." Where this seems to lead is toward a decentralized Jeffersonian polity of relatively small, intimate, locally

---

* The empirical evidence again supports the impressionistic judgment that when it comes to culture, bigger is not necessarily better. Some countries with no more than half the U.S. per capita consumption of energy actually outrank the United States statistically in important indicators of the quality of life, such as the rate of infant mortality and the number of persons per hospital bed, the number of books published per year per million persons, and even public expenditures for education as a percentage of national income (Watt 1974, Chap. 11). Of course, some minimum level of wealth is necessary for a reasonable level of comfort, but the level of production with appropriate technology in a steady-state society of reasonable population should be high enough to support moderate and judicious cultural aspirations.

autonomous, and self-governing communities rooted in the land (or other local ecological resources) and affiliated at the federal level only for a few clearly defined purposes. It leads, in other words, back to the original American vision of politics. Unlike mass society, such a minimal polity can place primary reliance on the inherent virtue of the citizen (or on the power of local public opinion to recall a straying citizen to her or his civic duty). This minimizes the perceived restrictions on individual freedom (in accordance with the principle of macro-constraint and micro-freedom described in Chapter 4, as well as the preference for design over planning expressed in Box 29). Of course, as is unfortunately true of all forms of political association, such a polity also has its attendant dangers. Local tyranny is first among them. However, the tyranny currently exercised over our lives by impersonal forces beyond any individual's capacity to comprehend, much less control, is far greater; we are largely at the mercy of market forces, efficiency, technological change, radical monopoly (that is, our almost total dependence on the ministrations of doctors, lawyers, teachers, and other professionals), and so on. By contrast, local tyrants are highly visible and few in number, so that at least one would know against whom to revolt. Cities would still exist within this basically Jeffersonian polity, but they should be less of an instrument for exacting an economic surplus from the countryside than they are now. Eventually they would probably come to resemble the pre-modern city state in size and spirit, a highly desirable development if the countless historians and political philosophers who have praised this organic form of political and social community are to be believed.* The minimal, frugal steady state would thus predominantly consist of medium sized communities and rural areas, but through modern communications, what Karl Marx called the idiocy (the political, social, and cultural unconsciousness) of rural life should be avoidable.

---

* As we noted before, an alternative to the city state as a primary model of political association is the agrarian empire, which has certain undeniable virtues and some correspondingly large drawbacks, as Maoist China seems to have shown. Nevertheless, given the large numbers of people in the world and the realities of international politics, a degree of international decentralization, decoupling, and autarky sufficient to support Jeffersonian politics at the local level may simply be unattainable. However, any form of minimal steady-state society would have to be supported by a large measure of international decentralization, decoupling, and autarky, for the current degree of interdependence is politically destabilizing and economically disruptive. It amplifies and universalizes problems instead of solving them, and it generates strong pressures toward political centralization.

In economics, too, less is better. The goal is frugality, which means neither poverty nor abundance, but rather an ample sufficiency. The governing principle of economic life in a minimal, frugal steady state would be "right livelihood" (Schumacher 1973, pp. 50-58). Honest work from which one can derive satisfaction (not simply a wage), a sense of working in community with and for one's fellows, and an opportunity to develop one's native talents for the benefit of self and others are just as important as enough income for a decent and dignified material exist-ence. This view of economics does not reject productivity or technology in itself, but it does demand that the value and dignity of human labor be restored and that the economy be run "as if people mattered." Following these prescriptions would inevitably promote small-scale, self-sufficient, virtually self-administering, locally oriented and controlled enterprise that depends on simple, inexpensive, more labor-intensive means of production that are ecologically appropriate. All of this should put in-dividuals back in charge of their own economic destiny and produce a frugal economy compatible with the minimal polity we have described.

Although in our search for a suitable civil religion we certainly ought to cast our net as widely as possible, it by no means follows that we must convert to Taoism or other seemingly alien faiths political, economic, or religious. As we have seen, the political philosophy of Thomas Jefferson can supply a large part of the ideological foundation for a minimal, frugal steady state. What is lacking may be found in the ideas of Thoreau and all the other "literary" critics of American civilization, such as Melville and Whitman, who chided us for following a path that must eventually lead to the betrayal of our basic principles.* Moreover, although much in Christianity has rightly been found by critics (for example, Roszak 1973) to be ecologically objectionable (in that nature is not viewed as sacred and humans are given dominion over creation), others point out that stewardship and other Christian virtues could easily form the basis of an ecological ethic. The historian Lynn White, for example, though generally critical of Christianity, nevertheless sees St. Francis of Assisi, who wor-shiped nature and preached absolute identification and harmonious equality with the rest of creation, as a potential "patron saint of ecology" (1967). The ecological philosopher René Dubos, on the other hand, prefers St. Benedict of Nursia, because he did not merely love nature but also founded an order of monks who worked with the natural environ-

---

* Historian Leo Marx's excellent essay "American Institutions and Ecological Ideals" (1970) shows how the literary and ecological critiques of American society have merged.

ment to create beautiful, productive, and harmonious landscapes, thus translating the ideal of stewardship into physical actuality (1972, Chap. 8). By contrast, economist E. F. Schumacher prefers to focus not on a particular figure but on the "Four Cardinal Virtues" of Christianity— *prudentia, justitia, fortitudo,* and *temperantia*—which would, if observed, almost automatically produce a minimal, frugal steady-state society (1974). Christian leaders in the United States are even today developing a new "green gospel"; the Baptist, United Methodist, Congregational, and Presbyterian churches have produced policy statements on the environment. Individual leaders of the Roman Catholic, Jewish, and Protestant faiths have been developing links between ecology and theology. Thus self-renewal or self-transformation based primarily on native American and Western principles is eminently possible, for the minimal and frugal steady state is in complete accords with the best in our own tradition.

The Grand Opportunity

Other visions of the minimal, frugal steady state are possible, but the foregoing should suggest that feelings of despair and impotence are not appropriate responses to the crisis of ecological scarcity. True, the transition to any conceivable form of steady-state society is likely to be wracking and painful, but some measure of destruction is a precondition of rebirth, and the industrial era was a necessary but in many respects ugly and disagreeable phase in human history that we should rejoice to put behind us. Moreover, if we act wisely and soon, the transition need not involve unbearable sacrifices or frightful turmoil. Indeed, we are confronted not with the end of the world (although it will surely be the end of the world as we have known it) but with an unparalleled opportunity to share in the creation of a new and potentially higher, more humane form of post-industrial civilization. But we must not delay, for unless we begin soon, an ugly and desperate transition to tyrannical version of the steady state may become almost inevitable.

A Politics of Transformation

Seizing this grand opportunity will require a politics of transformation. Metanoia is tantamount to religious conversion and is therefore not easily achieved. As in the revolutionary eras of the past, inspirational leadership will be needed to steer us clear of anarchy and chaos during the transition. The critical question is whether such leadership will be provided, on the one hand, by a "man on horseback" or Big Brother's Ministry of

Propaganda or, on the other, by a Gandhi or a group of Jeffersonian "natural aristocrats" resembling those who founded the American Republic. Unfortunately, the breadth of mind and nobility of character typical of the latter are hard to find these days, for our institutions are designed to turn out experts and other brilliant mediocrities whose distinguishing characteristic is what Thorstein Veblen called a "trained incapacity" to see beyond their professional blinders. Even those who avoid this pitfall often cling to the past. Few even entertain the idea that many of the Enlightenment values central to modern civilization, such as reliance on reductionist information acquired through endless schooling, might have to be discarded. What therefore typically emerges is a call for change in general that ignores most of the critical issues or, what is worse, a call for change in the other fellow that implies little real change in, or commitment from, the would-be leader. But this cannot be effective; only leaders who have themselves fully embraced the future can provide inspirational leadership. Next to the sheer lack of time in the face of onrushing events, the paucity of genuine leaders is probably our most serious obstacle to a better and more humane future.

People of Intemperate Minds Cannot Be Free

Leadership is only part of the politics of transformation, for even the most inspired leaders can do only so much. We as individuals must also stop clinging to the past and embrace the future, accepting our personal responsibility for helping to make this vision of a more beautiful and joyful steady-state future come true. Like charity, transformation begins at home.

Above all, we must somehow learn the essential lesson of the crisis of ecological scarcity. In the words of Edmund Burke, "men of intemperate minds cannot be free," for their passions do indeed "forge their fetters." It is not that nature has made scanty provision for our wants; nature's economy is generous and plentiful for those who would live modestly within its circle of interdependence. It is our numbers and our wants that have outrun nature's bounty. If we will not freely and joyfully place "moral chains" on our will and appetite, then we shall abdicate to the brute forces of nature or to a political Leviathan what should be our own moral duty. Because even nature's bounty can be exhausted by the infinitude of human wants, only a life of self-restraint and simple sufficiency in natural harmony with the earth will allow us and our descendants to continue to enjoy life, liberty, and estate. Having freely chosen such a life, we shall find that it has its own richness, for we become rich precisely to the degree to which we eliminate violence, greed, and pride from our lives. When we have rediscovered this primordial wealth, we shall see something the wise have always known: The greatest value of the earth is, always has been, and always will be not that it is useful but that it is beautiful—and that it simply is.

Afterword

Rereading *Ecology and the Politics of Scarcity* 14 years after its initial publication brings a sad awareness: Despite the existence of a vast literature on ecology and environmental problems, several decades of political activism in support of environmental causes, and considerable government activity to redress environmental wrongs, both domestically and internationally, very little has changed in humankind's relationship to its natural milieu. Nature is still seen as either a mine or a dump and is treated accordingly. The basic laws of ecology are ignored, denied, and flouted, and humanity continues to hasten down the path to ecological perdition. Thus the concerns of Earthday 1990 differed hardly at all from those of Earthday 1970. Everywhere on the planet, we can observe more and more people making more and more demands on their environment—demands that lead to an inexorable drawdown of finite resources, an acceleration of biological extinction and habitat destruction, and an increase in environmental pollution and stress. In short, the planet has grown steadily more crowded, and its physical condition continues to deteriorate alarmingly. Many large areas, such as Eastern Europe, have become acknowledged environmental disaster zones. And the social degradation that results when human demands outrun natural capacities has increased markedly.

Nor has there been a significant shift in the appetite of modern civilization for more material wealth defined in terms of "affluence." Populations continue to demand "prosperity" and "progress" above all. Political leaders who try to talk ecological sense to their constituents are likely to suffer the consequences (witness the anger and disdain that greeted President Carter's mention of limits). Thus, although the increased media attention focused on environmental matters during the past two decades has certainly promoted greater ecological awareness, along with the nagging suspicion that we must one day mend our profligate ways, the watchword of industrial civilization continues to be what it has been since its inception: "Après nous le déluge."

Rereading the previous edition after a lapse of many years also makes me aware that much has changed. Many of the facts recorded there are now out of date or even erroneous, and certain of my specific opinions were clearly mistaken. At the time of writing, for example, it seemed that an age of massive dependence on nuclear power was virtually inevitable, but a number of factors (principally energy conservation) have sharply reduced the need for new generating capacity. Of course, the warnings of environmental critics, myself included, were instrumental in mobilizing public opinion against nuclear power, so our prophecies of doom by radiation became, in effect, self-denying—an occupational risk that all who deal with the future willingly run. I am therefore deeply grateful to Professor Boyan for preparing this revised edition and to the publishers for making the work available to a new generation of readers.

Nevertheless, with all due respect for the labor that such updating entails, what is important in this work is not particular facts but rather the essence of the argument, which is unchanged. It may be that the supply of fossil fuels is more ample now than it was 15 years ago—though at what political and economic cost?—or that population growth has declined in some areas of the world faster than originally projected. But for each fact I viewed too pessimistically before, there is another I saw with too much optimism. For example, the extent and rate of rain-forest destruction has accelerated, both from commercial logging and land hunger, and we now understand in more depth the high price we will pay for its loss. Similarly acid rain, which was scarcely mentioned in the first edition, now looms as a very large and intractable problem noxious to the health of both natural environments and human populations. In other words, the changes in particulars over the past 20 years have at best canceled each other out, allowing the pervasive trend toward ecological scarcity to continue unchecked.

In addition, the human mind tends to assign too much importance to the latest news from the environmental front, especially when it is dramatic or threatening. It thereby overlooks or underestimates the real dangers and their root causes. Thus, for instance, the Exxon Valdez oil spill in Prince William Sound generated far more concern than it really warranted in the overall ecological scheme of things. Nature copes relatively well with isolated insults, but it succumbs over the long term to the slow and steady assault of chronic pollution, which lacks the high visibility of an oil spill.

Moreover, certain apparently grave problems may well be taken care of by natural feedback mechanisms. For example, many believe that the extraordinary fauna of Africa is all but doomed by population pressure, which is the underlying cause of poaching, fencing, and habitat destruction. It now appears, however, that a tragic and classically Malthusian combination of famine and diseases, such as malaria and AIDS, will apocalyptically reduce this continent's population, with humanity's loss being the animals' gain. Similarly, although the threat of global warming is real, and the eventual outcome may be as grim as some predict, this is by no means certain. Hidden feedback mechanisms of which we are currently unaware may sop up the excess carbon dioxide, resulting in no discernible change in climate despite greatly increased emissions, and it is even conceivable that these same feedback mechanisms could, by increasing the formation of clouds, induce significant global cooling in the next century. (Various acts of God, such as extensive vulcanism, or of man, such as nuclear war, would also result in cooling.) In short, basing the ecological case on particulars—no matter how dramatic, menacing, or "scientific"—is bad strategy.

The appropriate strategy is to place particular environmental events in the context of the fundamental ecological dynamics that underlie them. Facts and opinions may change, but basic laws and principles, such as the laws of thermodynamics, do not. These laws tell us clearly that the planet has a limited carrying capacity and that, like every other creature, we will soon have to find some way to survive within that capacity. In other words, we must learn to live on our biological income, for we cannot continue to squander capital without becoming ecological paupers. We need to pay primary attention to the forces that create ecological scarcity. It will then be clear why seeming improvements in our situation, such as the discovery of larger supplies of fossil fuel, are strictly temporary and why almost all of our attempts to evade ecological limits are bound to be self-defeating in the long run.

Similar remarks apply to the political discussion in Parts II and III. The core of the argument is that because the basic premises of modern industrial civilization are anti-ecological, all its values, practices and institutions are grossly maladapted to the emerging age of scarcity. The transition from abundance to scarcity will consequently be extremely difficult and painful, for what is required is a revolution in thought and action. The suffering will be especially severe in the United States, because American politics is predicated on cornucopian abundance. Our political institutions were designed for the easy job of dividing the rich spoils of an almost virgin continent, not for the much harder task of allocating scarce resources. In this light, the details of environmental policy and of the political infighting that surrounds it are not of primary importance. What is crucial is to understand the fundamental nature of the underlying dynamic.

Unfortunately, my argument has sometimes been misinterpreted as Hobbesian in spirit, so a few words to guide the reader are in order. As I said in the Preface to the previous edition, my intention was never to offer solutions, much less *the* solution, to our political–ecological predicament. If I used Hobbes extensively in the analysis, it was because he is, to echo Marx's homage, "the father of us all"—that is, the author of the basic political theory underlying all forms of modern politics, capitalist and socialist alike. His thought therefore reveals most clearly the profound tensions and contradictions implicit in this theory of politics. In addition, Hardin's analysis of the tragedy of the commons unknowingly replicated Hobbes, and *Hardin's* solution was indeed explicitly Hobbesian. Thus, although I expanded on Hardin's argument and suggested various ways in which an ecological Leviathan might be tamed, I never offered this as my own solution. (In the same manner, I used Burke, Plato, Rousseau, Saint-Simon, and other theorists to *elucidate* the issues we now confront, not to provide ready answers.) When in the final chapter I did permit myself to suggest the direction in which we should look for humane long-term answers, the tenor of the discussion was explicitly anti-Hobbesian:

> The essential political message of this book is that we must learn ecological self-restraint before it is forced on us by a potentially monolithic and totalitarian regime or by the brute forces of nature. We are currently sliding by default in the direction of one (or both) of these two outcomes. Only the restoration of some measure of civic virtue (to use the traditional term) can forestall this fate....

Moreover, it can hardly be accidental that the book's epigraph, drawn from Burke, urges self-restrained control of will and appetite or that the final paragraph reprises Burke's theme. The overall spirit of the work is therefore far from Hobbesian.

In sum, far from being the solution, Hobbes is rather the essence of the problem. The current environmental problematique is a direct outgrowth of the system of individualistic and economic politics that evolved out of the social contract theory elaborated in *Leviathan*. Thus we shall not begin to deal with our problems constructively until we acknowledge that we must reassess our whole world view and way of life.

Reforms intended merely to sustain the current political system will only deepen the crisis in the long run and, what is worse, feed the forces that are already pushing us in the direction of Leviathan.*

Nevertheless, I now see that I could have made the reader's task easier by employing Alexis de Tocqueville's all-important distinction between government and administration. No human group can exist without government—that is, without a fundamental agreement among its members on how their communal life is to be regulated. On the other hand, human beings can get along quite well without administration. So-called primitive tribes dispense with it entirely (leading some early observers to conclude, erroneously, that these tribes had no politics). Conversely, it is possible to have a polity that is all administration and little or no government. Such was the unhappy plight of the former Soviet Union: A monstrous bureaucratic apparatus refused to die, while the political spirit that once gave it life and meaning was all but extinguished. Alas, we in the United States have also moved quite far in the direction of the administrative state, and our governing spirit too burns much less brightly than before. Thus it is not surprising that my critique, which could be misunderstood as providing the rationale for an even bigger and more coercive administrative state, would alarm some readers.

If we use de Tocqueville's language, however, matters become much clearer. We desperately need more government—that is, stronger checks on the competitive overexploitation of the ecological commons and therefore on human self-aggrandizement. But it does not necessarily follow from this that we need more administration. On the contrary, or so it seems to me, given the appalling record of the administrative state in this century, the better solution is to be found in the other direction. We need a form of government that is effective in obliging humankind to live within its ecological means but that does not require us to erect an ecological Leviathan (which, as many of my critics rightly pointed out, simply would not work in the long run). This is why I championed design over planning (Box 32) and macro-constraints in the service of micro-freedoms (Box 24). Similarly, I suggested (Box 25) that the new form of government would have to be based on an "ecological contract" to ensure that a basic harmony between man and nature was at the core

---

* I describe these forces in a work in progress entitled *Moribund Liberalism: The Tragedy of Enlightenment Politics,* which explores the contradictions and self-destructive tendencies of modern politics and traces them back to Hobbes's fundamental error, believing that politics could ever be separated from virtue.

of politics.* I also pointed out (Chapter 8) that the most critical need is for a change of heart, or "metanoia," because until we have embraced an ecological ethos, we cannot possibly have a genuinely ecological politics.

It should now be apparent why those among my critics who believe that democratic activism within the current system is the solution to our environmental problematique are almost certainly mistaken. Unless the system is fundamentally restructured, "more democracy" cannot be effective. This is not to say that environmental pressure groups cannot be effective in preventing this or that ecological atrocity, but isolated victories are not enough to deflect the drive for development at the expense of nature. Moreover, although people who see the answer in political activism may be noble champions of the democratic ideal, they do not seem to appreciate what they are up against. The trouble with interest-group politics is that, for all the reasons outlined in Chapters 5 and 6, special interests are bound to be victorious over the common interest in the long run. The prospect of the ecological interest somehow prevailing over the commercial, financial, and manufacturing interests whose money pays the media pipers and finances the electoral process is therefore remote, to say the least. (Paradoxically, the more successful environmental interest groups are, the more they begin to resemble other interest groups and the more they necessarily collaborate in an established political process that is both fundamentally anti-ecological and increasingly anti-democratic.) In short, as I said in the final chapter, "environmental politicking within the system can only be a rear-guard holding action designed to slow the pace of ecological retreat" (p. 282).

Those who rely on "more democracy" within the current political context also commit the grave error of confusing "interest-group liberalism" with genuine democracy. Nothing could be further from the truth. Our current political system is statist, not democratic. The federal government is a bureaucratic and electoral behemoth dedicated to one primary end: the satisfaction of human appetite at the expense of nature. Popular participation, such as it is, is token, minimal, symbolic; and the behemoth is largely beholden to organized and monied interests. Genuine democracy, by contrast, is *self*-government, and self-government

---

* In retrospect, even though it is an obvious play off *social contract,* the phrase *ecological contract* was not apt. Contract is a commercial concept, and by using it, Hobbes was (knowingly or unknowingly) creating a polity in which the economic motive would predominate to the detriment of other values. Thus the appropriate term would have been *ecological covenant;* both the denotation and the connotations of *covenant* suggest the very different spirit that must underlie an ecological political order.

requires two things: a population that is willing and able to restrain its own appetites for the sake of the common good and a social and economic structure that is amenable to, and indeed fosters, popular understanding and local control. Neither of these conditions now exists in the United States. Thus, I too believe that democracy can be part of the long-term solution to our ecological predicament, but only if it is genuine democracy—democracy that is fundamentally Jeffersonian and Thoreauvian in spirit and practice. It is, in fact, precisely toward such an ecological democracy that most of the final chapter points.

In sum, therefore, those who see in this work an anti-democratic justification for increased state power to enforce ecological imperatives have fundamentally misunderstood the argument. Instead, I raise and explore the political contradictions of the current system in the light of ecology and then point toward a political order that would be both more ecologically sound and more truly democratic than our current one. I trust that, read in this spirit, my work remains a useful and important critique of the American political system, as well as a significant contribution to the intensifying debate over man's place in nature and long-term future.

I regret to say, however, that I am now less inclined than before to believe that we will respond positively to the ecological challenge. Then, I saw us confronted with a grand opportunity to create a more humane future, and I believed that the transition to a more ecologically enlightened world view had already begun. But we have frittered away the two decades since the first Earthday without seizing this grand opportunity. To the extent that we have acted other than symbolically, we have spent the last 20 years doing all the easiest and least painful things. Now we must do the hard things: reshape basic attitudes and expectations, alter established lifestyles, and restructure the economy accordingly. But rather than adopt ecological principles for public policy, we seem to do everything we can to avoid facing up to the inevitability of limits and of changing our profligate way of life. In other words, time has grown shorter and the problems have become larger and more entrenched, but our resistance to dealing with them constructively has increased.

Worse, the end of the Cold War, far from bringing about an era of universal peace, has made the world a more complex, unstable, and dangerous place than it was a decade ago. History has reawakened from nearly a half-century of hibernation. Long-frozen political, ethnic, and religious passions have thawed, and the economic struggle both for markets and for resources has simultaneously heated up. The depressing prospect is for widespread conflict and turmoil in many areas of the globe. Preoccupation with geopolitical advantage, military realignments, and economic competitiveness seems likely to preempt the policy agenda in the coming decade—hardly the best political environment for making thoughtful and far-sighted decisions about a human future based on ecological harmony.

Nevertheless, I have by no means lost heart or hope, and I continue to work for the benefit of the Earth and the life it bears, as well as for the posterity that has never done anything for me. I urge the reader to do the same. Together we may make a difference.

William Ophuls
January 1992

Suggested Readings

Foreword

Bookchin, Murray
    1990 "Death of a Small Planet." in *Environment 90/91.* Reprinted from *The Progressive*, August 1989, pp. 19–23, ed. John Allen (Guilford, Connecticut: The Dushkin Publishing Group).
Borrelli, Peter
    1989 "Environmental Ethics—the Oxymoron of Our Time," *The Amicus Journal*, Summer, 39–43.
Global Tomorrow Coalition
    1990 *The Global Ecology Handbook,* ed. Walter H. Corson (Boston: Beacon Press).
McRuer, John D.
    1990 "Conventions vs. Greens." *World,* March-April, 5–6, 63.
Montagna, Donald
    1991 *Interview,* 27 July.

Introduction

Barker, Ernest, trans. and ed.
    1952 *The Politics of Aristotle* (New York: Oxford).
Boulding, Kenneth E.
    1961 *The Image* (Ann Arbor: Michigan).
    1964 *The Meaning of the Twentieth Century: The Great Transition* (New York: Harper and Row).
    1966 "Is Scarcity Dead?" *Public Interest* 5:36–44.
    1970 "The Economics of the Coming Spaceship Earth," in his *Beyond Economics: Essays on Society, Religion and Ethics* (Ann Arbor: Michigan), pp. 275–287.
Brown, Harrison
    1954 *The Challenge of Man's Future* (New York: Viking).

Cole, H. S. D., et al., eds.
  1973 *Models of Doom: A Critique of The Limits to Growth* (New York: Universe).
Dubos, Rene
  1969 "A Social Design for Science," *Science* 166:823.
  1972 *A God Within* (New York: Scribner's).
Durrenberger, Robert W.
  1970 *Environment and Man: A Bibliography* (Palo Alto: National).
Geertz, Clifford
  1966 *Agricultural Involution* (Berkeley: California).
Glacken, Clarence J.
  1956 "Changing Ideas of the Habitable World", in Thomas 1956, pp. 70–92.
Goldsmith, Edward, et al.
  1972 "A Blueprint for Survival," *Ecologist* 2(l):1–43.
Hardin, Garrett
  1959 *Nature and Man's Fate* (New York: Holt, Rinehart and Winston).
———, ed.
  1969 *Population, Evolution, and Birth Control: A Collage of Controversial Ideas* (2nd ed.;
         New York: W. H. Freeman and Co.).
Hume, David
  1739 *A Treatise of Human Nature, in Theory of Politics,* ed. Frederick Watkins (New
         York: Nelson, 1951).
Jacob, Francois
  1974 *The Logic of Life: A History of Heredity,* trans. Betty E. Spillman (New York:
         Pantheon).
Kuhn, Thomas S.
  1970 *The Structure of Scientific Revolutions* (2nd ed.; University of Chicago Press).
Malthus, Thomas Robert
  1798 *Essay on the Principle of Population As It Affects the Future Improvement of Society,*
         reprinted as *First Essay on Population, 1798* (New York: Kelley, 1965).
  1830 "A Summary View of the Principle of Population," in *Three Essays,on Popula-
         tion,* ed. Frank W. Notestein (New York: New American Library).
Marsh, George Perkins
  1864 *Man and Nature,* ed. David Lowenthal (Cambridge: Harvard, 1965).
Meadows, Donella H., et al.
  1972 *The Limits to Growth* (New York: Universe).
  1982 *Groping in the Dark: The First Decade of Global Modelling* (New York:
         Wiley).
Mesarovic, Mihajlo, and Eduard Pestel
  1974 *Mankind at the Turning Point: The Second Report to the Club of Rome* (New York:
         Dutton/Reader's Digest).
National Research Council
  1989 *Alternative Agriculture* (Washington, D.C.: National Academy Press).
Ornstein, Robert E.
  1972 *The Psychology of Consciousness* (New York: W. H. Freeman and Co.).
Osborn, Fairfield
  1948 *Our Plundered Planet* (Boston: Little, Brown).
Pearce, Joseph C.
  1973 *The Crack in the Cosmic Egg: Challenging Constructs of Mind and Reality* (New
         York: Simon and Schuster).

Polak, Frederik L.
  1961 *The Image of the Future* (New York: Oceana).
Seaborg, Glenn T.
  1970 "The Birthpangs of a New World," *The Futurist* 4:205–208.
Sears, Paul B.
  1935 *Deserts on the March* (Norman: Oklahoma).
  1971 Letter to *Science* 174:263.
Thomas, William L., Jr., ed.
  1956 *Man's Role in Changing the Face of the Earth* (University of Chicago Press).
Vogt, William
  1948 *Road to Survival* (New York: William Sloane).
Wolin, Sheldon S.
  1968 "Paradigms and Political Theories," in *Politics and Experience,* ed. Preston King and B. C. Parekh (Cambridge, Eng.: University Press), pp. 125–152.
  1969 "Political Theory as a Vocation," *American Political Science Review* 63:1062–1082.
Woodhouse, Edward J.
  1972 "Re-visioning the Future of the Third World: An Ecological Perspective on Development," *World Politics* 25:1–33.
World Commission on Environment and Development.
  1987 *Our Common Future* (New York: Oxford University Press).

Chapter 1

Adams, M. W., A. H. Ellingboe, and E. C. Rossman
  1971 "Biological Uniformity and Disease Epidemics," *BioScience* 21:1067–1070.
Anon.
  1968 "Ecology: The New Great Chain of Being," *Natural History* 77(10):8–16, 60–69.
Armillas, Pedro
  1971 "Gardens on Swamps," *Science* 174:653–661.
Bates, Marston
  1960 *The Forest and the Sea* (New York: Vintage).
  1969 "The Human Ecosystem," in *Resources and Man,* ed. Preston Cloud for National Academy of Sciences-National Research Council (New York: W. H. Freeman and Co.), pp. 21–30.
Benson, Robert L.
  1971 "On the Necessity of Controlling the Level of Insecticide Resistance in Insect Populations," *BioScience* 21:1160–1165.
Blackburn, Thomas R.
  1973 "Information and the Ecology of Scholars," *Science* 181:1141–1146 [contains an excellent summary of ecosystem thermodynamics with references to original sources].
Cloud, Preston
  1974 "Evolution of Ecosystems," *American Scientist* 62:54–66.
Colinvaux, Paul A.
  1973 *Introduction to Ecology* (New York: Wiley).
Commoner, Barry
  1971 *The Closing Circle: Nature, Man, and Technology* (New York: Knopf).

Dansereau, Pierre
    1966 "Ecological Impact and Human Behavior," in *Future Environments of North America,* ed. F. Fraser Darling and John P. Milton (Garden City: Natural History Press), pp. 425–461.
Dasmann, Raymond F., John P. Milton, and Peter H. Freeman
    1973 *Ecological Principles for Economic Development* (London: Wiley).
Davis, James Sholto
    1973 "Forest-Farming: An Ecological Approach to Increase Nature's Food Productivity," *Impact of Science on Society* 23(2):117–132.
Dixon, Bernard
    1974 "Lethal Resistance," *New Scientist* 61:732.
Egerton, Frank N.
    1973 "Changing Concepts of the Balance of Nature," *Quarterly Review of Biology* 48:322–350 [a historical review showing that nature is both very stable and ever-changing].
Emlen, J. Merritt
    1973 *Ecology: An Evolutionary Approach* (Reading, Mass.: Addison-Wesley).
Farvar, M. Taghi, and John P. Milton, eds.
    1968 *The Careless Technology: Ecology and International Development* (Garden City: Natural History Press).
Flawn, Peter T.
    1970 *Environmental Geology: Conservation, Land-Use Planning, and Resource Management* (New York: Harper and Row).
Gomez-Pompa, A., C. Vazquez-Yanes, and S. Guevara
    1972 "The Tropical Rain Forest: A Nonrenewable Resource," *Science* 177:762–765.
Hardin, Garrett
    1966 *Biology: Its Principles and Implications* (2nd ed.; New York: W. H. Freeman and Co.).
Hirst, Eric
    1974 "Food-Related Energy Requirements," *Science* 184:134–138.
Kolata, Gina Bari
    1974 "Theoretical Ecology: Beginnings of a Predictive Science," *Science* 183:400–401, 450.
Kormoridy, Edward J.
    1969 *Concepts of Ecology* (Englewood Cliffs: Prentice-Hall).
Kucera, Clair L.
    1973 *The Challenge of Ecology* (St. Louis: Mosby).
McHarg, Ian
    1971 *Design with Nature* (Garden City: Natural History Press).
Margalef, Ramon
    1968 *Perspectives in Ecological Theory* (University of Chicago Press).
Menard, H. W.
    1974 *Geology, Resources, and Society: An Introduction to Earth Science* (New York: W. H. Freeman and Co.).
"Monsanto Experiment Seeks Herbicide-Resistant Plant" 1988 *The Washington Post,* May 17, C1.
Mott, Lawrie
    1988 "Pesticide Alert," *The Amicus Journal,* Spring.
Odum, Eugene P.
    1971 *Fundamentals of Ecology* (3rd ed.; Philadelphia: Saunders).

Odum, Howard T.
   1971 *Environment, Power and Society* (New York: Wiley).
Pimental, David, et al.
   1973 "Food Production and the Energy Crisis," *Science* 182:443–449.
Rappaport, Roy A.
   1971 "The Flow of Energy in an Agricultural Society," *Scientific American* 224(3):121–132.
Reichle, David E.
   1975 "Advances in Ecosystem Analysis," *BioScience* 25:257–264.
Richards, Paul W.
   1973 "The Tropical Rain Forest," *Scientific American* 229(6):58–67.
Ricklefs, Robert E.
   1973 *Ecology* (New York: W. H. Freeman and Co.).
Scientific American
   1970 *The Biosphere* (New York: W. H. Freeman and Co.).
Shepard, Paul, and Daniel McKinley, eds.
   1969 *The Subversive Science: Essays Toward an Ecology of Man* (Boston: Houghton Mifflin) [many fine articles on ecological science].
Siever, Raymond
   1974 "The Steady State of the Earth's Crust, Atmosphere and Oceans," *Scientific American* 230(6):72–79 [excellent on basic ecological cycles].
Steinhart, John S., and Carol E. Steinhart
   1974 "Energy Use in the U.S. Food System," *Science* 184:307–316.
Thurston, H. David
   1969 "Tropical Agriculture: A Key to the World Food Crisis," *BioScience* 19:29–34.
Watt, Kenneth E. F.
   1973 *Principles of Environmental Science* (New York: McGraw-Hill).
Woodwell, G. M.
   1967 "Toxic Substances and Ecological Cycles," *Scientific American* 220(3):24–31.
   1970 "Effects of Pollution on the Structure and Physiology of Ecosystems," *Science* 168:429–433.
   1974 "Success, Succession, and Adam Smith," *BioScience* 24:81–87.

Chapter 2

Abelson, Philip H.
   1974 "Water Pollution Abatement: Goals and Costs," *Science* 184:1333.
————, et al.
   1975 Special issue on "Food and Nutrition," *Science* 188:501–653 [many useful articles tending toward a rather optimistic conclusion that technology is capable of expanding production markedly].
   1976 Special issue on "Materials," *Science* 191:631–776 [excellent discussion of many important issues, including some that are neglected in other sources; generally optimistic about the ability of technology to cope with impending shortages, but the opposite point of view is represented].
Abert, James G., Harvey Alter, and J. Frank Bernheisel
   1974 "The Economics of Resource Recovery from Municipal Solid Waste," *Science* 183:1052–1058.

Albers, John P.
      1973 "Seabed Mineral Resources: A Survey," *Bulletin of the Atomic Scientists* 29(8):33–38 [optimistic].
Alexander, M.
      1973 "Microorganisms and Chemical Pollution," *BioScience* 23:509–515.
Allen, Jonathan
      1973 "Sewage Farming: Science Races Forward to the Eighteenth Century," *Environment* 15(3):36–41.
Allen, Robert
      1974 "Turning Platitudes into Policy," *New Scientist* 64:400–402 [by promoting the development of traditional agricultural techniques for expanding production].
Almqvist, Ebbe
      1974 "An Analysis of Global Air Pollution," *Ambio* 3:161–167.
American Chemical Society
      1969 *Cleaning Our Environment—The Chemical Basis for Action* (Washington: American Chemical Society).
Anon.
      1970 "Environmental Repairs," *Sierra Club Bulletin* 60(3):22 [cost of environmental repairs from OECD study].
      1973 "The BEIR Report: Effects on Populations of Exposure to Low Levels of Ionizing Radiation," *Bulletin of the Atomic Scientists* 29(3):47–49.
Bair, W. J., and R. C. Thompson
      1974 "Plutonium: Biomedical Research," *Science* 183:715–722.
Bardach, John E., John H. Ryther, and William O. McLarney
      1972 *Aquaculture: The Farming and Husbandry of Freshwater and Marine Organisms* (New York: Wiley).
Barnett, Harold J., and Chandler Morse
      1963 *Scarcity and Growth: The Economics of Natural Resource Availability* (Baltimore: Johns Hopkins).
Berg, Alan
      1973 *The Nutrition Factor: Its Role in National Development* (Washington: Brookings Institution).
Bergstrom, Georg
      1967 *The Hungry Planet: The Modern World at the Edge of Famine* (New York: Collier).
      1971 *Too Many: An Ecological Overview of the Earth's Limitations* (New York: Collier).
Bernarde, Melvin A.
      1970 *Our Precarious Habitat* (New York: Norton).
Berry, R. Stephen
      1971 "The Option for Survival," *Bulletin of the Atomic Scientists* 27(5):22–27 [recycling and pollution control].
Bevington, Rick, and Arthur H. Rosenfeld
      1990 "Energy for Buildings and Homes," *Scientific American* 263(3): 76–89.
Björkman, Olle, and Joseph Berry
      1973 "High-Efficiency Photosynthesis," *Scientific American* 229(4):80–93.
Bleviss, Deborah L., and Peter Walzer
      1990 "Energy for Motor Vehicles," *Scientific American* 263(3): 102–109.
Bohn, Hinrich I., and Robert C. Cauthom
      1971 "Pollution: The Problem of Misplaced Waste," *American Scientist* 60:561–565.

Bonner, James, and John Weir
    1963 *The Next Hundred Years* (New York: Viking).
Booth, William
    1990 "Warm Seas Killing Coral Reefs," *The Washington Post,* 12 October, A8.
Borlaug, Norman E.
    1972 "Mankind and Civilization at Another Crossroad: In Balance with Nature—A Biological Myth," *BioScience* 22:41–44.
Boughey, Arthur S., ed.
    1973 *Readings in Man, the Environment, and Human Ecology* (New York: Macmillan).
Brooks, David B., and P. W. Andrews
    1974 "Mineral Resources, Economic Growth, and World Population," *Science* 185:13–19.
Brown, Harrison
    1954 *The Challenge of Man's Future* (New York: Viking).
    1970 "Human Materials Production as a Process in the Biosphere," *Scientific American* 223(3): 19–5208.
Brown, Lester R.
    1972 *World Without Borders* (New York: Random House).
    1974 *By Bread Alone* (New York: Praeger).
    1989 "Reexamining the World Food Prospect." In *State of the World, 1989,* ed. Linda Starke (New York: W. W. Norton & Company).
    1990a "The Illusion of Progress" In *State of the World, 1990.* ed. Linda Starke (New York: W. W. Norton & Company).
    1990b "Feeding Six Billion," in *Environment 90/91.* Reprinted from World Watch September/October 1989, pp. 32–40, ed. John Allen, Annual Editions (Guilford, Connecticut: Dushkin Publishing Group).
    1991 "The New World Order," in *State of the World, 1991,* ed. Linda Starke (New York: W. W. Norton & Company).
———, and John E. Young
    1990 "Feeding the World in the Nineties," in *State of the World, 1990,* ed. Linda Starke (New York: W. W. Norton & Company).
Bryson, Reid A.
    1974 "A Perspective on Climatic Change," *Science* 184:753–760.
Carson, Rachel
    1962 *Silent Spring* (Boston: Houghton Mifflin).
Carter, Luther J.
    1974 "Cancer and the Environment (I): A Creaky System Grinds On," *Science* 186:239–242.
Chandler, William U., Alexei A. Makarov, and Zhou Dadi
    1990 "Energy for the Soviet Union, Eastern Europe, and China," *Scientific American* 263(3):120–127.
Chapman, Duane
    1973 "An End to Chemical Farming?" *Environment* 15(2):12–17.
Chasis, Sarah, and Lisa Speer
    1991 "Congressional Coastal Watch," *Amicus Journal,* Winter, 21.
Christy, Francis T., Jr., and Anthony Scott
    1965 *The Common Wealth in Ocean Fisheries* (Baltimore: Johns Hopkins).
Clawson, Marion, Hans H. Landsberg, and Lyle T. Alexander
    1969 "Desalted Water for Agriculture: Is It Economic?" *Science* 164:1141–1148.

Cloud, Preston E., Jr.
    1968 "Realities of Mineral Distribution," *Texas Quarterly* 2(2):103–126.
    ———, ed.
    1969 *Resources and Man* (New York: W. H. Freeman and Co.).
Coale, Ansley J.
    1970 "Man and His Environment," *Science* 170:132–136.
    1974 "The History of the Human Population," *Scientific American* 231(3):41–51.
Commission on Population Growth and the American Future
    1972 *Population and the American Future* (New York: New American Library).
Commoner, Barry
    1971 *The Closing Circle: Nature, Man, and Technology* (New York: Knopf).
    1990 *Making Peace With the Planet* (New York: Pantheon Books).
Conney, A. H., and J. J. Burns
    1972 "Metabolic Interactions Among Environmental Chemicals and Drugs,"
        Science 178:576–586.
Conservation Foundation
    1973a "Is Man Facing a Chronic Food Supply Problem?" *Conservation Foundation
        Letter,* October, pp. 1–8.
    1973b "How Far Can Man Push Nature in Search of Food?" *Conservation Founda-
        tion Letter,* November, pp. 1–8.
    1974 "Public Health: Still the Crux of Pollution Fights," *Conservation Foundation
        Letter,* May, pp. 1–8.
Cook, Earl
    1975 "The Depletion of Geological Resources," *Technology Review* 77(7):15–27.
Council on Environmental Quality, et al.
    1972 *The Economic Impact of Pollution Control: A Summary of Recent Studies*
        (Washington: Government Printing Office).
Council on Environmental Quality and Department of State, Gerald O. Barney, Study
    Director
    1980 "The Global 2000 Report to the President" (Washington, D.C.).
Critical Mass Energy Project
    1989 *Factsheet #5* (Washington, D.C.: Public Citizen).
Daly, Herman E., and John B. Cobb, Jr.
    1989 *For The Common Good* (Boston: Beacon Press).
Darnell, Rezneat M.
    1971 "The World Estuaries–Ecosystems in Jeopardy," *INTECOL Bulletin* 3:3–
        20.
Davis, Ged R.
    1990 "Energy for Planet Earth," *Scientific American* 263(3): 54–63.
DeBach, Paul
    1974 *Biological Control by Natural Enemies* (New York: Cambridge University Press)
        [an important book on a critical topic].
Dorst, Jean
    1971 *Before Nature Dies* (Baltimore: Penguin).
Dubos, Rene
    1965 *Man Adapting* (New Haven: Yale University Press).
Eckholm, Erik P.
    1976 *Losing Ground: Environmental Stress and World Food Prospects* (New York:
        Norton).

Ehrenfeld, David W.

  1974 "Conserving the Edible Sea Turtle: Can Mariculture Help?" *American Scientist* 62:23–31.

Ehrlich, Anne H. and John P. Holdren

  1973 *Human Ecology: Problems and Solutions* (New York: W. H. Freeman and Co.).

——, and John P. Holdren

  1969 "Population and Panaceas: A Technological Perspective," *BioScience* 19:1065–1071.

——, John P. Holdren, and Richard W. Holm, eds.

  1971 *Man and the Ecosphere* (New York: W. H. Freeman and Co.).

Ehrlich, Paul R.

  1968 *The Population Bomb* (New York: Ballantine).

——, and Anne H. Ehrlich

  1972 *Population, Resources, Environment: Issues in Human Ecology* (2nd ed.; New York: W. H. Freeman and Co.).

——, and Anne H. Ehrlich

  1990 "The Population Explosion," *The Amicus Journal,* Winter, 22–29.

Environmental Protection Agency

  1972 *The Economics Of Clean Air: Annual Report of the Administrator of The Enviromental Protection Agency to the Congress of the United States, February 1972* (Washington: Government Printing Office).

Fickett, Arnold P., Clark W. Gellings, and Amory B. Lovins

  1990 "Efficient Use of Electricity," *Scientific American* 263(3): 64–75.

Flavin, Christopher

  1987 "Reassessing Nuclear Power: The Fallout from Chernobyl," in *Worldwatch Paper 75,* March (Washington, D.C.: Worldwatch Institute).

  1990a "Ten Years of Fallout," in *Environment 90/91.* Reprinted from World Watch March/April 1989, pp. 30–37, ed. John Allen (Guilford, Connecticut: Dushkin Publishing Group).

  1990b "Slowing Global Warming," in *State of the World, 1990,* ed. Linda Starke (New York: W. W. Norton & Company).

——, and Nicholas Lenssen

  1991 "Designing a Sustainable Energy System," in *State of the World, 1991,* ed. Linda Starke (New York: W. W. Norton & Company).

Flawn, Peter T.

  1966 *Mineral Resources: Geology, Engineering, Economics, Politics, Law* (Chicago: Rand McNally).

Foster, G. G., et al.

  1972 "Chromosome Rearrangements for the Control of Insect Pests," *Science* 176:875–880.

Frejka, Tomas

  1968 "Reflections on the Demographic Conditions Needed to Establish a U. S. Stationary Population Growth," *Population Studies* 22:379-397.

  1973a *The Future of Population Growth: Alternative Paths to Equilibrium* (New York: Wiley).

  1973b "The Prospects for a Stationary World Population," *Scientific American* 228(3):15–23.

Fulkerson, William, Roddie R. Judkins, and Manoj K. Sanghvi

  1990 "Energy From Fossil Fuels," *Scientific American* 263(3): 128–135.

Furon, Raymond

  1967 *The Problem of Water* (New York: American Elsevier).

Gibbons, John, Peter Blair, and Holly Gwin
    1989 "Strategies for Energy Use," *Scientific American* 261(3):136–143.
Gillette, Robert
    1972 "Radiation Standards: The Last Word or at Least a Definitive One," *Science*
        178:966–967, 1012.
    1974 "Cancer and the Environment (II): Groping for New Remedies," *Science*
        186:242–245.
Gladwell, Malcolm
    1990 "Consumers' Choices About Money Consistently Defy Common Sense," *The
        Washington Post,* 12 February, A3.
Global Tomorrow Coalition
    1990 *The Global Ecology Handbook,* ed. Walter H. Corson (Boston: Beacon Press).
Grahn, Douglas
    1972 "Genetic Effects of Low Level Irradiation," *BioScience* 22:535–540.
Groth, Edward, III
    1975 "Increasing the Harvest," *Environment* 17(1):28–39 [an excellent summary of
        the key issues, amply documented].
Hafele, Wolf
    1990 "Energy from Nuclear Power," *Scientific American* 263(3)136–145.
Hammond, Allen L.
    1974a "Manganese Nodules (I): Mineral Resources on the Deep Seabed," *Science*
        183:502–503.
    1974b "Manganese Nodules (II): Prospects for Deep Sea Mining," *Science* 183:644–
        646.
Hannon, Bruce M.
    1972 "Bottles, Cans, Energy," *Environment* 14(2):11–21.
Harte, John, and Robert H. Socolow, eds.
    1971 *Patient Earth* (New York: Holt, Rinehart and Winston).
Haub, Carl
    1988 "Trial by Numbers," *Sierra* 73: 40–42.
Heichel, G. H.
    1974 "Energy Needs and Food Yields," *Technology Review* 76(8):19–25.
Hirst, Eric
    1973 "The Energy Cost of Pollution Control," *Environment* 15(8):37–44.
    1974 "Food-Related Energy Requirements," *Science* 184:134–138.
Hoff, Johan E., and Jules Janick, eds.
    1973 *Food* (New York: W. H. Freeman and Co.).
Hoffman, Allen R., and David Rittenhouse Inglis
    1972 "Radiation and Infants," *Bulletin of the Atomic Scientists* 28(10):45–52.
Holdren, John P., and Paul R. Ehrlich
    1971 *Global Ecology: Toward a Rational Strategy for Man* (New York: Harcourt Brace
        Jovanovich).
Holing, Dwight
    1991 "America's Energy Plan," *The Amicus Journal,* Winter, 12–20.
Holmberg, Bo, et al.
    1975 Special issue on "The Work Environment," *Ambio* 4(1):I–65 [an excellent
        review of an important problem].
Howland, H. Richard
    1975 "The Helium Conservation Question," *Technology Review* 77(7):42–49.

Huffaker, Carl B.
1971 "Biological Control and a Remodeled Pest Control Strategy," *Technology Review* 73(8):31–37.
Inman, Douglas L., and Birchard M. Brush
1973 "The Coastal Challenge," *Science* 181:20–31.
Janzen, Daniel H.
1973 "Tropical Agroecosystems," *Science* 182:1212–1219.
Kenward, Michael
1972 "Fighting for the Clean Car," *New Scientist* 51:553–555.
1989 "'Killer' Trees To The Rescue," *Newsweek* 114(14): 59.
Kramer, Eugene
1973 "Energy Conservation and Waste Recycling: Taking Advantage of Urban Congestion," *Bulletin of the Atomic Scientists* 29(4):13–18.
Laing, David
1974 "The Phosphate Connection," *Not Man Apart* 4(13):I, 10.
Lee, Douglas H. K.
1973 "Specific Approaches to Health Effects of Pollutants," *Bulletin of the Atomic Scientists* 29(8):45–47.
Lichtenstein, E. P., T. T. Liang, and B. N. Anderegg
1973 "Synergism of Insecticides by Herbicides," *Science* 181:847–849.
Likens, Gene E., and F. Herbert Bormann
1974 "Acid Rain: A Serious Regional Environmental Problem," *Science* 184:1176–1179.
Loosli, J. K.
1974 "New Sources of Proteins for Human and Animal Feeding," *BioScience* 24:26–31.
Lowe, Marcia
1991 "Rethinking Urban Transport," in *State of the World, 1991,* ed. Linda Starke (New York: W. W. Norton & Company).
McHale, John
1970 *The Ecological Context* (New York: Braziller).
1971 *The Future of the Future* (New York: Ballantine).
MacIntyre, Ferren
1974 "The Top Millimeter of the Ocean," *Scientific American* 230(5):62–77.
McKelvey, Vincent E.
1972 "Mineral Resource Estimates and Public Policy," *American Scientist* 60:32–40.
1974 "Approaches to the Mineral Supply Problem," *Technology Review* 76(5):13–23.
Maddox, John
1972 *The Doomsday Syndrome* (London: Macmillan).
Malenbaum, Wilfred
1973 "World Resources for the Year 2000," *Annals of the American Academy of Political and Social Science* 408:30–46.
Marx, Jean L.
1974 "Nitrogen Fertilizer," *Science* 185:133.
Mathews, Jessica Tuchman
1990 "Rescue Plan for Africa," in *Environment 90/91,* reprinted from World Monitor May 1989, pp. 28–36., ed. John Allen (Guilford, Connecticut: Dushkin Publishing Group).

Maugh, Thomas H., II
    1974 "Chemical Carcinogenesis: A Long-Neglected Field Blossoms," *Science*
        183:940–944.
Meadows, Dennis L., et al.
    1974 *The Dynamics of Growth in a Finite World* (Cambridge: Wright-Allen).
Meadows, Donella H., et al.
    1972 *The Limits to Growth* (New York: Universe).
Meier, Richard L.
    1966 *Science and Economic Development: New Patterns of Living* (2d ed.; Cambridge:
        MIT).
Metz, William D., and Allen L. Hammond
    1974a "Geodynamics Report: Exploiting the Earth Sciences Revolution," *Science*
        183:735–738, 769.
    1974b "Helium Conservation Program: Casting It to the Winds," *Science*
        183:59–63.
Meyer, Alden
    1990a "The 'White House Effect': Bush Backs Off Carbon Dioxide Stabilization,"
        Nucleus, Spring, 3.
    1990b "United States Increasingly Isolated on Global Warming," *Nucleus,* Sum-
        mer, 3.
Meyer, Judith E.
    1972 "Renewing the Soil," *Environment* 14(2):22–24, 29–32.
Murdoch, William W., ed.
    1971 *Environment: Resources, Pollution and Society* (Stamford, Conn.: Sinauer).
National Academy of Sciences, Office of the Foreign Secretary, ed.
    1971 *Rapid Population Growth* (Baltimore: Johns Hopkins).
National Research Council
    1989 *Alternative Agriculture* (Washington, D.C.: National Academy Press).
NCMP (National Commission on Materials Policy)
    1972 *Towards a National Materials Policy: Basic Data and Issues, An Interim Report*
        (Washington: Government Printing Office).
    1973 *Toward a National Materials Policy: World Perspective, Second Interim Report*
        (Washington: Government Printing Office).
Newill, Vaun A.
    1973 "Pollution's Price—The Cost in Human Health," *Bulletin of the Atomic Scien-
        tists* 29(8):47–49.
Newman, James E., and Robert C. Pickett
    1974 "World Climates and Food Supply Variations," *Science* 186:877–881.
Nogee, Alan
    1986 "Chernobyl: It Can Happen Here," *Environmental Action,* July-August, 12–14.
"Ocean Thermal Energy: Sunny Side Up."
    1987 *The Economist,* 20 June, 94.
Odell, Rice
    1974 "Water Pollution: The Complexities of Control," *Conservation Foundation
        Letter,* December.
Odum, Eugene P.
    1971 *Fundamentals of Ecology* (3d ed.; Philadelphia: Saunders).
Odum, Howard T.
    1971 *Environment, Power, and Society* (New York: Wiley).

Odum, William E.

1974 "Potential Effects of Aquaculture on Inshore Coastal Waters," *Environmental Conservation* 1:225–230.

Othmer, Donald F., and Oswald A. Roels

1973 "Power, Fresh Water, and Food from Cold, Deep Sea Water," *Science* 182:121–125.

Park, Charles F., Jr.

1968 *Affluence in Jeopardy: Minerals and the Political Economy* (San Francisco: Freeman, Cooper).

Payne, Philip

1974 "Protein Deficiency or Starvation?" *New Scientist* 64:393–398 [an excellent overview of the whole malnutrition-starvation syndrome].

Penney, Terry R., and Desikan Bharathan

1987 "Power from the Sea," *Scientific American* 256(1):86–92.

Perelman, Michael J.

1972 "Fanning with Petroleum," *Environment* 14(8):8–13.

Pimental, David, et al.

1973 "Food Production and the Energy Crisis," *Science* 182:443–449.

1975 "Energy and Land Constraints in Food Protein Production," *Science* 190:754–761.

Pinchot, Gifford B.

1970 "Marine Farming," *Scientific American* 223(6):15–21.

1974 "Ecological Aquaculture," *BioScience* 24:265.

Pollock, Cynthia

1986, April "Decommissioning: Nuclear Power's Missing Link," in *Worldwatch Paper 69* (Washington, D.C.: The Worldwatch Institute).

Postel, Sandra

1987, September "Defusing the Toxics Threat: Controlling Pesticides and Industrial Waste," in *Worldwatch Paper 79* (Washington, D. C.: Worldwatch Institute).

1990 "Saving Water for Agriculture," in *State of the World, 1990,* ed. Linda Starke (New York: W. W. Norton & Company).

Probstein, Ronald F.

1973 "Desalination," *American Scientist* 61:280–293.

Rauber, Paul

1991 "Better Nature Through Chemistry," *Sierra,* July/August, 32–34.

Revelle, Roger

1974 "Food and Population," *Scientific American* 231(3):161–170 [optimistic].

Roberts, Leslie

1989 "Does the Ozone Hole Threaten Antarctic Life?" *Science* 244: 288–289.

Ross, Marc H., and Daniel Steinmeyer

1990 "Energy for Industry," *Scientific American* 263(3): 88–101.

Russell, Dick

1987 "Rush to Market, Biotechnology and Agriculture," *The Amicus Journal,* Winter, 16–37.

Russett, Bruce M.

1967 "The Ecology of Future International Politics," *International Studies Quarterly* 11(l):14–19 [a good discussion of the use of exponential growth in making predictions about the future].

Ryther, John H.

1969 "Photosynthesis and Fish Production in the Sea," *Science* 166:72–76.

Sagan, L. A.
   1972 "Human Costs of Nuclear Power," *Science* 177:487–493.
Salk, Jonas
   1973 *The Survival of the Wisest* (New York: Harper and Row).
SCEP (Report of the Study of Critical Environment Problems)
   1970 *Man's Impact on the Global Environment* (Cambridge: MIT).
Schmidt-Perkins, Drusilla
   1989 "Are Alternative Fuels the Answer?" *Environmental Action*, July/August, 21–22.
Shapley, Deborah
   1973a "Auto Pollution: EPA Worrying That the Catalyst May Backfire," *Science* 182:368–371.
   1973b "Ocean Technology: Race to Seabed Wealth Disturbs More Than Fish," *Science* 180:849–851, 893.
Shepard, Paul, and Daniel McKinley, eds.
   1969 *The Subversive Science: Essays Toward an Ecology of Man* (Boston: Houghton Mifflin) [many excellent articles, especially on pollution].
Singer, S. Fred
   1971 "Environmental Quality—When Does Growth Become Too Expensive?" in *Is There an Optimum Level of Population?*, ed. S. Fred Singer (New York: McGraw-Hill).
Skinner, Brian J.
   1969 *Earth Resources* (2d ed.; Englewood Cliffs: Prentice-Hall).
Small, William E.
   1971 "Agriculture: The Seeds of a Problem," *Technology Review* 73(6):48–53.
Smith, Roger H., and R. C. von Borstel
   1972 "Genetic Control of Insect Populations," *Science* 178:1164–1174.
Spurgeon, David
   1973 "The Nutrition Crunch: A World View," *Bulletin of the Atomic Scientists* 29(8):50–54.
Staines, Andrew
   1974 "Digesting the Raw Materials Threat," *New Scientist* 61:609–611.
Starr, Roger, and James Carlson
   1968 "Pollution and Poverty," *Public Interest* 10:104–131 [pollution-control costs].
Steinhart, John S., and Carol E. Steinhart
   1974 "Energy Use in the U.S. Food System," *Science* 184:307–316.
Sterling, Theodor D.
   1971 "Difficulty of Evaluating the Toxicity and Teratogenicity of 2,4,5-T from Existing Animal Experiments," *Science* 174:1358–1359.
Summers, Claude M.
   1971 "The Conversion of Energy," in *Energy and Power*, ed. Scientific American (New York: W. H. Freeman and Co.), pp. 93–106.
Taylor, Theodore B., and Charles C. Humpstone
   1973 *The Restoration of the Earth* (New York: Harper and Row) [a "containment" pollution-control strategy].
Teitelbaum, Michael S.
   1975 "Relevance of Demographic Transition Theory for Developing Countries," *Science* 188:420
Valery, Nicholas
   1972 "Place in the Sun for Helium," *New Scientist* 56:496–500.

Wade, Nicholas

    1972 "A Message from Corn Blight: The Dangers of Uniformity," *Science* 177:678–679.

    1974a "Green Revolution (I): A Just Technology, Often Unjust in Use." *Science* 186:1093–1096.

    1974b "Green Revolution (II); Problems of Adapting a Western Technology," *Science* 186:1186–1192.

    1974c "Raw Materials: U.S. Grows More Vulnerable to Third World Cartels," *Science* 183:185–186.

    1974d "Sahelian Drought: No Victory for Western Aid," *Science* 185:234–237.

    1975 "New Alchemy Institute: Search for an Alternative Agriculture," *Science* 187:727–729.

Waldbott, George L.

    1973 *Health Effects of Environmental Pollutants* (St. Louis: Mosby).

Wallace, Bruce

    1974 "Commentary: Radioactive Wastes and Damage to Marine Communities," *BioScience* 24: 164–167.

Ward, Barbara, and Rene Dubos

    1972 *Only One Earth: The Care and Maintence of a Small Planet* (New York: Norton).

Weeks, W. F., and W. J. Campbell

    1973 "Towing Icebergs to Irrigate Arid Lands: Manna or Madness?" *Bulletin of the Atomic Scientists* 29(5):35–39.

Weinberg, Alvin M.

    1972 "Science and Trans–Science," *Minerva* 10(2):209–222.

Weinberg, Carl J., and Robert H. Williams

    1990 "Energy from the Sun" *Scientific American* 263(3): 146–155.

Weisskopf, Michael

    1987 "Pesticides in 15 Common Foods May Cause 20,000 Cancers a Year," *Washington Post,* 21 May, A33.

Westman, Walter E.

    1972 "Some Basic Issues in Water Pollution Control Legislation," *American Scientist* 60:767–773.

de Wilde, Jan

    1975 "Insect Population Management and Integrated Pest Control," *Ambio* 4:105–111.

Wilkes, H. Garrison, and Susan Wilkes

    1972 "The Green Revolution," *Environment* 14(8):32–39.

Wittwer, S. H.

    1974 "Maximum Production Capacity of Food Crops," *BioScience* 24:216–224.

Wood, J. M.

    1974 "Biological Cycles for Toxic Elements in the Environment," *Science* 183:1049–1052.

Woodwell, G. M.

    1969 "Radioactivity and Fallout: The Model Pollution," *BioScience* 19:884–887.

World Resources Institute

    1990 *World Resources, 1990–91* in collaboration with the U.N. Environmental Programme and the U.N. Development Program.

Wysham, Daphne

    1991 "Fueling the Fantasy" *Greenpeace,* May/June, 12–15.

Young, Gale
    1970 "Dry Lands and Desalted Water," *Science* 167:339–343.
Young, John E.
    1991 "Reducing Waste, Saving Materials," in *State of the World, 1991,* ed. Linda Starke (New York: W. W. Norton & Company).
Zwick, David, and Mary Benstock
    1971 *Water Wasteland: Ralph Nader's Study Group Report on Water Pollution* (New York: Grossman).

Chapter 3

Aaronson, Terri
    1971 "The Black Box," *Environment* 13(10):10–18 [on fuel cells].
Abelson, Philip H., ed.
    1974 "Energy," special issue of *Science* 184:245–389.
Ahmed, A. Karim
    1975 "Unshielding the Sun: Human Effects," *Environment* 17(3):6–14.
Alfvén, Hannes
    1972 "Energy and Environment," *Bulletin of the Atomic Scientists* 28(5):5–8.
    1974 "Fission Energy and Other Sources of Energy," *Bulletin of the Atomic Scientists* 30(1):4–8.
Allen, John, ed.
    1989 "The Planet Strikes Back" in *Environment 90/91,* reprinted from *National Wildlife,* February/March 1989, pp. 33–40, Annual Editions (Guilford, Connecticut: The Dushkin Publishing Group).
American Lung Association
    1990 "Air Pollution Health Costs Calculated," *The Washington Post,* 21 January, A12.
Anon.
    1972a "No Small Difference of Opinion," *World Environment Newsletter* in *World,* August 15, pp. 30–31.
    1972b "120 Million Mw. for Nothing," *Technology Review* 74(7):58.
    1975 "What the Shuttle Might Do to Our Environment," *New Scientist* 66:300 [the atmospheric and climatic dangers of the Space Shuttle program].
Anthrop, Donald F.
    1970 "Environmental Side Effects of Energy Production," *Bulletin of the Atomic Scientists* 26(8):39–41.
Armstead, H. C. H., ed.
    1973 *Geothermal Energy: Review of Research and Development* (New York: UNESCO).
Atwood, Genevieve
    1975 "The Strip–Mining of Western Coal," *Scientific American* 233(6):23–29.
Axtmann, Robert C.
    1975 "Environmental Impact of a Geothermal Plant," *Science* 187:795–803.
Ayres, Eugene
    1950 "Power from the Sun," *Scientific American* 183(2):16–21.
Baldwin, Pamela L., and Malcolm F. Baldwin
    1974 "Offshore Oil Heats Up as Energy Issue," *Conservation Foundation Letter,* November.
Bamberger, C. E., and J. Braunstein
    1975 "Hydrogen: A Versatile Element," *American Scientist* 63:438–447.

Barnaby, Frank, et al.
1975 Symposium on "Can We Live with Plutonium?" in *New Scientist* 66:494–506.

Barnea, Joseph
1972 "Geothermal Power," *Scientific American* 226(1):70–77.

Barraclough, Geoffrey
1974 "The End of an Era," *New York Review of Books*, June 27, pp. 14–20 [on the current economic disarray and its causes].

de Bell, Garrett, ed.
1970 *The Environmental Handbook* (New York: Ballantine).

Berg, Charles A.
1973 "Energy Conservation through Effective Utilization," *Science* 181:128–138.
1974 "A Technical Basis for Energy Conservation," *Technology Review* 76(4):15–23.

Berg, George G.
1973 "Hot Wastes from Nuclear Power," *Environment* 15(4):36–44.

Berry, R. Stephen
1971 "The Option for Survival," *Bulletin of the Atomic Scientists* 27(5):22–27.
———, and Margaret F. Fels
1973 "The Energy Cost of Automobiles," *Bulletin of the Atomic Scientists* 29(10):11–17, 58–60.
———, and Hiro Makino
1974 "Energy Thrift in Packaging and Marketing," *Technology Review* 76(4):33–43.

Bezdek, Roger, and Bruce Hannon
1974 "Energy, Manpower, and the Highway Trust Fund," *Science* 185:669–675.

Bockris, J. O'M.
1974 "The Coming Energy Crisis and Solar Sources," *Environmental Conservation* 1:241–249.

Boffey, Philip M.
1975 "Rasmussen Issues Revised Odds on a Nuclear Catastrophe," *Science* 190:640.

Bolin, Bert
1974 "Modelling the Climate and Its Variations," *Ambio* 3:180–188.

Booth, William
1990 "Carbon Dioxide Curbs May Not Halt Global Warming," *The Washington Post,* 10 March, A1.
1991a "Tropical Forests Disappearing at Faster Rate," *The Washington Post*, 9 September, A18.
1991b "Global Warming Continues, but Cause is Uncertain," *The Washington Post,* 10 January, A3.

Boulding, Kenneth E.
1964 *The Meaning of the Twentieth Century: The Great Transition* (New York: Harper and Row).
1973 "The Economics of the Coming Spaceship Earth," in *Daly* 1973, pp. 121–132.

Brinworth, B. J.
1973 *Solar Energy for Man* (New York: Wiley).

Broecker, Wallace S.
1975 "Climatic Change: Are We on the Brink of a Pronounced Global Warming?" *Science* 189:460–463.

Brooks, Harvey
1973 "The Technology of Zero–Growth," *Daedalus* 102(4):139–152.

Brown, Harrison, James Bonner, and John Weir
    1963 *The Next Hundred Years* (New York: Viking).
Browne, Malcolm W.
    1991 "Modern Alchemists Transmute Nuclear Waste," *The New York Times,* 29
        October, C1.
Bryson, Reid A.
    1973 "Drought in Sahelia: Who or What Is to Blame?" *The Ecologist* 3:366–371.
    1974 "A Perspective on Climatic Change," *Science* 184:753–760.
Bupp, Irvin C., and Jean–Claude Derian
    1974 "The Breeder Reactor in the U.S.: A New Economic Analysis," *Technology
        Review* 76(8):27–36.
Burnet, Macfarlane
    1971 "After the Age of Discovery?" *New Scientist* 52:96–100.
Bury, J. B.
    1955 *The Idea of Progress.: An Inquiry into Its Origin and Growth* (New York: Dover).
Callahan, Daniel
    1973 *The Tyranny of Survival* (New York: Macmillan) [esp. Chap. 3, which discusses
        ways of reexamining technology].
Calvin, Melvin
    1974 "Solar Energy by Photosynthesis," *Science* 184:375–381.
Carter, Luther J.
    1973 "Deepwater Ports: Issue Mixes Supertanker, Land Policy," *Science* 181:825–828.
    1974 "Floating Nuclear Plants: Power from the Assembly Line," *Science* 183:1063–
        1065.
Chapman, Peter
    1974 "The Ins and Outs of Nuclear Power," *New Scientist* 64:966–969 [net energy
        analysis].
Chedd, Graham
    1974 "Colonisation at Lagranges," *New Scientist* 64:247–249.
Cheney, Eric S.
    1974 "U.S. Energy Resources: Limits and Future Outlook," *American Scientist*
        62:14–22.
Clark, Wilson
    1974 *Energy for Survival: The Alternatives to Extinction* (Garden City, N.Y.: Doubleday).
Clarke, Arthur C.
    1962 *Profiles of the Future: An Inquiry into the Limits of the Possible* (New York: Harper
        and Row).
Cloud, Preston, ed.
    1969 *Resources and Man* (New York: W. H. Freeman and Co.).
Cochran, Thomas B.
    1974 *The Liquid Metal Fast Breeder Reactor: An Economic and Environmental Critique*
        (Baltimore: Johns Hopkins).
Cohen, Bernard L.
    1974 "Perspectives on the Nuclear Debate: An Opposing View," *Bulletin of the
        Atomic Scientists* 30(8):35–39 [nuclear power as the lesser evil].
Comey, David D.
    1974 "Will Idle Capacity Kill Nuclear Power?" *Bulletin of the Atomic Scientists*
        30(9):23–28.
    1975 "The Legacy of Uranium Tailings," *Bulletin of the Atomic Scientists* 31(7):43–45.

Commoner, Barry

1990 *Making Peace With the Planet* (New York: Pantheon Books).

———, Howard Boksenbaum, and Michael Corr, eds.

1975 *Energy and Human Welfare: A Critical Analysis* (3 vols; Riverside, N.J.: Macmillan Information).

Conservation Foundation

1970 "Can We Have All the Electricity We Want and a Decent Environment Too?" *CF Newsletter,* No. 3–70.

1973 "The Land Pinch: Where Can We Put Our Wastes?" *Conservation Foundation Letter,* May.

1974a "Carrying Capacity Analysis Is Useful—But Limited," *Conservation Foundation Letter,* June [a useful discussion of the multiple factors that have to be taken into account in thinking about carrying capacity for human use].

1974b "U.S. Coastline Is Scene of Many Energy Conflicts," *Conservation Foundation Letter,* January.

Cook, C. Sharp

1973 "Energy: Planning for the Future," *American Scientist* 61:61–65.

Cook, Earl

1971 "The Flow of Energy in an Industrial Society," *Scientific American* 224(3):134–144.

1976 *Man, Energy, Society* (New York: W. H. Freeman and Co.).

Cottrell, Fred

1955 *Energy and Society: The Relation Between Energy, Social Change, and Economic Development* (New York: McGraw–Hill).

Craven, Gwyneth

1975 "The Garden of Feasibility," *Harper's,* August, pp. 66–75 [a proposal for space colonization].

Crossland, Janice

1974 "Ferment in Technology," *Environment* 16(10):17–30 [fermenting organic materials for fuels and other useful products].

Dahlberg, Kenneth A.

1973 "Towards a Policy of Zero Energy Growth," *The Ecologist* 3:338–341.

Daly, Herman E., ed.

1973 *Toward a Steady State Economy* (New York: W. H. Freeman and Co.).

Daniels, Farrington

1964 *Direct Use of the Sun's Energy* (New Haven: Yale).

1971 "Direct Use of the Sun's Energy," *American Scientist* 55:5–47 [updates and summarizes his book, still a standard work in the field].

David, Edward E., Jr.

1973 "Energy: A Strategy of Diversity," *Technology Review* 75(7):26–31.

Day, M. C.

1975 "Nuclear Energy: A Second Round of Questions," *Bulletin of the Atomic Scientists* 31(10):52–59 [fuel supply problems].

DeNike, L. Douglas

1974 "Radioactive Malevolence," *Bulletin of the Atomic Scientists* 30(2):16–20 [security risks].

Dials, George E., and Elizabeth C. Moore

1974 "The Cost of Coal," *Environment* 16(7):18–37.

Dickson, David

1974 *Alternative Technology: And the Politics of Technical Change* (London: Fontana).

Dinneen, Gerald U., and Glenn L. Cook
   1974 "Oil Shale and the Energy Crisis," *Technology Review* 76(3):27–33.
Djerassi, Carl, et al.
   1974 "Insect Control of the Future: Operational and Policy Aspects," *Science* 186:596–607.
Dreschhoff, Gisela, D. F. Saunders, and E. J. Zeller
   1974 "International High Level Nuclear Waste Management," *Bulletin of the Atomic Scientists* 30(l):28–33.
Drucker, Daniel C.
   1971 "The Engineer in the Establishment," *Bulletin of the Atomic Scientists* 27(10):31–34.
Dudley, H. C.
   1975 "The Ultimate Catastrophe," *Bulletin of the Atomic Scientists* 31(9):21–34 [the remote possibility of a runaway chain reaction following a nuclear explosion].
Edsall, John T.
   1974 "Hazards of Nuclear Fission Power and the Choice of Alternatives," *Environmental Conservation* 1(l):21–30 [fossil fuel the lesser risk].
Ehricke, Kraffi A.
   1971 "Extraterrestrial Imperative," *Bulletin of the Atomic Scientists* 27(9):18–26 [escape to space].
Ehrlich, Paul R.,and Anne Ehrlich
   1972 *Population, Resources, Environment: Issues in Human Ecology* (2nd ed.; New York: W. H. Freeman and Co.).
EIC (Environment Information Center)
   1973 *The Energy Index* (New York: EIC).
Eigner, Joseph
   1975 "Unshielding the Sun: Environmental Effects,"*Environment* 17(3):15–18.
Ellis, A. J.
   1975 "Geothermal Systems and Power Development," *American Scientist* 63:510–521.
Emmett, John L., John Nuckolls, and Lowell Wood
   1974 "Fusion Power by Laser Implosion," *Scientific American* 230(6):24–37.
Enviro/Info
   1973 *Energy/Environment/Economy: An Annotated Bibliography of Selected U.S. Government Publications Concerning United States Energy Policy* (April) and *Supplement* (September) (mimeo; Green Bay: Enviro/Info).
Environmental Action
   1991 "U.S. Lacks National Plan for Packaging Reduction"*Environmental Action,* March–April, 25–29.
Ewell, Raymond
   1975 "Food and Fertilizer in the Developing Countries, 1975–2000," *BioScience* 25:771.
Ewing, Maurice, and W. L. Donn
   1956 "A Theory of Ice Ages," *Science* 123:1061–1066.
Ferkiss, Victor C.
   1969 *Technological Man: The Myth and the Reality* (New York: Braziller).
Fisher, John C.
   1974 *Energy Crises in Perspective* (New York: Wiley).
Fletcher, J. O.
   1970 "Polar Ice and the Global Climate Machine," *Bulletin of the Atomic Scientists* 26(10):40–47.

French, Hillary F.
  1990 "A Most Deadly Trade" *World Watch*, July–August, 11–17.
Frisken, W. R.
  1971 "Extended Industrial Revolution and Climate Change," *EOS* 52:500–507.
Gabel, Medard, ed.
  1975 *Energy, Earth & Everyone* (San Francisco: Straight Arrow) [an unconventional and stimulating treatment of energy by followers of Buckminster Fuller].
Georgescu–Roegen, Nicholas
  1971 *The Entropy Law and the Economic Process* (Cambridge: Harvard).
  1973 "The Entropy Law and the Economic Problem," in Daly 1973, pp. 37–49.
  1975 "Energy and Economic Myths," *The Ecologist* 5:164–174, 242–252.
Giddings, J. Calvin
  1973 "World Population, Human Disaster and Nuclear Holocaust," *Bulletin of the Atomic Scientists* 29(7):21–24, 45–50.
Gillette, Robert
  1973a "Energy R & D: Under Pressure, a National Policy Takes Form," *Science* 182:898–900.
  1973b "NAS: Water Scarcity May Limit Use of Western Coal," *Science* 181:525.
  1973c "Radiation Spill at Hanford: The Anatomy of An Accident," *Science* 181:728–730.
  1974a "Budget Review: Energy," *Science* 183:636–638.
  1974b "Oil and Gas Resources: Did USGS Gush Too High?" *Science* 185:127–130.
  1974c "Synthetic Fuels: Will Government Lend the Oil Industry a Hand?" *Science* 183:641–643.
  1975 "Geological Survey Lowers Its Sights," *Science* 189:200.
Gilliland, Martha W.
  1975 "Energy Analysis and Public Policy," *Science* 189:1051–1056.
Glaser, Peter E.
  1968 "Power from the Sun: Its Future," *Science* 162:857–861 [gathering solar energy in space].
Glass, Bentley
  1971 "Science: Endless Horizons or Golden Age?" *Science* 171:23–29.
Global Tomorrow Coalition
  1990 *The Global Ecology Handbook*, ed. Walter H. Corson (Boston: Beacon Press.)
Gofman, John
  1972 "Is Nuclear Fission Acceptable?" *Futures* 4:211–219.
Goldstein, Irving S.
  1975 "Potential for Converting Wood into Plastics," *Science* 189:847–852.
Gough, William C., and Bernard J. Eastlund
  1971 "The Prospects of Fusion Power," *Scientific American* 224(2):50–64.
Green, Harold P.
  1971 "Radioactive Waste and the Law," *Natural Resources Journal* 11:281–295.
Green, Leon, Jr.
  1967 "Energy Needs vs. Environmental Pollution—A Reconciliation," *Science* 156:1448–1450 [using ammonia as a fuel].
Greenhill, Basil
  1972 "The Sailing Ship in a Fuel Crisis," *The Ecologist* 2(9):8–10.
Gregory, Derek P.
  1973 "The Hydrogen Economy," *Scientific American* 228(i):13–21.

Gustafson, Philip F.
    1970 "Nuclear Power and Thermal Pollution: Zion, Illinois," *Bulletin of the Atomic
        Scientists* 26(3):17–23.
Hafele, Wolf
    1974 "A Systems Approach to Energy," *American Scientist* 62:438–447.
Hammond, Allen L.
    1974a "Academy Says Energy Self–Sufficiency Unlikely," *Science* 184:964 [reporting
        conclusions of National Academy of Engineering study].
    1974b "Energy: Ford Foundation Study Urges Action on Conservation," *Science*
        186:426–428.
    1974c "Individual Self–Sufficiency in Energy," *Science* 184:278–282.
    1974d "Modeling the Climate: A New Sense of Urgency," *Science* 185:1145–1147.
    1975a "Geothermal Resources: A New Look," *Science* 190:370.
    1975b "Ozone Destruction: Problem's Scope Grows, Its Urgency Recedes," *Science*
        187:1181–1183.
    1975c "Solar Energy Reconsidered: ERDA Sees Bright Future," *Science* 189:538–539.
    1976 "Lithium: Will Short Supply Constrain Energy Technologies?" *Science*
        191:1037–1038.
————, and Thomas H. Maugh, II
    1974 "Stratospheric Pollution: Multiple Threats to Earth's Ozone," *Science*
        186:335–338.
————, William Metz, and Thomas H. Maugh, II
    1973 *Energy and the Future* (Washington: AAAS).
Hammond, Ogden, and Martin B. Zimmerman
    1975 "The Economics of Coal–Based Synthetic Gas," *Technology Review* 77(8):43–51.
Hammond, R. Philip
    1974 "Nuclear Power Risks," *American Scientist* 62:155–160.
Hannon, Bruce
    1974 "Options for Energy Conservation," *Technology Review* 76(4):24–31.
    1975 "Energy Conservation and the Consumer," *Science* 189:95–102 [a first–rate
        discussion of the need for an energy standard of value].
Harleman, Donald R. F.
    1971 "Heat—The Ultimate Waste," *Technology Review* 74(2):45–51.
Harte, John, and Robert H. Socolow, eds.
    1971 *Patient Earth* (New York: Holt, Rinehart and Winston).
Hein, R. A.
    1974 "Superconductivity: Large–Scale Applications," *Science* 185:211–222.
Heronemus, William E.
    1975 "The Case for Solar Energy," *Center Report* 8(1):6–9.
Hirsch, Robert L., and William L. R. Rice
    1974 "Nuclear Fusion Power and the Environment," *Environmental Conservation*
        1:251–262.
Hirst, Eric
    1973 "Transportation Energy Use and Conservation Potential," *Bulletin of the
        Atomic Scientists* 29(9):36–42.
————, and John C. Moyers
    1973 "Efficiency of Energy Use in the United States," *Science* 179:1299–1304.
Hobbs, P. V., H. Harrison, and E. Robinson
    1974 "Atmospheric Effects of Pollutants," *Science* 183:909–915.

Hohenemser, Kurt H.
  1975 "The Failsafe Risk," *Environment* 17(l):6–10.
Holdren, John P.
  1974 "Hazards of the Nuclear Fuel Cycle," *Bulletin of the Atomic Scientists* 30(8):14–23.
  ——— , and Paul R. Ehrlich
  1974 "Human Population and the Global Environment," *American Scientist* 62:282–292.
House Committee on Energy and Commerce, United States Congress, Subcommittee
      on Health and the Environment.
  1989 *Hearing on Air Toxins Control Act of 1989,* 100th Cong., 1st sess., 22 June
        Testimony of Bruce K. Maillet, on behalf of the State and Territorial Air
        Pollution Program Administrators and the Association of Local Air Pollution
        Control Officials.
House Committee on Energy and Commerce, United States Congress, Subcommittee
      on Health and the Environment.
  1989 *Hearing on Air Toxins Control Act of 1989,* 100th Cong., 1st sess. , 22 June
        Statement of Henry A. Waxman, Chairman.
Hubbert, M. King
  1969 "Energy Resources," in *Cloud* 1969, pp. 157–242.
Hueckel, Glenn
  1975 "A Historical Approach to Future Economic Growth," *Science* 187:925–931
        [an article on technological growth that begs almost all the questions raised
        in this chapter].
Indonesia, Government of
  1989 *U. S. News and World Report,* 18 December, 80–81.
Inglis, David R.
  1973 Nuclear Energy—Its Physics and Its Social Challenge (Reading, Mass.: Ad-
        dison–Wesley).
Johnston, Harold S.
  1974 "Pollution of the Stratosphere," *Environmental Conservation* 1:163–176.
Kantrowitz, Arthur
  1969 "The Test: Meeting the Challenge of New Technology," *Bulletin of the Atomic
        Scientists* 25(9):20–22, 48 [even more Panglossian than Rabinowitch 1969].
Kariel, Pat
  1974 "The Athabasca Tar Sands," *Sierra Club Bulletin* 59(8):8–10, 32.
Kates, Robert W., et al.
  1973 "Human Impact of the Managua Earthquake," *Science* 182:981–990.
Kellogg, W. W., and S. H. Schneider
  1974 "Climate Stabilization: For Better or for Worse?" *Science* 186:1163–1172 [the
        perils of attempting climate control].
Kolb, Charles E.
  1975 "The Depletion of Stratospheric Ozone," *Technology Review* 78(1):39–47.
Krieger, David
  1975 "Terrorists and Nuclear Technology," *Bulletin of the Atomic Scientists*
        31(6):28–34.
Kubo, Arthur S., and David J. Rose
  1973 "Disposal of Nuclear Wastes, " *Science* 182:1205–1211.
Kuhn, Thomas S.
  1970 *The Structure of Scientific Revolutions* (2nd ed.; University of Chicago Press)
        [diminishing returns in scientific discovery].

Kukla, George J., and Helena J. Kukla
  1974 "Increased Surface Albedo in the Northern Hemisphere," *Science* 183:709–714.
Lamb, Hubert H.
  1974 "Is the Earth's Climate Changing?" *The Ecologist* 4:10–15.
Landsberg, Hans H.
  1974 "Low–Cost, Abundant Energy: Paradise Lost?" *Science* 184:247–253.
Landsberg, Helmut E., and Lester Machta
  1974 "Anthropogenic Pollution of the Atmosphere: Whereto?" *Ambio* 3:146–150.
Lapp, Ralph E.
  1972 "One Answer to the Atomic-Energy Puzzle—Put the Atomic Power Plants in the Ocean," *New York Times Magazine,* June 4, pp. 20–21, 80–90.
  1973 "The Chemical Century," *Bulletin of the Atomic Scientists* 29(7):8–14.
Lewis, Richard S.
  1972 *The Nuclear Power Rebellion: Citizens vs. the Atomic Industrial Establishment* (New York:Viking).
Lieberman, M. A.
  1976 "United States Uranium Resources—An Analysis of Historical Data," *Science* 192:431–436.
Lincoln, G. A.
  1973 "Energy Conservation," *Science* 180:155–162.
Lindop, Patricia J., and J. Rotblat
  1971 "Radiation Pollution of the Environment," *Bulletin of the Atomic Scientists* 27(7):17–24.
Lippman, Thomas W.
  1991 "Risk Found in Low Levels of Radiation," *The Washington Post,* 20 March, A3.
Lovins, Amory B.
  1975 *World Energy Strategies: Facts, Issues, and Options* (Cambridge: Friends of the Earth/Ballinger).
——— , and John H. Price
  1975 *Non-Nuclear Futures: The Case for an Ethical Energy Strategy* (Cambridge: Friends of the Earth/Ballinger).
McCaull, Julian
  1973 "Windmills," *Environment* 15(l):6–17.
  1974 "Wringing Out the West," *Environment* 16(7):10–17.
McCloughlin, Merrill
  1989 "Our Dirty Air," *U. S. News and World Report,* 12 June, 48–54.
McIntyre, Hugh C.
  1975 "Natural–Uranium Heavy–Water Reactors," *Scientific American* 233(4):17–27 [the CANDU system].
McKelvey, V. E.
  1972 "Mineral Resource Estimates and Public Policy," *American Scientist* 60:32–40.
Makhijani, A.B., and A.J. Lichtenberg
  1972 "Energy and Well–Being," *Environment* 14(5):11–18.
Manuel, Frank E.
  1962 *The Prophets of Paris* (Cambridge: Harvard) [the Enlightenment ideology of progress].

Maraniss, David, and Michael Weisskopf
    1988 "Jobs and Illness in Petrochemical Corridor," *The Washington Post,* 22 December, A1.
Margen, Peter, et al.
    1975 "The Capacity of Nuclear Power Plants," *Bulletin of the Atomic Scientists* 31(8):38–46.
Martin, S., and W. J. Campbell
    1973 "Oil and Ice in the Arctic Ocean: Possible Large–Scale Interactions," *Science* 181:56–58.
Marx, Wesley
    1973 "Los Angeles and Its Mistress Machine," *Bulletin of the Atomic Scientists* 29(4):4–7, 44–48.
Massumi, Brian
    1974 "Oil Shale Country," *Not Man Apart* 4(6):12.
Mathews, Jay
    1991 "Southern California Clean Air Agency is Criticized," *The Washington Post,* 1 May, A21.
Mazur, Allan, and Eugene Rosa
    1974 "Energy and Life–Style," *Science* 186:607–610.
Meadows, Dennis L., and Jorgen Randers
    1972 "Adding the Time Dimension to Environmental Policy," in *World Eco–Crisis: International Organizations in Response,* ed. David A. Kay and Eugene B. Skolnikoff (Madison: Wisconsin), pp. 47–66.
Meadows, Dennis L., et al.
    1974 *The Dynamics of Growth in a Finite World* (Cambridge: Wright–Allen).
Medawar, Peter
    1969 "On 'The Effecting of All Things Possible,' " *Technology Review* 72(2):30–35 [a modern descendant of Francis Bacon emotionally defends the hope of progress].
Meinel, Aden B., and Marjorie P. Meinel
    1971 "Is It Time for a New Look at Solar Energy?" *Bulletin of the Atomic Scientists* 27(8):32–37.
Mesarovic, Mihajlo, and Eduard Pestel
    1974 *Mankind at the Turning Point: The Second Report to the Club of Rome* (New York: Duttord Reader's Digest).
Metz, William D.
    1972 "Magnetic Containment Fusion: What Are the Prospects?" *Science* 178:291–292.
    1973 "Ocean Temperature Gradients: Solar Power from the Sea," *Science* 180:1266–1267.
    1974 "Oil Shale: A Huge Resource of Low–Grade Fuel," *Science* 184:1271–1275.
    1975a "Energy Conservation: Better Living through Thermodynamics," *Science* 188:820–821.
    1975b "Energy: ERDA Stresses Multiple Sources and Conservation," *Science* 189:369–370.
Metzger, H. Peter
    1972 *The Atomic Establishment* (New York: Simon and Schuster).
Meyer, Alden
    1990 "The 'White House Effect': Bush Backs Off Carbon Dioxide Stabilization," *Nucleus,* Spring, 3.
    1990 "United States Increasingly Isolated on Global Warming," *Nucleus,* Summer, 3.

Michaelis, Anthony R.
    1973 "Coping with Disaster," *Bulletin of the Atomic Scientists* 29(4):24–29.
Micklin, Philip P.
    1974 "Environmental Hazards of Nuclear Wastes," *Bulletin of the Atomic Scientists* 30(4):36–42 [a first-rate non-polemical review].
Miles, Rufus., Jr.
    1976 *Awakening from the American Dream* (New York: Universe).
Mishan, E. J.
    1974 "The New Inflation: Its Theory and Practice," *Encounter* 42(5):12–24.
Moore, Curtis
    1990 "Revenge of the Killer Trees," *The Washington Post,* 29 July, C3.
Mostert, Noel
    1974 *Supership* (New York: Knopf).
Mumford, Lewis
    1970 *The Pentagon of Power* (New York: Harcourt Brace Jovanovich).
Murdoch, William W., ed.
    1971 *Environment: Resources, Pollution and Society* (Stamford: Sinauer).
Mussett, Alan
    1973 "Discovery: A Declining Asset?" *New Scientist* 60:886–889.
Naill, Roger F., et al.
    1975 "The Transition to Coal," *Technology Review* 78(1):19–29.
Nash, Hugh
    1974 "Nader, UCS Release Suppressed AEC Report on Reactor Safety," *Not Man Apart* 4(2):8–9.
National Academy of Sciences
    1975 *Mineral Resources and the Environment* (Washington: National Academy of Sciences) [critical of the U.S. Geological Survey oil and gas estimates as too high].
National Research Council
    1989 *Alternative Agriculture* (Washington, D.C.: National Academy Press.)
Nelson, Saul
    1974 "The Looming Shortage of Primary Processing Capacity," *Challenge* 16(6):45–48.
de Nevers, Noel
    1973 "Enforcing the Clean Air Act of 1970," *Scientific American* 228(6): 14–21.
Newell, Reginald E.
    1974 "The Earth's Climatic History," *Technology Review* 77(2):31–45.
Nilsson, Sam
    1974 "Energy Analysis—A More Sensitive Instrument for Determining Costs of Goods and Services," *Ambio* 3:222–224.
Novick, Sheldon
    1969 *The Careless Atom* (Boston: Houghton Mifflin).
    1975 "A Troublesome Brew," *Environment* 17(4):8–11 [critique of AEC's final environmental impact statement on the breeder].
Odell, Rice
    1975 "Net Energy Analysis Can Be Illuminating," *Conservation Foundation Letter,* October [a very useful brief summary of the issues].
O'Donnell, Sean
    1974 "Ireland Turns to Peat," *New Scientist* 63:18–19 [the USSR and other countries also have substantial supplies].

Odum, Howard T.
  1971 *Environment, Power and Society* (New York: Wiley).
  1973 "Energy, Ecology and Economics," *Ambio* 2:220–227.
Okie, Susan
  1990 "Cancer Rates in Industrial Countries Rise," *The Washington Post,* 10 December, A1.
O'Neill, Gerard K.
  1975 "Space Colonies and Energy Supply to the Earth," *Science* 190:943–947.
Organization for Economic Co–Operation and Development
  1990 *The State of the Environment.* (Paris: OECD Publication Service).
Osborn, Elburt F.
  1974 "Coal and the Present Energy Situation," *Science* 183:477–481.
Page, James K., Jr.
  1974 "Growing Pains in Energy," *Smithsonian* 5(6):12–15.
Park, Charles F., Jr.
  1968 *Affluence in Jeopardy: Minerals and the Political Economy* (San Francisco: Freeman, Cooper).
Patterson, Walter C.
  1972 "The British Atom," *Environment* 14(10):2–9.
Pearl, Arthur, and Stephanie Pearl
  1971 "Toward an Ecological Theory of Value," *Social Policy* 2(1):30–38 [thermodynamic economics].
Perry, Harry
  1974 "The Gasification of Coal," *Scientific American* 230(3):19–25.
Peterson, James T.
  1973 "Energy and the Weather," *Environment* 15(8):4–9.
Plau, John R.
  1966 *The Step to Man* (New York: Wiley) [limits of technological scale].
Pollard, William G.
  1976 "The Long–range Prospects for Solar Energy," *American Scientist* 64:424–429 [why centralized generation of electricity using solar energy will be impractical, if not impossible].
Polunin, Nicholas
  1974 "Thoughts on Some Conceivable Ecodisasters," *Environmental Conservation* 1:177–189 [all the small but potentially lethal risks, especially in combination].
Post, Richard F.
  1971 "Fusion Power: The Uncertain Certainty," *Bulletin of the Atomic Scientists* 27(8):42–48.
———, and Stephen F. Post
  1973 "Flywheels," *Scientific American* 229(6):17–23.
———, and F. L. Ribe
  1974 "Fusion Reactors as Future Energy Sources," *Science* 186:397–407.
Postel, Sandra
  1990 "Trouble On Tap" in *Environment 90/91,* reprinted from World Watch, September/October, 1989, pp. 12–20, ed. John Allen, Annual Editions (Guilford, Connecticut: The Dushkin Publishing Group).
Price, Derek J. de Solla
  1961 *Science Since Babylon* (New Haven: Yale) ["diseases" of "big science"].

Primack, Joel, and Frank von Hippel
  1974 "Nuclear Reactor Safety: The Origins and Issues of a Vital Debate," *Bulletin of the Atomic Scientists* 30(8):5–12.
Prud'homme, Robert K.
  1974 "Automobile Emissions Abatement and Fuels Policy," *American Scientist* 62:191–199.
Pryde, Philip R., and Lucy T. Pryde
  1974 "Soviet Nuclear Power: A Different Approach to Nuclear Safety," *Environment* 16(3):26–34.
Rabinowitch, Eugene
  1969 "Responsibility of Scientists in Our Age," *Bulletin of the Atomic Scientists* 25(9):2–3, 26 [an argument—very typical in its rationale—that science and technology have abolished scarcity].
Ramseier, Rene O.
  1974 "Oil on Ice: How to Melt the Arctic and Warm the World," *Environment* 16(4):7–14.
RAND Corporation
  1973 *California's Electric Quandary* (3 vols; Santa Monica: RAND).
RANN (Research Applied to National Needs Program)
  1972 *Summary Report of the Cornell Workshop on Energy and the Environment* (Washington: Government Printing Office).
Reed, T. B., and R. M. Lemer
  1973 "Methanol: A Versatile Fuel for Immediate Use," *Science* 182:1299–1304.
Renner, Michael
  1991 "Assessing the Military War on the Environment," in *State of the World, 1991*, ed. Linda Starke (New York: W. W. Norton & Company).
Rex, Robert W.
  1971 "Geothermal Energy—The Neglected Energy Option," *Bulletin of the Atomic Scientists* 27(8):52–56.
RFF (Resources for the Future)
  1973 *Energy Research and Development—Problems and Prospects* (Washington: Government Printing Office).
Rhodes, Richard
  1974 "Los Alamos Revisited," *Harper's,* March, pp. 57–64.
Rice, Richard A.
  1974 "Toward More Transportation with Less Energy," *Technology Review* 76(4):45–53.
Ritchie–Calder, Peter R.
  1970 "Mortgaging the Old Homestead," *Foreign Affairs* 48:207–220 [supertanker problems].
Roberts, Marc J.
  1973 "Is There an Energy Crisis?" *Public Interest* 31:17–37.
Robinson, Arthur L.
  1974 "Energy Storage (II): Developing Advanced Technologies," *Science* 184:884–887.
Robson, Geoffrey
  1974 "Geothermal Electricity Production," *Science* 184:371–375.
Rose, David J.
  1974a "Energy Policy in the U.S.," *Scientific American* 230(l):20–29.
  1974b "Nuclear Eclectic Power," *Science* 184:351–359.

Rubin, Milton D.
  1974 "Plugging the Energy Sieve," *Bulletin of the Atomic Scientists* 30(10):7–17.
Russell, W. M. S.
  1971 "Population and Inflation," *The Ecologist* 1(8):4–8.
SCEP (Study of Critical Environmental Problems)
  1970 *Man's Impact on the Global Environment* (Cambridge: MIT).
Schneider, Stephen H.
  1974 "The Population Explosion: Can It Shake the Climate?" *Ambio* 3:150–155.
————, and Roger D. Dennett
  1975 "Climatic Barriers to Long-Term Energy Growth," *Ambio* 4:65–74.
Schumacher, E. F.
  1974 *Small Is Beautiful: Economics as if People Mattered* (New York: Harper and Row).
Seaborg, Glenn T., and William R. Corliss
  1971 *Man and Atom: Building a New World Through Nuclear Technology* (New York: Dutton).
Shea, Cynthia Pollock
  1989 "Protecting the Ozone Layer" in *State of the World, 1989*, ed. Linda Starke (New York: W. W. Norton & Company).
Shen–Miller, J.
  1970 "Some Thoughts on the Nuclear Agro–Industrial Complex," *BioScience* 20:98–100.
Shogren, Elizabeth
  1990 "4 Years Later, Chernobyl's Ills Widen," *The Washington Post*, 27 April, A1.
Skinner, Brian J.
  1969 *Earth Resources* (Englewood Cliffs: Prentice–Hall).
Slesser, Malcolm
  1973 "Energy Analysis in Policy Making," *New Scientist* 60:328–330.
  1974 "The Energy Ration," *The Ecologist* 4:139–140.
SMIC (Study of Man's Impact on Climate)
  1971 *Inadvertent Climate Modification* (Cambridge: MIT).
Smith, R. Jeffrey
  1989 "Low-Level Radiation Causes More Deaths Than Assumed, Study Finds," *The Washington Post*, 20 December, A3.
Snowden, Donald P.
  1972 "Superconductors for Power Transmission," *Scientific American* 226(4):84–91.
Sorensen, Bent
  1975 "Energy and Resources," *Science* 189:255–260 [a solar energy economy for Denmark].
Spurgeon, David
  1973 "Natural Power for the Third World," *New Scientist* 60:694–697.
Squires, Arthur M.
  1974 "Coal: A Past and Future King," *Ambio* 3:1–14.
Starr, Chauncey
  1971 "Energy and Power," *Scientific American* 225(3)134–144.
————, and Richard Rudman
  1973 "Parameters of Technological Growth," *Science* 182:358–364.
Stein, Richard G.
  1972 "A Matter of Design," *Environment* 14(8):17–20, 25–29.

Stent, Gunther S.
1969 *The Coming of the Golden Age: A View of the End of Progress* (New York: Natural History Press).

Stevens, William K.
1991 "As Nations Meet on Global Warming, U.S. Stands Alone," *The New York Times,* 10 September, C1.

————, 1991 "Danes Link Sunspot Intensity to Global Temperature Rise," *The New York Times,* 5 November, C4.

Stever, H. Guyford
1975 "Whither the NSF?—The Higher Derivatives," *Science* 189:264–267 [the growing capital intensity of research and development].

Strong, Maurice F.
1973 "One Year After Stockholm: An Ecological Approach to Management," *Foreign Affairs* 51(4):690–707.

Stunkel, Kenneth R.
1973 "The Technological Solution," *Bulletin of the Atomic Scientists* 29(7):42–44.

Tamplin, Arthur R.
1973 "Solar Energy," *Environment* 15(5):16–20, 32–34 [one of the best short reviews; extensive citations].

Taylor, Theodore B., and Charles C. Humpstone
1973 *The Restoration of the Earth* (New York: Harper and Row).

UNESCO
1973 "Appropriate Technology," a special issue of *Impact of Science on Society* 23:251–352.

United Nations
1961 *Proceedings of the Conference on New Sources of Energy* (6 vols; Rome: United Nations) [extensive discussions of wind, tide, and sun as sources of power].

Vacca, Roberto
1973 *The Coming Dark Age,* trans. J. S. Whale (Garden City, N.Y.: Doubleday) [an alarmist view of the industrial system's intrinsic instability].

Wade, Nicholas
1974 "Windmills: The Resurrection of an Ancient Energy Technology," *Science* 184:1055–1058.

Walsh, John
1974 "Uranium Enrichment: Both the Americans and Europeans Must Decide Where to Get the Nuclear Fuel of the 1980's," *Science* 184:1160–1161.

Wanniski, Jude
1975 "The Mundell–Laffer Hypothesis—A New View of the World Economy," *Public Interest* 39:31–52 [scarcity and inflation].

Wasserman, Harvey
1991 "Bush's Pro–Nuke Energy Strategy." *The Nation,* 20 May, 656–660.

Waters, W. G., II
1973 "Landing a Man Downtown," *Bulletin of the Atomic Scientists* 29(9):34–35 [how environmental management differs from space programs].

Watt, Kenneth E. F.
1974 *The Titanic Effect: Planning for the Unthinkable* (Stamford, Conn.: Sinauer).

Weinberg, Alvin M.
1972 "Science and Trans-Science," *Minerva* 10:209–222.

1973 "Technology and Ecology—Is There a Need for Confrontation?" *BioScience* 23:41–45.
1974 "Global Effects of Man's Production of Energy," *Science* 186:205.
Weir, David, and Constance Matthiessen
1990 "Will the Circle Be Unbroken?" in *Environment 90/91,* reprinted from *Mother Jones,* June, 1989, pp. 20–27, Annual Editions (Guildford, Connecticut: The Dushkin Publishing Group).
Weisskopf, Michael
1991 "Ozone Layer over U.S. Thinning Swiftly," *The Washington Post,* 5 April, A1.
Wentorf, R. H., Jr., and R. E. Hanneman
1974 "Thermochemical Hydrogen Generation," *Science* 185:311–319.
Westman, Walter E., and Roger M. Gifford
1973 "Environmental Impact: Controlling the Overall Level," *Science* 181:819–825 [with an energy currency].
Whittemore, F. Case
1973 "How Much in Reserve?" *Environment* 15(7):16–20, 31–35.
Wilkinson, John
1974 "A Modest Proposal for Recycling Our Junk Heap Society," *Center Report* 7(3):7–12 [a computer simulation suggests that future living standards will resemble those of the early 1900s].
Willrich, Mason, and Theodore B. Taylor
1974 *Nuclear Theft: Risks and Safeguards* (Cambridge: Ballinger).
Wilson, Richard
1973 "Natural Gas Is a Beautiful Thing?" *Bulletin of the Atomic Scientists* 29(7):35–40.
Wind Energy Weekly
1991 4 June, #403.
Winsche, W. E., et al.
1973 "Hydrogen: Its Future Role in the Nation's Energy Economy," *Science* 180:1325–1332.
Wolf, Martin
1974 "Solar Energy Utilization by Physical Methods," *Science* 184:382–386.
Wood, Lowell, and John Nuckolls
1972 "Fusion Power," *Environment* 14(4):29–33.
World Resources Institute.
1990 *World Resources, 1990–91* in collaboration with the U.N. Environmental Programme and the U.N. Development Program.
Wright, John, and John Syrett
1975 "Energy Analysis of Nuclear Power," *New Scientist* 65:66–67.
Young, Louise B., and H. Peyton Young
1974 "Pollution by Electrical Transmission: The Environmental Impact of High Voltage Lines," *Bulletin of the Atomic Scientists* 30(10):34–38.

Chapter 4

Attah, Ernest B.
1973 "Racial Aspects of Zero Population Growth," *Science* 180:1143–115.
Barker, Ernst, trans.
1962 *The Politics of Aristotle* (New York: Oxford).

Barnett, Larry D.
    1971 "Zero Population Growth, Inc.," *BioScience* 21:759–765.
Bell, Daniel
    1973 *The Coming of Post–Industrial Society: A Venture in Social Forecasting* (New York: Basic Books).
Berlin, Isaiah
    1969 *Four Essays on Liberty* (New York: Oxford).
Brown, Harrison
    1954 *The Challenge of Man's Future* (New York: Viking).
Buchanan, James
    1969 *The Demand and Supply of Public Goods* (Chicago: Rand McNally).
Burch, William R., Jr.
    1971 *Daydreams and Nightmares: A Sociological Essay on the American Environment* (New York: Harper and Row).
Butler, Samuel
    1872 *Erewhon* (New York: Signet, 1960).
Callahan, Daniel J.
    1973 *The Tyranny of Survival; and Other Pathologies of Civilized Life* (New York: Macmillan).
Carney, Francis
    1972 " Schlockology, " *New York Review of Books,* June 1, pp. 26–29.
Chamberlin, Neil W.
    1970 *Beyond Malthus: Population and Power* (New York: Basic Books).
Christy, Francis T., Jr., and Anthony Scott
    1965 *The Common Wealth in Ocean Fisheries* (Baltimore: Johns Hopkins).
Cohen, David
    1973 "Chemical Castration," *New Scientist* 57:525–526.
Comford, Francis M., trans.
    1945 *The Republic of Plato* (New York: Oxford).
Crowe, Beryl L.
    1969 "The Tragedy of the Commons Revisited," *Science* 166:1103–1107.
Dahl, Robert A.
    1970 *After the Revolution?: Authority in a Good Society* (New Haven: Yale).
Delgado, Jose Manuel R.
    1969 *Physical Control of the Mind: Toward a Psychocivilized Society* (New York: Harper and Row).
Eisner, Thomas, Ari van Tienhaven, and Frank Rosenblatt
    1970 "Population Control, Sterilization, and Ignorance," *Science* 167:337.
Ellul, Jacques
    1967 *The Technological Society* (rev.; New York: Knopf).
Elmer–Dewitt, Phillip
    1989 "A Drastic Plan to Banish Smog," *Time,* 27 March, 65.
Fife, Daniel
    1971 "Killing the Goose," *Environment* 13(3):20–27 [the logic of the commons].
Forester, E. M.
    1928 *The Eternal Moment* (New York: Harcourt, Brace).
Fuller, R. Buckminster
    1968 "An Operating Manual for Spaceship Earth" in *Environment and Change: The Next Fifty Years,* ed. William R. Ewald, Jr. (Bloomington: Indiana).

1969 "Vertical Is to Live, Horizontal Is to Die," *American Scholar* 39(1):27–47.

Geesaman, Donald P., and Dean E. Abrahamson

1974 "The Dilemma of Fission Power," *Bulletin of the Atomic Scientists* 30(9):37–41 [the extreme security measures a nuclear power economy will require].

Global Tomorrow Coalition

1990 *The Global Ecology Handbook,* ed. Walter H. Corson (Boston: Beacon Press).

Haefele, Edwin T., ed.

1975 *The Governance of Common Property Resources* (Baltimore: Johns Hopkins).

Hardin, Garrett

1968 "The Tragedy of the Commons," *Science* 162:1243–1248.

1972 *Exploring New Ethics for Survival* (New York: Viking).

———, ed.

1969 *Population, Evolution, and Birth Control: A Collage of Controversial Ideas* (2nd ed.; New York: W. H. Freeman and Co.).

Heilbroner, Robert L.

1974 *An Inquiry into the Human Prospect* (New York: Norton).

Hobbes, Thomas

1651 *Leviathan, or the Matter, Form and Power of a Commonwealth, ecclesiastical and civil,* ed. H. W. Schneider (Indianapolis: Bobbs–Merrill, 1958).

Holden, Constance

1973 "Psychosurgery: Legitimate Therapy or Laundered Lobotomy?" *Science* 179:1109–1114.

Huxley, Aldous L.

1932 *Brave New World* (New York: Modern Library, 1956).

1958 *Brave New World Revisited* (New York: Harper).

Illich, Ivan

1973 *Tools for Conviviality* (New York: Harper and Row).

Kahn, Alfred E.

1966 "The Tyranny of Small Decisions: Market Failures, Imperfections, and the Limits of Economics," *Kyklos* 19(1):23–47 [the logic of the commons].

Kahn, Herman, and Anthony J. Wiener

1968 "Faustian Powers and Human Choice: Some Twenty-First Century Technological and Economic Issues" in *Environment and Choice,* ed. William R. Ewald, Jr. (Bloomington: Indiana), pp. 101–131.

Kass, Leon R.

1971 "The New Biology: What Price Relieving Man's Estate?" *Science* 174:779–788.

1972 "Making Babies—The New Biology and the 'Old' Morality," *Public Interest* 26:18–56.

Lewis, C. S.

1965 *The Abolition of Man* (New York: Macmillan).

Locke, John

1690 *Second Treatise, in Two Treatises of Government,* ed. Peter Laslett (New York: New American Library, 1965).

McDermott, John

1969 "Technology: The Opiate of the Intellectuals," *New York Review of Books,* July 31, pp. 25–35.

Michael, Donald N.

1970 *The Unprepared Society: Planning for a Precarious Future* (New York: Harper and Row).

Morrison, Denton E., Kenneth E. Horseback, and W. Keith Warner
    1974 *Environment: A Bibliography of Social Science and Related Literature* (Washington: GPO).
    1975 *Energy: A Bibliography of Social Science and Related Literature* (New York: Garland).
Myers, Norman
    1975 "The Whaling Controversy," *American Scientist* 63:448–455 [an excellent case study of the kinds of pressures that promote overexploitation].
Odell, Rice
    1975 "How Will We React to an Age of Scarcity?" *Conservation Foundation Letter,* January [a review of many different opinions].
Olson, Mancur, Jr.
    1968 *The Logic of Collective Action: Public Goods and the Theory of Groups* (New York: Schocken).
———, and Hans Landsberg, eds.
    1973 *The No-Growth Society* (New York: Norton).
Ophuls, William
    1973 "Leviathan or Oblivion?" in *Toward Steady–State Economy,* ed. Herman E. Daly (New York: W. H. Freeman and Co.), pp. 215–230.
Orwell, George
    1963 *Nineteen Eighty–Four: Text, Sources, Criticism,* ed. Irving Howe (New York: Harcourt, Brace and World).
Pirages, Dennis C., and Paul R. Ehrlich
    1974 *Ark II: Social Response to Environmental Imperatives* (New York: W. H. Freeman and Company).
Popper, Karl R.
    1966 *The Open Society, and Its Enemies* (2 vols, 5th ed., rev.; Princeton University Press).
Reich, Charles A.
    1971 *The Greening of America* (New York: Random House).
Rousseau, Jean-Jacques
    1762 *The Social Contract,* ed. Charles Frankel (New York: Hafner, 1947).
Russett, Bruce M., and John D. Sullivan
    1971 "Collective Goods and International Organization," *International Organization* 25:845–865.
Schelling, Thomas C.
    1971 "On the Ecology of Micromotive,," *Public Interest* 25:61–98.
Skinner, B. F.
    1971 *Beyond Freedom and Dignity* (New York: Knopf).
Smith, Adam
    1776 *An Inquiry into the Nature and Causes of the Wealth of Nations,* ed. Edwin Cannan (New York: Modern Library, 1937).
Speth, J. Gustave, Arthur R. Tamplin, and Thomas B. Cochran
    1974 "Plutonium Recycle: The Fateful Step," *Bulletin of the Atomic Scientists* 30(9):15–22.
Stillman, Peter G.
    1975 "The Tragedy of the Commons: A Re-Analysis," *Alternatives* 4(2):12–15.
Stone, Christopher D.
    1974 *Should Trees Have Standing?: Toward Legal Rights for Natural Objects* (Los Altos, Calif.: William Kaufmann).

Susskind, Charles
   1973 *Understanding Technology* (Baltimore: Johns Hopkins).
Tuan, Yi-Fu
   1970 "Our Treatment of the Environment in Ideal and Actuality," *American Scientist*
        58:244–249 [the Chinese and their environment through history].
Wade, Nicholas
   1974 "Sahelian Drought: No Victory for Western Aid," *Science* 185:234–237 [how
        an aid program destroyed the traditional controls on a common—with
        catastrophic results].
Webb, Walter Prescott
   1952 *The Great Frontier* (Boston: Houghton Mifflin).
Weinberg, Alvin M.
   1972a "Social Institutions and Nuclear Energy," *Science* 177:27–34.
   1972b Review of John Holdren and Philip Herrers, *Energy: A Crisis in Power* in
         American Scientist 60:775–776.
   1973 "Technology and Ecology—Is There a Need for Confrontation?" *BioScience*
        23:41–46.
White, Lynn, Jr.
   1967 "The Historical Roots of Our Ecologic Crisis,"*Science* 155:1203–1207.
Wilkinson, Richard G.
   1973 *Poverty and Progress: An Ecological Perspective on Economic Development* (New
        York: Praeger).
Willrich, Mason
   1975 "Terrorists Keep Out!: The Problem of Safeguarding Nuclear Materials in a World
        of Malfunctioning People," *Bulletin of the Atomic Scientists* 31(5):12–16.
Wynne–Edwards, V. C.
   1970 "Self–Regulatory Systems in Populations of Animals," in *The Subversive
        Science*, ed. Paul Shepard and Daniel McKinley (Boston: Houghton Mif-
        flin), pp. 99–111 [valuable biological perspective on the tragedy of the
        commons].

Chapter 5

Abrahamson, Dean E.
   1974a "Energy: All in the Family," *Environment* 16(7):50–52.
   1974b "Energy: Sidestepping NEPA Reviews," *Environment* 16(9):39.
Anderson, Frederick R., and Robert H. Daniels
   1973 *NEPA in the Courts: A Legal Analysis of the National Environmental Policy Act*
        (Baltimore: Johns Hopkins).
Ayres, Robert U., and Allen V. Kneese
   1969 "Production, Consumption, and Externalities," *American Economic Review*
        59:282–297 [Kneese et al. 1970 in a nutshell].
Barnett, Harold J., and Chandler Morse
   1963 *Scarcity and Growth: The Economics of Natural Resource Availability* (Baltimore:
        Johns Hopkins).
Beckerman, Wilfred
   1974 *In Defence of Economic Growth* (London: Cape).
Bell, Daniel
   1971 "The Corporation and Society in the 1970's," *Public Interest* 24:5–32.

Boguslaw, Robert
    1965 *The New Utopians* (Englewood Cliffs, N.J.: Prentice–Hall).
Boulding, Kenneth E.
    1949 "Income or Welfare?" *Review of Economic Studies* 17:77–86.
    1966 "The Economics of the Coming Spaceship Earth," in *Environmental Quality in a Growing Economy*, ed. Henry Jarrett (Baltimore: Johns Hopkins), pp. 3–14.
    1967 "Fun and Games with the Gross National Product—The Role of Misleading Indicators in Social Policy," in *The Environmental Crisis*, ed. Harold W. Helfrich, Jr. (New Haven: Yale), pp. 157–170.
    1970 *Economics as a Science* (New York: McGraw–Hill), Chap. 7.
Brooks, Harvey, and Raymond Bowers
    1971 "The Assessment of Technology" in *Man and the Ecosphere*, ed. Paul R. Ehrlich, John P. Holdren, and Richard W. Holm (New York: W. H. Freeman and Co.).
Carter, Luther J.
    1973 "Alaska Pipeline: Congress Deaf to Environmentalists," *Science* 179:1310–1312, 1350.
Clark, Colin W.
    1973 "The Economics of Overexploitation," *Science* 181:630–634.
Commoner, Barry
    1973 "Trains into Flowers," *Harper's*, December. pp. 78–86 [why trains cannot compete with the auto].
Conservation Foundation
    1971 "Indiscriminate Economic Growth, Measured with Little Regard for Environmental Costs and Social Well–Being, Is Challenged," *CF Letter*, May.
    1972 "NEPA Challenges the Nation's Plans and Priorities—But Progress Is Slow, and Some Are Reacting Against It," *CF Letter*, May.
Culbertson, John M.
    1971 *Economic Development: An Ecological Approach* (New York: Knopf).
Dales, J. H.
    1968 *Pollution, Property and Prices: An Essay in Policy–Making and Economics* (University of Toronto Press).
Daly, Herman E., ed.
    1973 *Toward a Steady–State Economy* (New York: W. H. Freeman and Co.).
Dolan, Edwin G.
    1971 *TANSTAAFL: The Economic Strategy for Ecologic Crisis* (New York: Holt, Rinehart and Winston).
Edel, Matthew
    1973 *Economics and the Environment* (Englewood Cliffs, N.J.: Prentice–Hall).
Freeman, A. Myrick, and Robert H. Haveman
    1972 "Clean Rhetoric and Dirty Water," *Public Interest* 28:51–65.
————, Robert H. Haveman, and Allen V. Kneese
    1973 *The Economics of Environmental Policy* (New York: Wiley).
Gabor, Dennis
    1972 *The Mature Society* (London: Secker and Warburg).
Galbraith, John K.
    1958 *The Affluent Society* (Boston: Houghton Mifflin).
    1967 *The New Industrial State* (Boston: Houghton Mifflin).

Garvey, Gerald
   1972 *Energy, Ecology, Economy: A Framework for Environmental Policy* (New York: Norton).

Gillette, Robert
   1972 "National Environmental Policy Act: Signs of Backlash Are Evident," *Science* 176: 30–33.

Hagevik, George
   1971 "Legislating for Air Quality Management," in *The Politics of Ecosuicide*, ed. Leslie L. Roos, Jr. (New York: Holt, Rinehart and Winston), pp. 311–345. [Excellent on the difficulties of internalizing costs.]

Hardesty, John, Norris C. Clement, and Clinton E. Jencks
   1971 "The Political Economy of Environmental Disruption," in *Economic Growth vs. the Environment*, ed. Warren E. Johnson and John Hardesty (Belmont, Calif.: Wadsworth), pp. 85–106.

Hardin, Garrett
   1972 *Exploring New Ethics for Survival* (New York: Viking).

Harnik, Peter
   1973 "The Biggest Going-Out-of-Business Sale of All Time," *Environmental Action*, September 1, pp. 9–12.

Hays, Samuel P.
   1959 *Conservation and the Gospel of Efficiency* (Cambridge: Harvard).

Heller, Walter W.
   1973 *Economic Growth and Environmental Quality: Collision, or Co–Existence* (Morristown, N.J.: General Learning Press).

Henderson, Hazel
   1976 "The End of Economics," *The Ecologist* 6:137–146 [a first-rate critique by an important radical economist; a valuable supplement to the argument of this chapter, with useful references to her own previous work and to the work of others].

Hirschman, Albert O.
   1967 *Development Projects Observed* (Washington: Brookings Institute) [the hidden costs of development].

Kapp, K. William
   1950 *The Social Costs of Private Enterprise* (New York: Schocken, 1971).

Klausener, Samuel Z.
   1971 *On Man and His Environment* (San Francisco: Jossey–Bass) [an attempt to come to terms with some of the sociological externalities of development].

Kneese, Allen V.
   1973 "The Faustian Bargain: Benefit–Cost Analysis and Unscheduled Events in the Nuclear Fuel Cycle," *Resources* 44:1–5.
   ————, Robert U. Ayres, and Ralph C. d'Arge
   1970 *Economics and Environment: A Materials Balance Approach* (Baltimore: Johns Hopkins)

Kraus, James
   1974 "American Environmental Case Law: An Update," *Alternatives* 3(2):25–30.

Krieger, Martin H.
   1973 "What's Wrong with Plastic Trees," *Science* 179:446–455 [the perversities of pure economic analysis].

Krieth, Frank
   1973 "Lack of Impact," *Environment* 15(1):26–33.

Miller, G. Tyler
    1990 *Living in the Environment* (6th ed.; Belmont, California: Wadsworth Publishing
        Company).
Mishan, Ezra J.
    1969 *Technology and Growth: The Price We Pay* (New York: Praeger).
    1971 "On Making the Future Safe for Mankind," *Public Interest* 24:33–61.
Novick, Sheldon
    1974 "Nuclear Breeders," *Environment* 16(6):6–15.
Odell, Rice
    1973 "Environmental Politicking—Business as Usual," *Conservation Foundation Let-
        ter,* August.
Passell, Peter, and Leonard Ross
    1973 *The Retreat from Riches: Affluence and Its Enemies* (New York: Viking).
Pearce, David
    1973 "Is Ecology Elitist?" *The Ecologist* 3:61–63.
Polanyi, Karl
    1944 *The Great Transformation* (Boston: Beacon).
Ridker, Ronald G.
    1972 "Population and Pollution in the United States," *Science* 176:1085–1090.
Rothman, Harry
    1972 *Murderous Providence: A Study of Pollution in Industrial Societies* (New York:
        Bobbs–Merrill).
Ruff, Larry E.
    1970 "The Economic Common Sense of Pollution," *Public Interest* 19:69–85.
Sachs, Ignacy
    1971 "Approaches to a Political Economy of Environment," *Social Science Informa-
        tion* 5(5):47–58 [a very perceptive brief overview of the clash between market
        traditionalists and the new economic holists].
Stone, Richard
    1972 "The Evaluation of Pollution: Balancing Gains and Losses," *Minerva* 10:412–
        425.
Tribe, Lawrence H.
    1971 "Legal Frameworks for the Assessment and Control of Technology," *Minerva*
        9:243–255.
Tsuru, Shigeto
    1971 "In Place of GNP," *Social Science Information* 10(4):7–21 [an especially good
        discussion of the drawbacks of GNP as an indicator].
UNESCO
    1973 "The Social Assessment of Technology," special issue *of International Social
        Science Journal* 25(3).
Weisskopf, Walter A.
    1971 *Alienation and Economics* (New York: Dutton).
Wildavsky, Aaron
    1967 "Aesthetic Power or the Triumph of the Sensitive Minority over the
        Vulgar Mass: A Political Analysis of the New Economics," *Daedalus*
        96:1115–1128.
Wilkinson, Richard G.
    1973 *Poverty and Progress: An Ecological Perspective on Economic Development* (New
        York: Praeger).

Winner, Langdon
   1972 "On Controlling Technology," *Public Policy* 20:35–59.
Wollman, Nathaniel
   1967 "The New Economics of Resources," *Daedalus* 96:1099–1114.

Chapter 6

Abelson, Philip H.
   1972a "Environmental Quality," *Science* 177:655.
   1972b "Federal Statistics," *Science* 175:1315.
Bachrach, Peter
   1967 *The Theory of Democratic Elitism: A Critique* (Boston: Little, Brown).
Bell, Daniel
   1974 "The Public Household—On 'Fiscal Sociology' and the Liberal Society," *Public Interest* 37:29–68.
Brown, Harrison, James Bonner, and John Weir
   1963 *The Next Hundred Years* (New York: Viking) [esp. Chaps. 14–17, which discuss manpower].
Bruce–Briggs, B.
   1974 "Against the Neo-Malthusians," *Commentary* July, pp. 25–29.
Burch, William R., Jr.
   1971 *Daydreams and Nightmares: A Sociological Essay on the American Environment* (New York: Harper and Row).
Caldwell, Lynton K.
   1971 *Environment: A Challenge to Modern Society* (Garden City, N.Y.: Doubleday).
——, and Toufiq A. Siddiqi
   1974 *Environmental Policy, Law and Administration: A Guide to Advanced Study* (Bloomington: University of Indiana School of Public and Environmental Affairs).
Carpenter, Richard A.
   1972 "National Goals and Environmental Laws," *Technology Review* 74(3):58–63.
Carter, Luther J.
   1973a "Environment: A Lesson for the People of Plenty," *Science* 182:1323–1324.
   1973b "Environmental Law (I): Maturing Field for Lawyers and Scientists," *Science* 179:1205–1209.
   1973c "Environmental Law (II): A Strategic Weapon Against Degradation?" *Science* 179:1310–1312, 1350.
   1973d "Pesticides: Environmentalists Seek New Victory in a Frustrating War," *Science* 181:143–145.
   1974a "Cancer and the Environment (I): A Creaky System Grinds On," *Science* 186:239–242.
   1974b "Con Edison: Endless Storm King Dispute Adds to Its Troubles," *Science* 194:1353–1358.
   1974c "The Energy Bureaucracy: The Pieces Fall into Place," *Science* 185:44–45.
   1974d "Energy: Cannibalism in the Bureaucracy," *Science* 186:511.
   1974e "Pollution and Public Health: Taconite Case Poses Major Test," *Science* 186:31–36.
   1975a "The Environment: A 'Mature' Cause in Need of a Lift," *Science* 187:45–48.

1975b *The Florida Experience: Land and Water Policy in a Growth State* (Baltimore: Johns Hopkins).

Cohn, Victor
1975 "The Washington Energy Show," *Technology Review* 77(3):8, 68.

Conservation Foundation
1972 "Wanted: A Coordinated, Coherent National Energy Policy Geared to the Public Interest," *CF Letter,* No. 6–72.

Cooley, Richard A., and Geoffrey Wandesforde–Smith, eds.
1970 *Congress and the Environment* (Seattle: Washington).

Crossland, Janice
1974 "Cars, Fuel, and Pollution," *Environment* 16(2):15–27.

Dahl, Robert A.
1970 *After the Revolution?: Authority in a Good Society* (New Haven: Yale).

Davies, Barbara S., and Clarence J. Davies, III
1975 *The Politics of Pollution* (2nd ed.; New York: Pegasus).

Davis, David H.
1974 *Energy Politics* (New York: St. Martin's).

Dexter, Lewis A.
1969 *The Sociology and Politics of Congress* (Chicago: Rand McNally).

Downs, Anthony
1972 "Up and Down with Ecology—The 'Issue-Attention Cycle,' " *Public Interest* 28:38–50.

Dror, Yehezkel
1968 *Public Policymaking Reexamined* (San Francisco: Chandler).

Edelman, Murray
1964 *The Symbolic Uses of Politics* (Urbana: Illinois).

Forrester, Jay W.
1971 *World Dynamics* (Cambridge: Wright–Allen) [esp. Chaps. 1 and 7 for a radical critique of nonsystematic, incremental decision making].

Forsythe, Dall W.
1974 "An Energy-Scarce Society: The Politics and Possibilities," *Working Papers for a New Society* 2(l):3–12 [an excellent short analysis].

Gillette, Robert
1973a "Energy: The Muddle at the Top," *Science* 182:1319–1321.
1973b "Western Coal: Does the Debate Follow Irreversible Commitment?" *Science* 182:456–458.
1975 "In Energy Impasse, Conservation Keeps Popping Up," *Science* 187:42–45.

Goldstein, Paul, and Robert Ford
1973 "On the Control of Air Quality: Why the Laws Don't Work," *Bulletin of the Atomic Scientists* 29(6):31–34.

Green, Charles S., III
1973 "Politics, Equality and the End of Progress," *Alternatives* 2(2):4–9.

Haefele, Edwin T.
1974 *Representative Government and Environmental Management* (Baltimore: Johns Hopkins).

Hartz, Louis
1955 *The Liberal Tradition in America: An Interpretation of American Political Thought Since the Revolution* (New York: Harcourt, Brace).

Henning, Daniel H.

1974 *Environmental Policy and Administration* (New York: American Elsevier).

Hirschman, Albert O.

1970 *Exit, Voice and Loyalty* (Cambridge: Harvard) [esp. Chap. 8 on frontier-style decision making and problem avoidance].

Horowitz, Irving L.

1972 "The Environmental Cleavage: Social Ecology versus Political Economy," *Social Theory and Practice* 2(1):125–134.

Jacobsen, Sally

1974 "Anti–Pollution Backlash in Illinois: Can a Tough Protection Program Survive?" *Bulletin of the Atomic Scientists* 30(1):39–44.

Jones, Charles O.

1975 *Clean Air: The Policies and Politics of Pollution Control* (University of Pittsburgh Press).

Kohlmeier, Louis M., Jr.

1969 *The Regulators: Watchdog Agencies and the Public Interest* (New York: Harper and Row).

Kraft, Michael

1972 "Congressional Attitudes Toward the Environment," *Alternatives* 1(4):27–37 [congressional avoidance of the environmental issue].

1974 "Ecological Politics and American Government: A Review Essay," in Nagel 1974, pp. 139–159 [the best critical review of the political science literature in the light of environmental problems].

Lecht, L. A.

1966 *Goals, Priorities and Dollars* (New York: Free Press).

1969 *Manpower Needs for National Goals in the 1970's* (New York: Praeger).

Lewis, Richard

1972 *The Nuclear Power Rebellion* (New York: Viking).

Lindblom, Charles E.

1965 *The Intelligence of Democracy: Decisionmaking Through Mutual Adjustment* (New York: Free Press).

1969 "The Science of 'Muddling Through,' " *Public Administration Review* 19(2):79–88.

Little, Charles E.

1973 "The Environment of the Poor: Who Gives a Damn?" *Conservation Foundation Letter*, July.

Loveridge, Ronald O.

1971 "Political Science and Air Pollution: A Review and Assessment of the Literature," in *Air Pollution and the Social Sciences*, ed. Paul B. Downing (New York: Praeger), pp. 45–85 [why we are not coping with the problem].

1972 "The Environment: New Priorities and Old Politics," in *People and Politics in Urban Society*, ed. Harlan Hahn (Los Angeles: Sage), pp. 499–529.

Lowi, Theodore

1969 *The End of Liberalism: Ideology, Policy, and the Crisis of Public Authority* (New York: Norton).

McConnell, Grant

1966 *Private Power and American Democracy* (New York: Knopf).

McLane, James

1974 "Energy Goals and Institutional Reform," *The Futurist* 8:239–242.

Michael, Donald N.
    1968 *The Unprepared Society: Planning for a Precarious Future* (New York: Harper and Row).
Miller, John C.
    1957 *Origins of the American Revolution* (Stanford University Press).
Moorman, James W.
    1974 "Bureaucracy v. The Law," *Sierra Club Bulletin* 59(9):7–10 [how agencies evade or flout their legal responsibilities].
Murphy, Earl F.
    1967 *Governing Nature* (Chicago: Quadrangle).
Nagel, Stuart S., ed.
    1974 *Environmental Politics* (New York: Praeger).
Nelkin, Dorothy
    1974 "The Role of Experts in a Nuclear Siting Controversy," *Bulletin of the Atomic Scientists* 30(9):29–36.
Neuhaus, Richard
    1971 *In Defense of People* (New York: Macmillan).
de Nevers, Noel
    1973 "Enforcing the Clean Air Act of 1970," *Scientific American* 228(6):14–21.
Odell, Rice
    1975a "Automobiles Keep Posing New Dilemmas," *Conservation Foundation Letter*, March.
    1975b "Should Americans Be Pried Out of Their Cars?" *Conservation Foundation Letter*, April.
Pirages, Dennis C., and Paul R. Ehrlich
    1974 *Ark II: Social Response to Environmental Imperatives* (New York: W. H. Freeman and Co.).
Platt, John
    1969 "What We Must Do," *Science* 166:1115–1121.
Potter, David M.
    1954 *People Of Plenty: Economic Abundance and the American Character* (University of Chicago).
Quarles, John
    1974 "Fighting the Corporate Lobby," *Environmental Action*, December 7, pp. 3–6 [how the political and other resources of corporations overwhelm the environmental regulators].
Quigg, Philip W.
    1974 "Energy Shortage Spurs Expansion of Nuclear Fission," *World Environment Newsletter* in *SIR World*, June 29, pp. 21–22.
Rauber, Paul
    1991 "O Say, Can You See," *Sierra*, July/August, 24–29.
Roos, Leslie L., Jr., ed.
    1971 *The Politics of Ecosuicide* (New York: Holt, Rinehart and Winston).
Rose, David J.
    1974 "Energy Policy in the U.S.," *Scientific American* 230(l):20–29.
Rosenbaum, Walter A.
    1973 *The Politics of Environmental Concern* (New York: Praeger).
Ross, Charles R.
    1970 "The Federal Government as an Inadvertent Advocate of Environmental

Degradation," in *The Environmental Crisis*, ed. Harold W. Helfiich, Jr. (New Haven: Yale), pp. 171–187.

Ross, Douglas, and Harold Wolman
  1971 "Congress and Pollution—The Gentleman's Agreement," in *Economic Growth vs. the Environment*, ed. Warren A. Johnson and John Hardesty (Belmont, Calif.: Wadsworth), pp. 134–144.

Schaeffer, Robert
  1990 "Car Sick," *Greenpeace*, May/June, 13–17.

Schick, Allen
  1971 "Systems Politics and Systems Budgeting," in Roos 1971, pp. 135–158.

Shapley, Deborah
  1973 "Auto Pollution: Research Group Charged with Conflict of Interest," *Science* 181:732–735.

Shubik, Martin
  1967 "Information, Rationality, and Free Choice in a Future Democratic Society," *Daedalus* 96:771–778.

Sills, David L.
  1975 "The Environmental Movement and Its Critics," *Human Ecology* 3:1–41.

Smith, Adam
  1776 *An Inquiry into the Nature and Causes of the Wealth of Nations*, ed. Edwin Cannan (New York: Modern Library, 1937).

Smith, James N., ed.
  1974 *Environmental Quality and Social Justice* (Washington, D.C.: Conservation Foundation).

Sprout, Harold, and Margaret Sprout
  1971 *Ecology and Politics in America: Some Issues and Alternatives* (New York: General Learning Press).
  1972 "National Priorities: Demands, Resources, Dilemmas," *World Politics* 24:293–317.

Weisskopf, Michael
  1991 "Rule-Making Process Could Soften Clean Air Act," *The Washington Post*, 21 September, A1.

White, Lawrence J.
  1973 "The Auto Pollution Muddle," *Public Interest* 32:97–112.

Wolff, Robert Paul
  1968 *The Poverty of Liberalism* (Boston: Beacon).

Chapter 7

Anon.
  1974 "Take Water and Heat from Third World," *New Scientist* 62:549.

d'Arge, Ralph C., and Allen V. Kneese
  1972 "Environmental Quality and International Trade," *International Organization* 26:419–465.

Banks, Fred
  1974 "Copper is Not Oil," *New Scientist* 63:255–257.

Barraclough, Geoffrey
  1975a "The Great World Crisis," *New York Review of Books*, January 23, pp. 20–30.

1975b "Wealth and Power: The Politics of Food and Oil," *New York Review of Books,*
August 7, pp. 23–30.

Baxter, William F., et al.
1973 Special issue on Stockholm Conference, *Stanford Journal of International Studies*
18:1–153.

Bennett, John W., Sukehiro Hasegawa, and Solomon B. Levine
1973 "Japan: Are There Limits to Growth?" *Environment* 15(10):6–13.

Bergsten, C. Fred
1974 "The New Era in World Commodity Markets," *Challenge* 17(4):34–42.

Boserup, Mogens
1975 "Sharing Is a Myth," *Development Forum* 3(2):1–2.

Brower, David, et al.
1972 "The Stockholm Conference," *Not Man Apart* 2(7):1–ll.

Brown, Lester R.
1972 *World Without Borders* (New York: Random House).

Brown, Seyom, and Larry L. Fabian
1974 "Diplomats at Sea," *Foreign Affairs* 52:301–321.

Caldwell, Lynton K.
1972 *In Defense of Earth: International Protection of the Biosphere* (Bloomington: Indiana).

Castro, Joao A. de A.
1972 "Environment and Development: The Case of the Developing Countries,"
International Organization 26:401–416.

CESI (Center for Economic and Social Information)
1974 "Oil and the Poor Countries," *Environment* 16(2):10–14.

Clawson, Marion
1971 "Economic Development and Environmental Impact: International Aspects,"
Social Science Information 10(4):23–43.

Connelly, Philip, and Robert Perlman
1975 *The Politics of Scarcity: Resource Conflicts in International Relations* (New York:
Oxford).

Cox, Richard H.
1960 Locke on War and Peace (Oxford: University Press).

Durning, Alan
1991 "Asking How Much is Enough," in *State of the World 1991,* ed. Linda Starke
(New York: W. W. Norton & Company).

Enloe, Cynthia
1975 The Politics of Pollution in Comparative Perspective: Ecology and Power in
Four Nations (New York: McKay).

Enviro/Info
1973 *Stockholm '72: A Bibliography of Selected Post-Conference Articles and Documents on
the United Nations Conference on the Human Environment* (Green Bay, Wisc.:
Enviro/Info).

Epstein, William
1975 "The Proliferation of Nuclear Weapons," *Scientific American* 232(4):18–33.

Falk, Richard A.
1971 *This Endangered Planet* (New York: Random House).
1975 "Toward a New World Order: Modest Methods and Drastic Visions," in *On
the Creation of a Just World Order: Preferred Worlds for the 1990's,* ed. Saul H.
Mendlovitz (New York: Free Press), pp. 253–300.

Farvar, M. Taghi, and Theodore N. Soule
    1973 *International Development and the Human Environment: An Annotated Bibliography* (Riverside, N.J.: Macmillan Information).

Finsterbusch, Gail W.
    1973 "International Cooperation Is Picking Up Steam," *Conservation Foundation Letter,* September.

Flavin, Christopher
    1990 "Last Road to Shangri-La," *World Watch,* July-August, 18–26.

Fyodorov, Yevgeny
    1973 "Against the Limits of Growth," *New Scientist* 57:431–432 [abridged from *Kommunist,* No. 14].

Goldman, Marshall I.
    1970 "The Convergence of Environmental Disruption," *Science* 170:37–42 [a synopsis of Goldman 1972].
    1972 *The Spoils of Progress: Environmental Pollution in the Soviet Union* (Cambridge: MIT).

Goldsmith, Edward, et al.
    1972 "Critique of the Stockholm Conference," *The Ecologist* 2(6):1–42.

Graubard, Stephen R., et al.
    1975 "The Oil Crisis: In Perspective," special issue of *Daedalus* 104(4).

Hardin, Garrett
    1974 "Living in a Lifeboat," *BioScience* 24:561–568 [a controversial proposal for American ecological autarky].

Harding, James A.
    1974 "Ecology as Ideology," *Alternatives* 3(4):18–22.

Heilbroner, Robert L.
    1974 *An Inquiry into the Human Prospect* (New York: Norton).

Holt, S. J.
    1974 "Prescription for the Mediterranean: International Cooperation for a Sick Sea," *Environment* 16(4):28–33.

Kay, David A., and Eugene B. Skolnikoff, eds.
    1972 *World Eco-Crisis: International Organizations in Response* (Madison: Wisconsin).

Kelley, Donald, Kenneth R. Stunkel, and Richard R. Wescott
    1976 *The Economic Superpowers and the Environment* (New York: W. H. Freeman and Co.).

Kiseleva, Galina
    1974 "A Soviet View: The Earth and Population," *Development Forum* 2(4):9.

Kristoferson, Lars, ed.
    1975 Special issue on "War and Environment," *Ambio* 4:178–244 [a first-rate treatment].

Laurie, Peter, et al.
    1975 "Towards a Self-Sufficient Britain?" symposium in *New Scientist* 65:690, 695–710.

MacDonald, Gordon J.
    1975 "Weather Modification as a Weapon," *Technology Review* 78(1):57–63.

Mikesell, Raymond F.
    1974 "More Third World Cartels Ahead?" *Challenge* 17(5):24–31.

Miller, Willard M.
    1972 "Radical Environmentalism," *Not Man Apart* 2(11):14–15 [socialism as the answer].

Nash, A. E. Keir
    1970 "Pollution, Population and the Cowboy Economy," *Journal of Comparative Administration* 2:109–128.
Omo–Fadaka, Jimoh
    1973 "The Tanzanian Way of Effective Development," *Impact of Science on Society* 23:107–116.
Packer, Arnold
    1975 "Living with Oil at $10 per Barrel," *Challenge* 17(6):17–25.
Powell, David E.
    1971 "The Social Costs of Modernization: Ecological Problems in the USSR," *World Politics* 23:618–634.
Pryde, Philip R.
    1972 *Conservation in the Soviet Union* (New York: Cambridge).
Quigg, Philip W.
    1974 "The Consumption Dilemma," *World Environment Newsletter* in SR/World, November 2, p. 49.
Ritchie–Calder, Peter R.
    1974 "Caracas—'Smash and Grab,' " *Center Magazine* 7(6):35–38.
Rothman, Harry
    1972 *Murderous Providence.: A Study of Pollution in Industrial Societies* (New York: Bobbs–Merrill).
Rotkirch, Holger
    1974 "Claims to the Ocean: Freedom of the Sea for Whom?" *Environment* 16(5):34–41.
Shapley, Deborah
    1973 "Ocean Technology: Race to Seabed Wealth Disturbs More than Fish," *Science* 180:849–851, 893.
    1975 "Now, a Draft Sea Law Treaty—But What Comes After?" *Science* 188:918.
Shields, Linda P., and Marvin C. Ott
    1974 "Environmental Decay and International Politics: The Uses of Sovereignty," *Environmental Affairs* 3:743–767.
Sigurdson, Jon
    1973 "The Suitability of Technology in Contemporary China," *Impact of Science on Society* 23:341–352.
    1975 "Resources and Environment in China," *Ambio* 4:112–119.
Sivard, Ruth L.
    1975 "Let Them Eat Bullets!" *Bulletin of the Atomic Scientists* 31(4):6–10.
Skolnikoff, Eugene B.
    1971 "Technology and the Future Growth of International Organizations," *Technology Review* 73(8):39–47.
Slocum, Marianna
    1974 "Soviet Energy: An Internal Assessment," *Technology Review* 77(1):17–33.
Spengler, Joseph J.
    1969 "Return to Thomas Hobbes?" *South Atlantic Quarterly* 68:443–453.
Sprout, Harold, and Margaret Sprout
    1971 *Toward a Politics of the Planet Earth* (New York: Van Nostrand Reinhold).
Staines, Andrew
    1974 "Digesting the Raw Materials Threat," *New Scientist* 61:609–611.
Syer, G. N.
    1971 "Marx and Ecology," *The Ecologist* 1(16):19–21.

Tinker, Jon, et al.

1975a "Cocoyoc: The New Economics," *New Scientist* 67:529–531.

1975b "Cocoyoc Revisited," *New Scientist* 67:480–483 [Third World demands for fundamental reform of the world system].

1975c "World Environment: What's Happening at UNEP?" *New Scientist* 66:600–613.

UNIPUB

1972 *United Nations Conference on the Human Environment: A Guide to the Conference Bibliography* (New York: UNIPUB).

Utton, Albert E., and Daniel H. Henning, eds.

1973 *Environmental Policy: Concepts and International Implications* (New York: Praeger).

Wade, Nicholas

1974 "Raw Materials: U.S. Grows More Vulnerable to Third World Cartels," *Science* 183:185–186.

Walsh, John

1974 "UN Conferences: Topping Any Agenda Is the Question of Development," *Science* 185:1143–1144, 1192–1193.

Westing, Arthur H.

1974 "Arms Control and the Environment: Proscription of Ecocide," *Bulletin of the Atomic Scientists* 30(1):24–27.

Wilson, Carroll L.

1973 "A Plan for Energy Independence," *Foreign Affairs* 51:657–675.

Wilson, Thomas W., Jr.

1971 *International Environmental Action: A Global Survey* (Cambridge: Dunellen).

Woodhouse, Edward J.

1972 "Re–Visioning the Future of the Third World: An Ecological Perspective on Development," *World Politics* 25:1–33.

Yablokov, Alexei, et al.

1991 "Russia: Gasping for Breath, Choking in Waste, Dying Young," trans. Klose, Eliza K., *The Washington Post,* 18 August, C3 [adapted from *Russian Gazette*].

Chapter 8

Barash, David P.

1973 "The Ecologist as Zen Master," *American Midland Naturalist* 89:214–217.

Bookchin, Murray

1971 *Post-Scarcity Anarchism* (Berkeley: Ramparts Press).

Boulding, Kenneth E.

1964 *The Meaning of the Twentieth Century: The Great Transition* (New York: Harper and Row).

Burch, William R., Jr.

1971 *Daydreams and Nightmares: A Sociological Essay on the American Environment* (New York: Harper and Row).

Callenbach, Ernest

1975 *Ecotopia* (Berkeley: Banyan Tree Books).

Churchman, C. West

1968 *The Systems Approach* (New York: Dell).

Colwell, Thomas B., Jr.
    1969 "The Balance of Nature: A Ground of Human Values," *Main Currents in Modern Thought* 26(2):46–52.

Dasmann, Raymond F.
    1974 "Conservation, Counter–culture, and Separate Realities," *Environmental Conservation* 1:133–137.

Doctor, Adi H.
    1975 "Gandhi's Political Philosophy," *The Ecologist* 5:300–321 [a succinct summary, with copious excerpts from Gandhi's own writings].

van Dresser, Peter
    1972 *A Landscape for Humans* (Albuquerque: Biotechnic Press).

Dubos, Rene
    1968 *So Human an Animal* (New York: Scribners).
    1972 *A God Within* (New York: Scribners).

Easterlin, Richard A.
    1973 "Does Money Buy Happiness?" *Public Interest* 30:3–10.

Goldsmith, Edward, et al.
    1972 "A Blueprint for Survival," *The Ecologist* 2(l):1-43 [a concrete plan for a minimal, frugal steady-state society].

Huxley, Aldous
    1962 *Island* (New York: Harper and Row).

Illich, Ivan
    1971 *Deschooling Society* (New York: Harper and Row).
    1973 *Tools for Conviviality* (New York: Harper and Row).
    1974a *Energy and Equity* (London: Calder and Boyars).
    1974b "Energy and Social Disruption," *The Ecologist* 4:49–52.

Iyer, Raghavan
    1973 *The Moral and Political Thought of Mahatma Gandhi* (New York: Oxford).

Kateb, George
    1973 *Utopia and Its Enemies* (New York: Schocken).

Keynes, John Maynard
    1971 "Economic Possibilities for Our Grandchildren," in *Economic Growth vs. the Environment,* ed. Warren A. Johnson and John Hardesty (Belmont, Calif.: Wadsworth), pp. 189–193.

Koch, Adrienne
    1964 *The Philosophy of Thomas Jefferson* (Chicago: Quadrangle).

Kozlovsky, Daniel G.
    1974 *An Ecological and Evolutionary Ethic* (Englewood Cliffs, N.J.: Prentice–Hall).

Kropotkin, Peter
    1899 *Fields, Factories and Workshops Tomorrow,* ed. Colin Ward (New York: Harper and Row, 1975).

Lao Tzu
    1958 *Tao Teh King,* ed. Archie J. Bahm (New York: Frederick Ungar).

Laszlo, Ervin
    1972 *The Systems View of the World: The Natural Philosophy of the New Developments in the Sciences* (New York: Braziller).

Leiss, William
    1972 *The Domination of Nature* (New York: Braziller).

Leopold, Aldo
  1968 *A Sand County Almanac* (New York: Oxford).
Lévi–Strauss, Claude
  1966 *The Savage Mind* (University of Chicago Press).
Lindner, Staffan B.
  1970 *The Harried Leisure Class* (New York: Columbia).
Livingston, John A.
  1973 *One Cosmic Instant: Man's Fleeting Supremacy* (Boston: Houghton Mifflin).
McKinley, Daniel
  1970 "Lichens—Mirror to the Universe," *Audubon* 72(6):51–54.
Maslow, Abraham H.
  1966 *The Psychology of Science: A Reconnaissance* (New York: Harper and Row).
  1971 *The Farther Reaches of Human Nature* (New York: Viking).
Marx, Karl
  1844 *The Economic and Philosophic Manuscripts of 1844,* ed. Dirk J. Struik (New York: International, 1964).
Marx, Leo
  1964 *The Machine in the Garden: Technology and the Pastoral Ideal in America* (New York: Oxford).
  1970 "American Institutions and Ecological Ideals," *Science 170:945–952.*
Meeker, Joseph W.
  1974 *The Comedy of Survival: Studies in Literary Ecology* (New York: Scribners).
Mill, J. S.
  1871 *Principles of Political Economy,* ed. W. J. Ashley (New York: Sentry, 1965).
More, Thomas
  1516 *Utopia,* ed. H. V. S. Ogden (New York: Meredith, 1949).
Mumford, Lewis
  1961 *The City in History: Its Origins, Its Transformations, and Its Prospects* (New York: Harcourt, Brace and World).
  1967 *The Myth of the Machine: Technics and Human Development* (New York: Harcourt, Brace and World).
  1970 *The Myth of the Machine: The Pentagon of Power* (New York: Harcourt Brace Jovanovich).
  1973 *Interpretations and Forecasts* (New York: Harcourt Brace Jovanovich).
Nasr, Seyyed Hossein
  1968 *The Encounter of Man and Nature: The Spiritual Crisis of Modern Man* (London: Allen and Unwin).
Passmore, John
  1974 *Man's Responsibility for Nature* (New York: Scribners).
Roszak, Theodore
  1969 *The Making of a Counter Culture: Reflections on the Technocratic Society and Its Youthful Opposition* (Garden City, N.Y.: Doubleday).
  1973 *Where the Wasteland Ends: Politics and Transcendence in Postindustrial Society* (Garden City, N.Y.: Doubleday).
Sahlins, Marshall
  1970 *Stone-Age Economics* (Chicago: Aldine–Atherton).
Schumacher, E. F.
  1973 *Small Is Beautiful: Economics As If People Mattered* (New York: Harper and Row).

1974 "Message from the Universe," *The Ecologist* 4:318–320.

Shepard, Paul
1973 *The Tender Carnivore and the Sacred Game* (New York: Scribners).
———, and Daniel McKinley, eds.
1969 *The Subversive Science: Essays Toward an Ecology of Man* (Boston: Houghton Mifflin).

Sibley, Mulford Q.
1973 "The Relevance of Classical Political Theory for Economy, Technology, and Ecology," *Alternatives* 2(2):14–35.

Slater, Philip E.
1970 *The Pursuit of Loneliness: American Culture at the Breaking Point* (Boston: Beacon).
1974 *Earthwalk* (Garden City, N.Y.: Doubleday).

Smith, Adam
1792 *The Theory of Moral Sentiments,* in *The Works of Adam Smith,* vol. I (Aalen, W. Germany: O. Zeller, 1963).

Snyder, Gary
1969 *Earth House Hold* (New York: New Directions).
1974 *Turtle Island* (New York: New Directions).

Stavrianos, L. S.
1976 *The Promise of the Coming Dark Age* (New York: W. H. Freeman and Co.).

Taylor, Gordon Rattray
1974 *Rethink: Radical Proposals to Save a Disintegrating World* (Baltimore: Penguin).

Thoreau, Henry David
1854 *Walden,* in *The Portable Thoreau,* ed. Carl Bode (New York: Viking, 1964).

Wagar, W. Warren
1971 *Building the City of Man: Outlines of a World Civilization* (New York: Grossman).

Watt, Kenneth E. F.
1974 *The Titanic Effect: Planning for the Unthinkable* (Stamford, Conn.: Sinauer).

White, Lynn, Jr.
1967 "The Historical Roots of Our Ecologic Crisis," *Science* 155:1203–1207.

Index

Biographical Sketches

———

William Ophuls graduated from Princeton University in 1955 with an A.B. in oriental studies, then served for four years as an officer in the U.S. Coast Guard. He joined the U.S. Foreign Service in 1959 and spent the next eight years at the Department of State and at U.S. Embassies in the Ivory Cast and Japan. In 1973, he received a Ph.D. in political science from Yale. He taught at Northwestern and Oberlin and has lectured and consulted widely. When *Ecology and the Politics of Scarcity* was first published in 1977, it was recognized as a pioneering work, winning the Sprout Prize from the International Studies Association for the best book on ecology and international politics and the Kammerer Award from the American Political Science Association for the best book on U.S. domestic politics. Dr. Ophuls is at work on a sequel to *Ecology and the Politics of Scarcity*. His new work focuses on the internal contradictions of modern politics.

A. Stephen Boyan, Jr., received an A.B. from Brown University in 1959 and a Ph.D. from the University of Chicago in 1966, both in political science. He taught at Pennsylvania State University for four years and since 1971 has been a member of the faculty of the University of Maryland Baltimore County. His special interests are environmental ethics and constitutional law. He is editor of the six-volume work, *Constitutional Aspects of Watergate*. He is a Leader in the Ethical Culture Movement and was organizer of the Earth Ethics project at the Washington Ethical Society.

Thomas E. Lovejoy received his B.S. and Ph.D. in biology from Yale University. From 1973 to 1987, he directed the program of the World Wildlife Fund-U.S. He is generally credited with having brought the problems of tropical forests to the fore as a public issue. He founded the public television series "Nature." He is president of the Society for Conservation Biology and chairman of the U.S. Man and Biosphere Program. Since 1987, he has been Assistant Secretary for External Affairs of the Smithsonian Institution.